Lectures on the Philosophy of Mathematics

Lectures on the Philosophy of Mathematics

Joel David Hamkins

The MIT Press
Cambridge, Massachusetts
London, England

This book was set in Times Roman by the author using LATEX and TikZ. Printed and bound in the United States of America.

Library of Congress Cataloging-in-Publication Data

Names: Hamkins, Joel David, author.
Title: Lectures on the philosophy of mathematics / Joel David Hamkins.
Description: Cambridge, Massachusetts : The MIT Press, [2020] | Includes
    bibliographical references and index.
Identifiers: LCCN 2020009876 | ISBN 9780262542234 (paperback)
Subjects: LCSH: Mathematics–Philosophy.
Classification: LCC QA8.6 .H35 2020 | DDC 510.1–dc23
LC record available at https://lccn.loc.gov/2020009876

10 9 8 7 6 5 4 3 2

To all and only those authors who do not dedicate books to themselves.

Contents

Preface

Philosophical conundrums pervade mathematics, from fundamental questions of mathematical ontology—What is a number? What is infinity?—to questions about the relations among truth, proof, and meaning. What is the role of figures in geometric argument? Do mathematical objects exist that we cannot construct? Can every mathematical question be solved in principle by computation? Is every truth of mathematics true for a reason? Can every mathematical truth be proved?

This book is an introduction to the philosophy of mathematics, in which we shall consider all these questions and more. I come to the subject from mathematics, and I have strived in this book for what I hope will be a fresh approach to the philosophy of mathematics—one grounded in mathematics, motivated by mathematical inquiry or mathematical practice. I have strived to treat philosophical issues as they arise organically in mathematics. Therefore, I have organized the book by mathematical themes, such as number, infinity, geometry, and computability, and I have included some mathematical arguments and elementary proofs when they bring philosophical issues to light.

The philosophical positions of platonism, realism, logicism, structuralism, formalism, intuitionism, type theorism, and others arise naturally in various mathematical contexts. The mathematical progression from ancient Pythagorean incommensurability and the irrationality of $\sqrt{2}$, for example, through to Liouville's construction of transcendental numbers, paralleling the discovery of nonconstructible numbers in geometry, is an opportunity to contrast platonism with structuralism and other accounts of what numbers and mathematical objects are. Structuralism finds its origin in Dedekind's arithmetic categoricity theorem, gaining strength with categorical accounts of the real numbers and our other familiar mathematical structures. The rise of rigor in the calculus is a natural setting to discuss whether the indispensability of mathematics in science offers grounds for mathematical truth. Zeno's paradoxes of motion and Galileo's paradoxes of infinity lead to the Cantor-Hume principle, and then to both Frege's number concept and Cantor's work on the transfinite. Thus, mathematical themes traverse millennia, giving rise again and again to philosophical considerations.

I therefore aim to present a mathematics-oriented philosophy of mathematics. Years ago, Penelope Maddy (1991) criticized parts of the philosophy of mathematics at that time as amounting to

> an intramural squabble between metaphysicians, and a squabble in which it is not clear what, if anything, is really at stake. (p. 158)

She sought to refocus the philosophy of mathematics on philosophical issues closer to mathematics:

> What I'm recommending is a hands-on sort of philosophy of mathematics, a sort relevant to actual practice, a sort sensitive to the problems, procedures, and concerns of mathematicians themselves. (p. 159)

I find that inspiring, and part of what I have aimed to do in this book is follow that advice— to present an introduction to the philosophy of mathematics that both mathematicians and philosophers might find relevant. Whether or not you agree with Maddy's harsh criticism, there are many truly compelling issues in the philosophy of mathematics, which I hope to share with you in this book. I hope that you will enjoy them.

Another aim I have with the book is to try to help develop a little the mathematical sophistication of the reader on mathematical topics important in the philosophy of mathematics, such as the foundations of number theory, non-Euclidean geometry, nonstandard analysis, Gödel's incompleteness theorems, and uncountability. Readers surely come to this subject from diverse mathematical backgrounds, ranging from novice to expert, and so I have tried to provide something useful for everyone, always beginning gently but still reaching deep waters. The allegory of Hilbert's Grand Hotel, for example, is an accessible entryway to the discussion of Cantor's results on countable and uncountable infinities, and ultimately to the topic of large cardinals. I have aimed high on several mathematical topics, but I have also strived to treat them with a light touch, without getting bogged down in difficult details.

This book served as the basis for the lecture series I gave on the philosophy of mathematics at the University of Oxford for Michaelmas terms in 2018, and again in 2019 and 2020. I am grateful to the Oxford philosophy of mathematics community for wide-ranging discussions that have helped me to improve this book. Special thanks to Daniel Isaacson, Alex Paseau, Beau Mount, Timothy Williamson, Volker Halbach, and especially Robin Solberg, who gave me extensive comments on earlier drafts. Thanks also to Justin Clarke-Doane of Columbia University in New York for comments. And thanks to Theresa Carcaldi for extensive help with editing.

This book was typeset using LATEX. Except for the image on page 89, which is in the public domain, I created all the other images in this book using TikZ in LATEX, specifically for this book, and in several instances also for my book *Proof and the Art of Mathematics* (2020), available from MIT Press.

About the Author

I am both a mathematician and a philosopher, undertaking research in the area of mathematical and philosophical logic, with a focus on the mathematics and philosophy of the infinite, especially set theory and the philosophy of set theory and the mathematics and philosophy of potentialism. My new book on proof-writing, *Proof and the Art of Mathematics* (2020), is available from MIT Press.

I have recently taken up a position in Oxford, after a longstanding appointment at the City University of New York. I have also held visiting positions over the years at New York University, Carnegie Mellon University, Kobe University, University of Amsterdam, University of Muenster, University of California at Berkeley, and elsewhere. My 1994 Ph.D. in mathematics was from the University of California at Berkeley, after an undergraduate degree in mathematics at the California Institute of Technology.

Joel David Hamkins
Professor of Logic and Sir Peter Strawson Fellow in Philosophy
Oxford University, University College
High Street, Oxford 0X1 4BH

joeldavid.hamkins@philosophy.ox.ac.uk
joeldavid.hamkins@maths.ox.ac.uk
Blog: http://jdh.hamkins.org
MathOverflow: http://mathoverflow.net/users/1946/joel-david-hamkins
Twitter: @JDHamkins

1 Numbers

Abstract. Numbers are perhaps the essential mathematical idea, but what are numbers? There are many kinds of numbers—natural numbers, integers, rational numbers, real numbers, complex numbers, hyperreal numbers, surreal numbers, ordinal numbers, and more—and these number systems provide a fruitful background for classical arguments on incommensurability and transcendentality, while setting the stage for discussions of platonism, logicism, the nature of abstraction, the significance of categoricity, and structuralism.

1.1 Numbers versus numerals

What is a number? Consider the number 57. What is it? We distinguish between the number and the numerals used to represent it. The notation 57—I mean literally the symbol 5 followed by the symbol 7—is a description of how to build the number: take five tens and add seven. The number 57 is represented in binary as 111001, which is a different recipe: start with a fresh thirty-two, fold in sixteen and eight, and garnish with one on top, chill and serve. The Romans would have written LVII, which is the following recipe: start with fifty, add five, and then two more.

In the delightful children's novel, *The Phantom Tollbooth*, Juster and Feiffer (1961), numbers come from the number mine in Digitopolis, and they found there *the largest number*! It was...*ahem*...a gigantic number 3—over 4 meters tall—made of stone. Broken numbers from the mine were used for fractions, like 5/3, when the number 5 has broken into three pieces. But of course, this confuses the number with the numeral, the object with its description. We would not confuse Hypatia, the person, with *Hypatia*, the string of seven letters forming her name; or the pecan pie (delicious!) with the written instructions (chewy, like cardboard) for how to prepare it.

There are many diverse kinds of natural numbers. The *square* numbers, for example, are those that can be arranged in the form of a square:

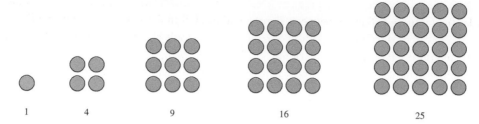

The *triangular* numbers, in contrast, are those that can be arranged in the form of a triangle:

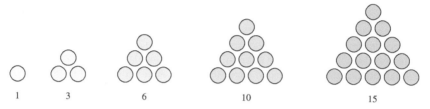

One proceeds to the *hexagonal* numbers, and so forth. The number zero is a degenerate instance in each case.

The *palindromic* numbers are the numbers, such as 121 or 523323325, whose digits read the same forward and backward, like the palindromic phrase, "I prefer pi." Whereas squareness and triangularity are properties of numbers, however, the question of whether a number is palindromic depends on the base in which it is presented (and for this reason, mathematicians sometimes find the notion unnatural or amateurish). For example, 27 is not a palindrome, but in binary, it is represented as 11011, which is a palindrome. Every number is a palindrome in any sufficiently large base, for it thereby becomes a single digit, a palindrome. Thus, palindromicity is not a property of numbers, but of number descriptions.

1.2 Number systems

Let us lay our various number systems on the table.

Natural numbers

The *natural numbers* are commonly taken to be the numbers:

$$0 \quad 1 \quad 2 \quad 3 \quad 4 \quad \cdots$$

The set of natural numbers is often denoted by $\mathbb{N}$. The introduction of zero, historically, was a difficult conceptual step. The idea that one might need a number to represent *nothing* or an absence is deeper than one might expect, given our current familiarity with this concept. Zero as a number was first fully explicit in the fifth century in India, although it was used earlier as a placeholder, but not with Roman numerals, which have no place values. Even today, some mathematicians prefer to start the natural numbers with 1; some computer programming languages start their indices with 0, and others with 1. And consider the cultural difference between Europe and the US in the manner of counting floors in a building, or the Chinese method of counting a person's age, where a baby is "one" at birth and during their first year, turning "two" after one year, and so on.

A rich number theory has emerged, built upon the natural numbers, described by Johann Carl Friedrich Gauss as the "queen of mathematics" and admired by G. H. Hardy (1940) for its pure, abstract isolation. Despite Hardy's remarks, number theory has nevertheless found key applications: advanced number-theoretic ideas form the core of cryptography, the basis of internet security, and therefore a foundational element in our economy.

Integers

The *integers*, usually denoted $\mathbb{Z}$ for the German word *Zahlen*, meaning "numbers," include all the natural numbers and their negations, the negative numbers:

$$\cdots \quad -3 \quad -2 \quad -1 \quad 0 \quad 1 \quad 2 \quad 3 \quad \cdots$$

The integers enjoy mathematical features not shared by the natural numbers, such as the fact that the sum *and difference* of any two integers is an integer. Every integer has an additive inverse, a number whose sum with the first number is zero, the additive identity; and multiplication distributes over addition. Thus, the integers form a *ring*, a kind of mathematical structure with addition and multiplication operations for which (among other familiar properties) every number should have an additive inverse.

Rational numbers

Rational numbers are formed by the ratios or quotients of integers, also known as fractions, such as $1/2$ or $5/7$ or $129/53$. The collection of all rational numbers, usually denoted by $\mathbb{Q}$ for "quotient," is a dense-in-itself collection of points on the number line, for the average of any two rational numbers is again rational. The rational numbers in turn enjoy structural features not shared by the natural numbers or the integers, such as the existence of multiplicative inverses for every nonzero element. This makes the rational numbers $\mathbb{Q}$ a *field*, and in fact an *ordered* field, since the order relation $<$ interacts with the arithmetic operations in a certain way, such as that $x < y$ if and only if $x + a < y + a$.

But what is a number, really? Consider a confounding case: the fraction $3/6$ is not in lowest terms, while $1/2$ is, and yet we say that they are equal:

$$\frac{3}{6} = \frac{1}{2}.$$

What gives? If a thing has a property that another does not, how can they be identical? That would violate Leibniz's law on the indiscernibility of identicals. The solution is that like the palindromic numbers, the property of being in lowest terms is not actually a property of the number itself, but a property of the way that we have described or represented that number. Again, we have confused number with numeral.

The distinctions between object and description, between semantics and syntax, between use and mention—these are core distinctions in mathematical logic, clarifying many issues. Often, distinct descriptions can characterize the very same object: the morning star is the evening star—both are the planet Venus—even though this took some time to learn.

But again, what is a number?

1.3   Incommensurable numbers

A gem of classical mathematics—and, as I see it, a pinnacle of human achievement—is the discovery of incommensurable numbers, quantities that cannot be represented by the ratio of integers. In the fifth century BC, the Pythagoreans discovered that the side and diagonal of a square have no common unit of measure; there is no smaller unit length of which they are both integral multiples. In modern terminology, this amounts to the claim that $\sqrt{2}$ is irrational, since this is the ratio of diagonal to side in any square. If you divide the side of a square into 10 units, then the diagonal will have a little more than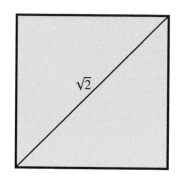
14 of them; if into 100 units, then a little more than 141; if 1000, then a little more than 1414; but it will never come out exact—the quantities are incommensurable. One sees these numbers in the initial digits of the decimal expansion:

$$\sqrt{2} = 1.414213\ldots$$

It was a shocking discovery for the Pythagoreans, and indeed heretical in light of their quasi-religious number-mysticism beliefs, which aimed to comprehend all through proportion and ratio, taking numbers as the foundational substance of reality. According to legend, the man who made the discovery of incommensurability was drowned at sea, punished by the gods for revealing the secret knowledge of irrationality. But the word has been out for two thousand years now, and so let us together go through a proof of this beautiful mathematical fact. We have dozens of proofs now, using diverse methods. My opinion is that every self-respecting, educated person should be able to prove that $\sqrt{2}$ is irrational.

Theorem 1. *$\sqrt{2}$ is irrational.*

Proof. A widely known, classical argument proceeds as follows. Suppose toward contradiction that $\sqrt{2}$ is rational, so it can be represented as a fraction:

$$\sqrt{2} = \frac{p}{q}.$$

We may assume that this fraction is in lowest terms, meaning that p and q are integers with no common factor. If we square both sides of the equation and multiply by q^2, we arrive at

$$2q^2 = p^2.$$

From this, we see that p^2 is a multiple of 2, and it therefore must be an even number. It follows that p also is even, because odd × odd is odd. So $p = 2k$ for some integer k, and therefore $2q^2 = p^2 = (2k)^2 = 4k^2$, which implies that $q^2 = 2k^2$. Thus, q^2 also is even, and so q must be even as well. So both p and q are even, which contradicts our assumption that the fraction p/q was in lowest terms. So $\sqrt{2}$ cannot be represented as a fraction and therefore is irrational. □

An alternative geometric argument

Here is another argument—one with a more geometric flavor, due to the logician Stanley Tennenbaum from New York City.

Proof. If $\sqrt{2} = p/q$ is rational, then let us represent the resulting equation $p^2 = q^2 + q^2$ geometrically, with a large, blue integer square having the same area as two medium, red integer squares:

$$p^2 \quad = \quad q^2 \quad + \quad q^2$$

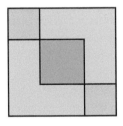

Let us consider the smallest possible integer instance of this relationship. Place the two red squares overlapping inside the blue square, as shown here. The two blue corners are uncovered, peeking through, while the red square of overlap in the center is covered twice. Since the original blue square had the same area as the two red squares, it follows that the area of the double-counted central square is exactly balanced by the two small, uncovered blue squares in the corners. Let us pull these smaller squares out of the figure and consider them separately. We have said that the red, central square has the same area as the two small blue squares in the corners.

Furthermore, these smaller squares also each have integer sides, since they arise as differences from the previous squares. Thus, we have found a strictly smaller integer square as the sum of two copies of another integer square. This contradicts our assumption that we had begun with the smallest instance. So there can be no such instance at all, and therefore $\sqrt{2}$ is irrational. □

1.4   Platonism

Truly, what is a number? Let us begin to survey some possible answers. According to the philosophical position known as *platonism*, numbers and other mathematical objects exist as abstract objects. Plato held them to exist in a realm of ideal forms. A particular line or circle that you might draw on paper is flawed and imperfect; in the platonic realm, there are perfect lines and circles—and numbers. From this view, for a mathematician to say, "There is a natural number with such-and-such property," means that in the platonic realm, there are such numbers to be found. Contemporary approaches to platonism assert that abstract objects exist—this is the core issue—but are less connected with Plato's idea of an ideal form or the platonic realm, a place where they are all gathered together.

What does it mean to say, "There is a function f that is continuous but not differentiable," "There is a solution to this differential equation," or "There is a topological space that is sequentially compact, but not compact"? This is not physical existence; we cannot hold these "objects" in our hands, as we might briefly hold a hot potato. What kind of existence is this?

According to platonism, mathematical objects are abstract but enjoy a real existence. For the platonist, ordinary talk in mathematics about the existence of mathematical objects can be taken literally—the objects do exist, but abstractly rather than physically. According to this perspective, the nature of mathematical existence is similar to the nature of existence for other abstractions, such as beauty or happiness. Does beauty exist? I believe so. Do parallel lines exist? According to platonism, the answers are similar. But what are abstract objects? What is the nature of this existence?

Consider a piece of writing: Henrik Ibsen's play *A Doll's House*. This exists, surely, but what is it specifically that exists here? I could offer you a printed manuscript, saying, "This is *A Doll's House*." But that would not be fully true, for if that particular manuscript were damaged, we would not say that the play itself was damaged; we would not say that the play had been taken in my back pocket on a motorcycle ride. I could see a performance of the play on Broadway, but no particular performance would seem to be the play itself. We do not say that Ibsen's play existed only in 1879 at its premiere, or that the play comes into and out of existence with each performance. The play is an abstraction, an idealization of its various imperfect instantiations in manuscripts and performances.

Like the play, the number 57 similarly exists in various imperfect instantiations: 57 apples in the bushel and 57 cards in the cheater's deck. The existence of abstract objects is mediated somehow through the existence of their various instantiations. Is the existence of the number 57 similar to the existence of a play, a novel, or a song? The play, as with other pieces of art, was created: Ibsen wrote *A Doll's House*. And while some mathematicians describe their work as an act of creation, doubtless no mathematician would claim to have created the number 57 in that sense. Is mathematics discovered or created? Part of the contemporary platonist view is that numbers and other mathematical objects have an *independent* existence; like the proverbial tree falling in the forest, the next number would exist anyway, even if nobody ever happened to count that high.

Plenitudinous platonism

The position known as *plenitudinous platonism,* defended by Mark Balaguer (1998), is a generous form of platonism, generous in its metaphysical commitments; it overflows with them. According to plenitudinous platonism, every coherent mathematical theory is realized in a corresponding mathematical structure in the platonic realm. The theory is true in an ideal mathematical structure, instantiating the subject matter that the theory is about. According to plenitudinous platonism, every conceivable coherent mathematical theory is true in an actual mathematical structure, and so this form of platonism offers us a rich mathematical ontology.

1.5 Logicism

Pursuing the philosophical program known as *logicism,* Gottlob Frege, and later Bertrand Russell and others at the dawn of the twentieth century, aimed to reduce all mathematics, including the concept of number, to logic. Frege begins by analyzing what it means to say that there are a certain number of things of a certain kind. There are exactly two things with property P, for example, when there is a thing x with that property and there is another distinct thing y with that property, and furthermore, anything with the property is either x or y. In logical notation, therefore, "There are exactly two Ps" can be expressed like this:

$$\exists x, y \,(Px \wedge Py \wedge x \neq y \wedge \forall z(Pz \to z = x \vee z = y)).$$

The quantifier symbol $\exists$ is read as "There exists" and $\forall$ as "For all," while $\wedge$ and $\vee$ mean "and" and "or," and $\to$ means "implies." In this way, Frege has expressed the concept of the number 2 in purely logical terms. You can have two sheep or two apples or two hands, and the thing that is common between these situations is what the number 2 is.

Equinumerosity

Frege's approach to cardinal numbers via logic has the effect that classes placed in a one-to-one correspondence with each other will fulfill exactly the same Fregean number assertions, because the details of the truth assertion transfer through the correspondence from one class to the other. Frege's approach, therefore, is deeply con-

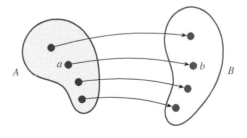

nected with the *equinumerosity* relation, a concept aiming to express the idea that two sets or classes have the same cardinal size. Equinumerosity also lies at the core of Georg Cantor's analysis of cardinality, particularly the infinite cardinalities discussed in chapter 3. Specifically, two classes of objects (or as Frege would say: two concepts) are *equinumerous*—they have the same cardinal size—when they can be placed into a one-to-one correspondence. Each object in the first class is associated with a unique object in the second class and conversely, like the shepherd counting his sheep off on his fingers.

So let us consider the equinumerosity relation on the collection of all sets. Amongst the sets pictured here, for example, equinumerosity is indicated by color: all the green sets are equinumerous, with two elements, and all the red sets are equinumerous, and the orange sets, and so on.

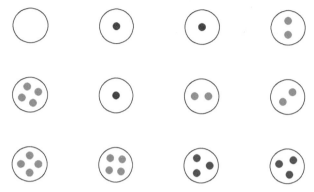

In this way, equinumerosity enables us systematically to compare any two sets and determine whether they have the same cardinal size.

The Cantor-Hume principle

At the center of Frege's treatment of cardinal numbers, therefore, is the following criterion for number identity:

Cantor-Hume principle. *Two concepts have the same number if and only if those concepts can be placed in a one-to-one correspondence.*

In other words, the number of objects with property P is the same as the number of objects with property Q exactly when there is a one-to-one correspondence between the class $\{\, x \mid P(x) \,\}$ and the class $\{\, x \mid Q(x) \,\}$. Expressed in symbols, the principle asserts

$$\#P = \#Q \qquad \text{if and only if} \qquad \{\, x \mid P(x) \,\} \quad \simeq \quad \{\, x \mid Q(x) \,\},$$

where $\#P$ and $\#Q$ denote the number of objects with property P or Q, respectively, and the symbol $\simeq$ denotes the equinumerosity relation.

The Cantor-Hume principle is also widely known simply as *Hume's principle*, in light of Hume's brief statement of it in *A Treatise of Human Nature* (1739, I.III.I), which Frege mentions (the relevant Hume quotation appears in section 4.5 of this book). Much earlier, Galileo mounted an extended discussion of equinumerosity in his *Dialogues Concerning Two New Sciences* (1638), considering it as a criterion of size identity, particularly in the confounding case of infinite collections, including the paradoxical observation that line segments of different lengths and circles of different radii are nevertheless equinumerous as collections of points. Meanwhile, to my way of thinking, the principle is chiefly to be associated with Cantor, who takes it as the core motivation underlying his foundational development of cardinality, perhaps the most successful and influential, and the first finally to be clear on the nature of countable and uncountable cardinalities (see chapter 3). Cantor treats equinumerosity in his seminal set-theoretic article, Cantor (1874), and states a version of the Cantor-Hume principle in the opening sentence of Cantor (1878). In an 1877 letter to Richard Dedekind, he proved the equinumerosity of the unit interval with the square, the cube, and indeed the unit hypercube in any finite dimension, saying of the discovery, "I see it, but I don't believe it!" (Dauben, 2004, p. 941) We shall return to this example in section 3.8, page 106.

The Cantor-Hume principle provides a criterion of number identity, a criterion for determining when two concepts have the same number. Yet it expresses on its face merely a necessary feature of the number concept, rather than identifying fully what numbers are. Namely, the principle tells us that numbers are classification invariants of the equinumerosity relation. A classification *invariant* of an equivalence relation is a labeling of the objects in the domain of the relation, such that equivalent objects get the same label and inequivalent objects get different labels. For example, if we affix labels to all the apples we have picked, with a different color for each day of picking, then the color of the label will be an invariant for the picked-on-the-same-day-as relation on these apples. But there are many other invariants; we could have written the date on the labels, encoded it in a bar code, or we could simply have placed each day's apples into a different bushel.

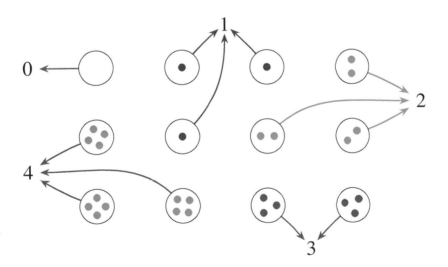

The Cantor-Hume principle tells us that numbers—whatever they are—are assigned to every class in such a way that equinumerous classes get the same number and nonequinumerous classes get different numbers. And this is precisely what it means for numbers to be a classification invariant of the equinumerosity relation. But ultimately, what are these "number" objects that get assigned to the sets? The Cantor-Hume principle does not say.

The Julius Caesar problem

Frege had sought in his logicist program an *eliminative* definition of number, for which numbers would be defined in terms of other specific concepts. Since the Cantor-Hume principle does not tell us what numbers are, he ultimately found it unsatisfactory to base a number concept solely upon it. Putting the issue boldly, he proclaimed

> we can never—to take a crude example—decide by means of our definitions whether any concept has the number Julius Caesar belonging to it, or whether that same familiar conqueror of Gaul is a number or not. (Frege, 1968 [1884], §57)

The objection is that although the Cantor-Hume principle provides a number identity criterion for identities of the form $\#P = \#Q$, comparing the numbers of two classes, it does not provide an identity criterion for identities of the form $\#P = x$, which would tell us which objects x were numbers, including the case $\#P =$ Julius Caesar.

On its face, this objection is strongly antistructuralist, a position we shall discuss at length in section 1.10, for it is concerned with what the numbers are, rather than with their structural features and roles. Yet, Greimann (2003) and others argue, nevertheless, that there is a nuanced structuralism in Frege.

Numbers as equinumerosity classes

In order to define a concept of number suitable for his program, Frege undertakes a process of abstraction from the equinumerosity relation. He realizes ultimately that *the equinumerosity classes themselves* can serve as numbers.

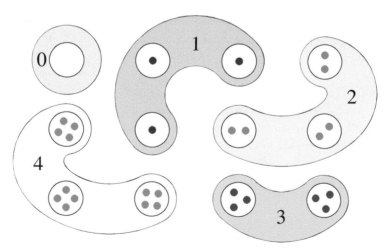

Specifically, Frege defines the cardinal "number" of a concept P to be the concept of being equinumerous with P. In other words, the cardinal number of a set is the class of all sets equinumerous with it, the equinumerosity class of the set:

$$\#P \qquad =_{\text{def}} \qquad \{\, x \mid x \simeq P \,\}.$$

For Frege, the number 2 *is* the class of all two-element collections; the number 3 *is* the class of all three-element collections, and so on. The number 0 is precisely $\{\varnothing\}$, since only the empty set has no elements. Frege's definition fulfills the Cantor-Hume principle because sets have the same equinumerosity class if and only if they are equinumerous.

Frege proceeds to develop his arithmetic theory, defining zero as the cardinal of the empty class and defining that a number *m* is the *successor* cardinal of another *n* if *m* is the cardinal of a set obtained by adding one new element to a set of cardinal *n*; the natural numbers are the cardinals generated from 0 by successive applications of the successor operation, or more precisely, the numbers having every property of 0 that also is passed from every cardinal to its successor. If one regards equinumerosity and these other notions as logical, then this approach seems ultimately to reduce arithmetic to logic.

At first independently of Frege, but later responding to and building upon Frege's work, Bertrand Russell also sought to found mathematics in logic. Russell was impressed by the Peano (1889) formalization of arithmetic, but he viewed Frege's foundation as still deeper in providing purely logical accounts of the arithmetic notions of "zero," "successor," and "natural number." In their subsequent monumental work, *Principia Mathematica*, Russell and Alfred North Whitehead wove both themes into their account of arithmetic.

Neologicism

The logicist programs of Frege and Russell came widely to be viewed as ultimately unsuccessful in their attempts to reduce mathematics to logic. Frege's foundational system was proved inconsistent by the Russell paradox (discussed in chapter 8), and Russell's system is viewed as making nonlogical existence assertions with the axiom of infinity and the axiom of choice. Nevertheless, logicism is revived by Bob Hale and Crispin Wright and others in the *neologicist* program, reconstruing Frege's approach to logicism in a way that avoids the pitfall of inconsistency, aiming once again to reduce mathematics to logic.

Specifically, Wright (1983) aims to found Frege's arithmetic directly upon the Cantor-Hume principle, establishing the Peano axioms without need for the problematic foundational system. Equinumerosity, expressed in second-order logic, is taken as logical. (Some philosophers object at this point, regarding second-order principles as inherently set-theoretic and mathematical, but let us proceed with the neologicist perspective.) The neologicists address the Julius Caesar problem by seeking a purely logical account of the number-assignment problem, building upon suggestions of Frege regarding the process of *abstraction*. Frege explains, for example, how we refer to the *direction* of a line, having initially only a criterion of same-directionality, namely, parallelism. Similarly, we might speak of the *order* of a sequence of chess moves, or the *pattern* of an arrangement of colored tiles, or perhaps the *value* of a certain item. In each case, we refer to an abstract entity with a functional expression f, defined only implicitly by a criterion of the form

$$f(a) = f(b) \quad \text{if and only if} \quad Rab,$$

with an equivalence relation R. The direction of line ℓ_1 is the same as the direction of line ℓ_2, for example, exactly when ℓ_1 is parallel to ℓ_2; one tiling pattern is the same as another if the tiles can be associated to match color and adjacency from one arrangement to the other. The Fregean *process of abstraction* takes such an abstraction principle to define an abstract function f. The main point, of course, is that the Cantor-Hume principle itself is such an abstraction principle, where we refer to the *number* of elements in a class, and two classes have the same number if and only if the classes are equinumerous. By the process of abstraction, we therefore achieve numbers as abstract objects.

Meanwhile, there are logical problems with taking every abstraction principle as legitimate, as some of them lead to inconsistency. The general comprehension principle, for example, can arise via Fregean abstraction. Thus, the principle of abstraction is said to consort with "bad company"; known-to-be-false principles arise as instances of it. This is merely a gentle way of saying, of course, that abstraction is wrong as a principle of functional definition. We do not usually say that the fallacy of denying the antecedent consorts with "bad company" because it has false instances; rather, we say that it is fallacious.

1.6 Interpreting arithmetic

Mathematicians generally seek to interpret their various theories within one another. How can we interpret arithmetic in other domains? Let us explore a few of the proposals.

Numbers as equinumerosity classes

We have just discussed how Frege, and later Russell, defined numbers as equinumerosity equivalence classes. According to this account, the number 2 is the class of all two-element sets, and the number 3 is the class of all three-element sets. These are proper classes rather than sets, which can be seen as set-theoretically problematic, but one can reduce them to sets via *Scott's trick* (due to Dana Scott) by representing each class with its set of minimal-rank instances in the set-theoretic hierarchy.

Numbers as sets

Meanwhile, there are several other interpretations of arithmetic within set theory. Ernst Zermelo represented the number zero with the empty set and then successively applied the singleton operation as a successor operation, like this:

$$
\begin{aligned}
0 &= \varnothing \\
1 &= \{\varnothing\} \\
2 &= \{1\} = \{\{\varnothing\}\} \\
3 &= \{2\} = \{\{\{\varnothing\}\}\} \\
&\vdots
\end{aligned}
$$

One then proceeds to define in purely set-theoretic terms the ordinary arithmetic structure on these numbers, including addition and multiplication.

John von Neumann proposed a different interpretation, based upon an elegant recursive idea: *every number is the set of smaller numbers.* On this conception, the empty set $\varnothing$ is the smallest number, for it has no elements and therefore, according to the slogan, there are no numbers smaller than it. Next is $\{\varnothing\}$, since the only element of this set (and hence the only number less than it) is $\varnothing$. Continuing in this way, one finds the natural numbers:

$$
\begin{aligned}
0 &= \varnothing \\
1 &= \{0\} \\
2 &= \{0,1\} \\
3 &= \{0,1,2\} \\
&\ \vdots
\end{aligned}
$$

The successor of any number n is the set $n \cup \{n\}$, because in addition to the numbers below n, one adds n itself below this new number. In the von Neumann number conception, the order relation $n < m$ on numbers is the same as the element-of relation $n \in m$, since m is precisely the set of numbers that are smaller than m.

Notice that if we unwind the definition of 3 above, we see that

$$3 = \{0,1,2\} = \{\varnothing, \{\varnothing\}, \{\varnothing, \{\varnothing\}\}\}.$$

But what a mess! And it gets worse with 4 and 5, and so on. Perhaps the idea is unnatural or complicated? Well, that way of looking at the von Neumann numbers obscures the underlying idea that every number is the set of smaller numbers, and it is this idea that generalizes so easily to the transfinite, and which smoothly enables the recursive definitions central to arithmetic. Does the fact that $1 = \frac{1}{\sqrt{\pi}} \int_{-\infty}^{\infty} e^{-x^2}\, dx$ mean that the number 1 is complicated? I do not think so; simple things can also have complex descriptions. The von Neumann ordinals are generally seen by set theorists today as a fundamental set-theoretic phenomenon, absolutely definable and rigid, forming a backbone to the cumulative hierarchy, and providing an interpretation of finite and transfinite arithmetic with grace and utility.

One should not place the Zermelo and von Neumann number conceptions on an entirely equal footing, since the Zermelo interpretation is essentially never used in set theory today except in this kind of discussion comparing interpretations, whereas the von Neumann interpretation, because of its convenient and conceptual advantages, is the de facto standard, used routinely without remark by thousands of set theorists.

Numbers as primitives

Some mathematicians and philosophers prefer to treat numbers as undefined primitives rather than interpreting them within another mathematical structure. According to this view, the number 2 is just that—a primitive mathematical object—and there is nothing more to say about what it is or its more fundamental composition. There is a collection of natural numbers 0, 1, 2, and so on, existing as urelements or irreducible primitives, and we may proceed to construct other, more elaborate mathematical structures upon them.

For example, given the natural numbers, we may construct the integers as follows: We would like to think of every integer as the difference between two natural numbers, representing the number 2 as $7 - 5$, for example, and -3 as $6 - 9$ or as $12 - 15$. The positive numbers arise as differences between a larger number and a smaller number, and negative numbers conversely. Since there are many ways to represent the same difference, we define an equivalence relation, the *same-difference* relation, on pairs of natural numbers:

$$(a, b) \sim (c, d) \quad \Longleftrightarrow \quad a + d = c + b.$$

Notice that we were able to express the same-difference idea by writing $a + d = c + b$ rather than $a - b = c - d$, using only addition instead of subtraction, which would have been a problem because the natural numbers are not closed under subtraction. Just as Frege defined numbers as equinumerosity classes, we now may define the integers simply to *be* the equivalence classes of pairs of natural numbers with respect to the same-difference relation. According to this view, the integer 2 simply *is* the set $\{(2, 0), (3, 1), (4, 2), \dots\}$, and -3 is the set $\{(0, 3), (1, 4), (2, 5), \dots\}$. We may proceed to define the operations of addition and multiplication on these new objects. For example, if $[(a, b)]$ denotes the same-difference equivalence class of the pair (a, b), then we define

$$[(a, b)] + [(c, d)] \quad = \quad [(a + c, b + d)] \qquad \text{and}$$
$$[(a, b)] \cdot [(c, d)] \quad = \quad [(ac + bd, ad + bc)].$$

If we think of the equivalence class $[(a, b)]$ as representing the integer difference $a - b$, these definitions are natural because we want to ensure in the integers that

$$(a - b) + (c - d) \quad = \quad (a + c) - (b + d) \qquad \text{and}$$
$$(a - b)(c - d) \quad = \quad (ac + bd) - (ad + bc).$$

There is a subtle issue with our method here; namely, we defined an operation on the equivalence classes $[(a, b)]$ and $[(c, d)]$, but in doing so, we referred to the particular representatives of those classes when writing $a + c$ and $b + d$. For this to succeed in defining an operation on the equivalence class, we need to show that the choice of representatives does not matter. We need to show that our operations are *well defined* with respect to the same-difference relation—that equivalent inputs give rise to equivalent outputs. If we have equivalent inputs $(a, b) \sim (a', b')$ and $(c, d) \sim (c', d')$, then we need to show that the corresponding outputs are also equivalent, meaning that $(a + c, b + d) \sim (a' + c', b' + d')$, and similarly for multiplication. And indeed our definition does have this well defined feature. Thus, we build a mathematical structure from the same-difference equivalence classes of pairs of natural numbers, and furthermore, we can prove that it exhibits all the properties that we expect in the ring of integers. In this way, we construct the integers from the natural numbers.

Similarly, we construct the rational numbers as equivalence classes of pairs of integers by the *same-ratio* relation. Namely, writing the pair (p, q) in the familiar fractional form $\frac{p}{q}$, and insisting that $q \neq 0$, we define the same-ratio equivalence relation by

$$\frac{p}{q} \equiv \frac{r}{s} \quad \Longleftrightarrow \quad ps = rq.$$

Note that we used only the multiplicative structure on the integers to do this. Next, we define addition and multiplication on these fractions:

$$\frac{p}{q} + \frac{r}{s} = \frac{ps + rq}{qs} \quad \text{and} \quad \frac{p}{q} \cdot \frac{r}{s} = \frac{pr}{qs},$$

and verify that these operations are well defined with respect to the same-ratio relation. This uses the fraction $\frac{p}{q}$ as a numeral, a mere representative of the corresponding rational number, which is the same-ratio equivalence class $\left[\frac{p}{q}\right]$. Thus, we construct the field of rational numbers from the integers, and therefore ultimately from the natural number primitives.

The program continues to the real numbers, which one may define by various means from the rational numbers, such as by the Dedekind cut construction or with equivalence classes of Cauchy sequences explained in section 1.11, and then the complex numbers as pairs of real numbers $a + bi$, as in section 1.13. And so on. Ultimately, all our familiar number systems can be constructed from the natural number primitives, in a process aptly described by the saying attributed to Leopold Kronecker:

> *"Die ganzen Zahlen hat der liebe Gott gemacht, alles andere ist Menschenwerk"*
> ("God made the integers, all the rest is the work of man.")
> Kronecker (1886), quoted in (Weber, 1893, p. 15)

Numbers as morphisms

Many mathematicians have noted the power of category theory to unify mathematical constructions and ideas from disparate mathematical areas. A construction in group theory, for example, might fulfill and be determined up to isomorphism by a universal property for a certain commutative diagram of homomorphisms, and the construction might be identical in that respect to a corresponding construction in rings, or in partial orders. Again and again, category theory has revealed that mathematicians have been undertaking essentially similar constructions in different contexts.

Because of this unifying power, mathematicians have sought to use category theory as a foundation of mathematics. In the elementary theory of the category of sets (ETCS), a category-theory-based foundational system introduced by F. William Lawvere (1963), one works in a *topos*, which is a certain kind of category having many features of the category of sets. The natural numbers in a topos are represented by what is called a *natural-numbers object*, an object $\mathbb{N}$ in the category equipped with a morphism $z : 1 \to \mathbb{N}$ that serves to pick out the number zero, where 1 is a terminal object in the category, and another morphism $s : \mathbb{N} \to \mathbb{N}$ that satisfies a certain universal free-action property, which ensures that it acts like the successor function on the natural numbers. The motivating idea is that every natural number is generated from zero by the successor operation, obtained from the composition of successive applications of s to z:

$$
\begin{aligned}
0 &= z : 1 \to \mathbb{N} \\
1 &= s \circ z : 1 \to \mathbb{N} \\
2 &= s \circ s \circ z : 1 \to \mathbb{N} \\
3 &= s \circ s \circ s \circ z : 1 \to \mathbb{N} \\
&\vdots
\end{aligned}
$$

In this conception, a natural number is simply a morphism $n : 1 \to \mathbb{N}$. The natural numbers object is unique in any topos up to isomorphism, and any such object is able to interpret arithmetic concepts into category theory.

Numbers as games

We may even interpret numbers as games. In John Conway's account, games are the fundamental notion, and one defines numbers to be certain kinds of games. Ultimately, his theory gives an account of the natural numbers, the integers, the real numbers, and the ordinals, all unified into a single number system, the *surreal* numbers. In Conway's framework, the games have two players, Left and Right, who take turns making moves. One describes a game by specifying for each player the move options; on their turn, a player selects one of those options, which in effect constitutes a new game starting from that position, with turn-of-play passing to the other player. Thus, games are hereditarily gamelike: every game is a pair of sets of games,

$$
G = \{ G_L \mid G_R \},
$$

where the games in G_L are the choices available for Left and those in G_R for Right. This idea can be taken as a foundational axiom for a recursive development of the entire theory. A player loses a game play when they have no legal move, which happens when their set of options is empty.

We may build up the universe of games from nothing, much like the cumulative hierarchy in set theory. At first, we have nothing. But then we may form the game known as *zero*,

$$0 = \{ \; | \; \},$$

which has no options for either Left or Right. This game is a loss for whichever player moves first. Having constructed this game, we may form the game known as *star*,

$$* = \{ 0 \, | \, 0 \},$$

which has the zero game as the only option for either Left or Right. This game is a win for the first player, since whichever player's turn it is will choose 0, which is then a loss for the other player. We can also form the games known as *one* and *two*:

$$1 \;\; = \;\; \{ 0 \, | \quad \}$$
$$2 \;\; = \;\; \{ 1 \, | \quad \},$$

which are wins for Left, and the games *negative one* and *negative two*:

$$-1 \;\; = \;\; \{ \quad | \, 0 \}$$
$$-2 \;\; = \;\; \{ \quad | \, {-1} \},$$

which are wins for Right. And consider the games known as *one-half* and *three-quarters*:

$$\frac{1}{2} \;\; = \;\; \{ \, 0 \, | \, 1 \, \} \qquad\qquad \frac{3}{4} \;\; = \;\; \{ \, \frac{1}{2} \, | \, 1 \, \}.$$

Can you see how to continue?

Conway proceeds to impose numberlike mathematical structure on the class of games: a game is *positive*, for example, if Left wins, regardless of who goes first, and *negative* if Right wins. The reader can verify that 1, 2, and $\frac{1}{2}$ are positive, while –1 and –2 are negative, and 0 and $*$ are each neither positive nor negative. Conway also defines a certain hereditary orderlike relation on games, guided by the idea that a game G might be greater than the games in its left set and less than the games in its right set, like a Dedekind cut in the rationals. Specifically, $G \leq H$ if and only if it never happens that $h \leq G$ for some h in the right set of H, nor $H \leq g$ for some g in the left set of G; this definition is well founded since we have reduced the question of $G \leq H$ to lower-rank instances of the order with earlier-created games. A *number* is a game G where every element of its left set stands in the $\leq$ relation to every element of its right set. When games are constructed transfinitely, this conception leads to the *surreal numbers*. Conway defines sums of games $G + H$ and products $G \times H$ and exponentials G^H and proves all the familiar arithmetic properties for his game conception of number. It is a beautiful and remarkable mathematical theory.

Junk theorems

Whenever one has provided an interpretation of one mathematical theory in another, such as interpreting arithmetic in set theory, there arises the *junk-theorem* phenomenon, unwanted facts that one can prove about the objects in the interpreted theory which arise because of their nature in the ambient theory rather than as part of the intended interpreted structure. One has junk theorems, junk properties, and even junk questions.

If one interprets arithmetic in set theory via the von Neumann ordinals, for example, then one can easily prove several strange facts:

$$2 \in 3 \qquad 5 \subseteq 12 \qquad 1 = P(0) \qquad 2 = P(1).$$

The P notation here means "power" set, the set of all subsets. Many mathematicians object to these theorems on the grounds that we do not want an interpretation of arithmetic, they stress, in which the number 2 is an element of the number 3, or in which it turns out that the number 2 is the same mathematical object as the set of all subsets of the number 1. These are "junk" theorems, in the sense that they are true of those arithmetic objects, but only by virtue of the details of this particular interpretation, the von Neumann ordinal interpretation; they would not necessarily be true of other interpretations of arithmetic in set theory. In the case of interpreting arithmetic in set theory, many of the objections one hears from mathematicians seem concentrated in the idea that numbers would be interpreted as sets at all, of any kind; many mathematicians find it strange to ask, "What are the elements of the number 7?" In an attempt to avoid this junk-theorem phenomenon, some mathematicians advocate certain alternative non-set-theoretic foundations.

To my way of thinking, the issue has little to do with set theory, for the alternative foundations exhibit their own junk. In Conway's game account of numbers, for example, it is sensible to ask, "Who wins 17?" In the arithmetic of ETCS, one may speak of the domain and codomains of the numbers 5 and 7, or form the composition of 5 with the successor operation, or ask whether the domain of 5 is the same as the domain of the real number π. This counts as junk to mathematicians who want to say that 17 is not a game or that numbers do not have domains and cannot be composed with morphisms of any kind. The junk theorem phenomenon seems inescapable; it will occur whenever one interprets one mathematical theory in another.

Interpretation of theories

Let us consider a little more carefully the process of interpretation in mathematics. One interprets an object theory T in a background theory S, as when interpreting arithmetic in set theory or in the theory of games, by providing a meaning in the language of the background theory S for the fundamental notions of the object theory T. We interpret arithmetic in games, for example, by defining which games we view as numbers and explaining how to add and multiply them. The interpretation provides a translation $\varphi \mapsto \varphi^*$

of assertions φ in the object theory T to corresponding assertions φ^* in the background theory S. The theory T is successfully interpreted in S if S proves the translations φ^* of every theorem φ proved by T. For example, Peano arithmetic (PA) can be interpreted in Zermelo-Fraenkel set theory (ZFC) via the von Neumann ordinals (and there are infinitely many other interpretations).

It would be a stronger requirement to insist on the biconditional—that S proves φ^* if and only if T proves φ—for this would mean that the background theory S was no stronger than the object theory T concerning the subject that T was about. In this case, we say that theory S is *conservative* over T for assertions in the language of T; the interpreting theory knows nothing more about the object theory than T does.

But sometimes we interpret a comparatively weak theory inside a strong universal background theory, and in this case, we might expect the background theory S to prove additional theorems about the interpreted objects, even in the language of the object theory T. For example, under the usual interpretation, ZFC proves the interpretation of arithmetic assertions not provable in PA, such as the assertion Con(PA) expressing the consistency of PA (see chapter 7). This is not a junk theorem, but rather reflects the fact that the background set theory ZFC simply has stronger arithmetic consequences than the arithmetic theory PA. Meanwhile, other nonstandard interpretations of PA in ZFC, such as those obtained by interpreting arithmetic via certain nonstandard models, are not able to prove the interpretation of Con(PA) in ZFC.

The objectionable aspect of junk theorems are not cases where the foundational theory S proves additional theorems beyond T in the language of T, but rather where it proves S-features of the interpreted T-objects. A junk theorem of arithmetic, for example, when interpreted by the von Neumann ordinals in set theory is a theorem about the set-theoretic features of these numbers, not a theorem about their arithmetic properties.

1.7   What numbers could not be

Paul Benacerraf (1965) tells the story of Ernie and Johnny, who from a young age study mathematics and set theory from first principles, with Ernie using the von Neumann interpretation of arithmetic and Johnny using the Zermelo interpretation (one wonders why Benacerraf did not use the names the other way around, so that each would be a namesake). Since these interpretations are isomorphic as arithmetic structures—there is a way to translate the numbers and arithmetic operations from one system to the other—naturally Ernie and Johnny agree on all the arithmetic assertions of their number systems. Since the sets they used to interpret the numbers are not the same, however, they will disagree about certain set-theoretic aspects of their numbers, such as the question of whether $3 \in 17$, illustrating the junk-theorem phenomenon.

Benacerraf emphasizes that whenever we interpret one structure in another foundational system such as set theory, then there will be numerous other interpretations, which disagree on their extensions, on which particular sets are the number 3, for example. Therefore, they cannot all be right, and indeed, at most one—possibly none—of the interpretations are correct, and all the others must involve nonnecessary features of numbers.

> Normally, one who identifies 3 with some particular set does so for the purpose of presenting some theory and does not claim that he has *discovered* which object 3 really is. (Benacerraf, 1965, III.B)

Pressing the point harder, Benacerraf argues that no single interpretation can be the necessarily correct interpretation of number.

> To put the point differently—and this is the crux of the matter—that any recursive sequence whatever would do suggests that what is important is not the individuality of each element but the structure which they jointly exhibit...[W]hether a particular "object"—for example, $\{\{\{\varnothing\}\}\}$—would do as a replacement for the number 3 would be pointless in the extreme, as indeed it is. "Objects" do not do the job of numbers singly; the whole system performs the job or nothing does. I therefore argue, extending the argument that led to the conclusion that numbers could not be sets, that numbers could not be objects at all; for there is no more reason to identify any individual number with any one particular object than with any other (not already known to be a number). (Benacerraf, 1965, III.C)

The epistemological problem

In a second influential paper, Benacerraf (1973) identifies an epistemological problem with the platonist approach of taking mathematical objects as being abstract. If mathematical objects exist abstractly or in an ideal platonic realm, totally separate in causal interaction from our own physical world, how do we interact with them? How can we know of them or have intuitions about them? How are we able even to refer to objects in that perfect platonic world?

W. D. Hart (1991) describes Benacerraf's argument as presenting a dilemma—a problem with two horns, one metaphysical and the other epistemological. Specifically, mathematics is a body of truths about its subject matter—numbers, functions, sets—which exist as abstract objects. And yet, precisely because they are abstract, we are causally disconnected from their realm. So how can we come to have mathematical knowledge?

> It is at least obscure how a person could have any knowledge of a subject matter that is utterly inert, and thus with which he could have no causal commerce. And yet by the first horn of the dilemma, the numbers, functions and sets have to be there for the pure mathematics of numbers, functions and sets to be true. Since these objects are very abstract, they are utterly inert. So it is at least obscure how a person could have any knowledge of the subject matter needed for the truth of the pure mathematics of numbers, functions and sets. As promised, Benacerraf's dilemma is that what seems necessary for mathematical truth also seems to make mathematical knowledge impossible. (Hart, 1991, p. 98)

Penelope Maddy (1992), grabbing the bull firmly by the horns, addresses the epistemological objection by arguing that we can gain knowledge of abstract objects through experience with concrete instantiations of them. You open the egg carton from the refrigerator and see three eggs; thus you have perceived, she argues at length, a set of eggs. Through this kind of experience and human evolution, humans have developed an internal set detector, a certain neural configuration that recognizes these set objects, much as we perceive other ordinary objects, just as a frog has a certain neural configuration, a bug detector, that enables it to perceive its next meal. By direct perception, she argues, we gain knowledge of the nature of these abstract objects, the sets we have perceived. However, see the evolution of her views expressed in Maddy (1997) and subsequent works.

Barbara Gail Montero (1999, 2020) deflates the significance of the problem of causal interaction with abstract objects by pointing out that we have long given up the idea that physical contact is required for causal interaction: consider the gravitational attraction of the Sun and the Earth, for example, or the electrical force pushing two electrons apart. But further, she argues, objections based on the difficulty of causal interactions with abstract objects lack force in light of our general failure to give an adequate account of causality of any kind. If we do not have a clear account of what it means to say that *A* causes *B* even in ordinary instances, then how convincing can an objection be based on the difficulty of causal interaction with abstract objects? And this does not even consider the difficulty of providing a sufficient account of what abstract objects are in the first instance, before causality enters the picture.

Sidestepping the issue of causal interaction, Hartry H. Field (1988) argues that the essence of the Benacerraf objection can be taken as the problem of explaining the reliability of our mathematical knowledge, in light of the observation that we would seem to have exactly the same mathematical beliefs, even if the mathematical facts had turned out to be different, and this undermines those beliefs. Justin Clarke-Doane (2017), meanwhile, argues that it is difficult to pin down exactly what the Benacerraf problem is in a way that exhibits all the features of the problem that are attributed to it.

1.8 Dedekind arithmetic

Richard Dedekind (1888) identified axioms that describe (and indeed characterize) the mathematical structure of the natural numbers. The main idea is that every natural number is ultimately generated from zero by repeated applications of the successor operation S, the operation that generates from any number the next number. From the initial number 0, we generate the successor $S0$, the successor of that $SS0$, the successor of *that SSS0*, and so on. More precisely, the axioms of *Dedekind arithmetic* assert (1) zero is not a successor; (2) the successor operation is one-to-one, meaning that $Sx = Sy$ exactly when $x = y$; and (3) every number is eventually generated from 0 by the application of the successor operation, in the sense that every collection A of numbers containing 0 and closed under the successor operation contains all the numbers. This is the second-order *induction axiom*, which is expressed in second-order logic because it quantifies over arbitrary sets of natural numbers.

Since zero is not a successor, the successor operation does not circle back to the origin; and since the successor operation is one-to-one, it never circles back from a later number to an earlier one. It turns out that one can derive all the familiar facts about natural numbers and their arithmetic just from Dedekind's principles. For example, one defines the order $n \leq m$ to hold exactly when every set A containing n and closed under successors also contains m. This is immediately seen to be reflexive and transitive, and one can easily verify that $x \leq Sx$ and that $x \leq 0$ only for $x = 0$. One can then prove (as in question 1.10) that $x \leq y \leftrightarrow Sx \leq Sy$ by induction, and using this, one can show that $\leq$ is antisymmetric ($x \leq y \leq x \implies x = y$) and linear ($x \leq y$ or $y \leq x$), both by induction on x. So it is a linear order. It is also discrete, in the sense that there are no numbers between x and Sx.

Giuseppe Peano (1889), citing Dedekind, gave an improved and simplified treatment of Dedekind's axiomatization, closer to the contemporary account described above, and furthermore used it to develop much of the standard theory of arithmetic. This work became influential, so that the axiomatization is also known as the second-order *Peano axiomatization* or as *Dedekind-Peano arithmetic*, and some of Peano's notation, notably $a \, \epsilon \, x$ for set membership, survives to this day. Bertrand Russell was particularly influenced by Peano's formalization.

Meanwhile, in current usage *Peano arithmetic* (PA) refers to the strictly first-order theory in the language with $+$, $\cdot$, 0, 1, $<$, asserting that $\mathbb{N}$ is a discretely ordered semiring with the principle of induction for any first-order definable set in this language. This weaker theory nevertheless still proves essentially all the familiar arithmetic facts.

Arithmetic categoricity

In a critical development of fundamental importance, Dedekind observed that his axioms determine a unique mathematical structure—in other words, that they are *categorical*, which means that all systems obeying these rules are isomorphic copies of one another.

Theorem 2. *Any two models of Dedekind arithmetic are isomorphic.*

Let me explain more precisely what this means and how to prove it. Suppose that you and I each have a Dedekind natural number system; you have your zero 0 and successor operation S while I have my alternative zero $\bar{0}$ and alternative successor operation $\bar{S}$, and both of our systems fulfill the full second-order induction axiom. We can immediately begin to match up our numbers, associating your 0 with my $\bar{0}$, your $S0$ with my $\bar{S}\bar{0}$, your $SS0$ with my $\bar{S}\bar{S}\bar{0}$, and so on. Thus, we begin to build a correspondence between initial segments of your system with initial segments of mine. To say that your system $\langle N, 0, S \rangle$ is *isomorphic* to my system $\langle \bar{N}, \bar{0}, \bar{S} \rangle$ is to say that there is a bijective correspondence $f : N \to \bar{N}$ defined fully on the entire domains, for which $f(0) = \bar{0}$ and $f(S(x)) = \bar{S}(f(x))$.

Let A be the collection of your numbers n for which the initial segment up to n in your system admits a partial isomorphism to an initial segment of my system. We have already observed that your 0 is in A; and it is easy to see that any partial isomorphism can be extended from n to the successor Sn, and so A is closed under successors. By induction, therefore, all your numbers are in A. Furthermore, by induction, the partial isomorphisms all agree with one another, providing the same counterparts, since they must agree on 0, and if they agree on n, then they will also agree on the successor Sn. Thus, the partial isomorphisms taken together form an isomorphism of your system to an initial segment of mine. And since the range of this isomorphism includes my $\bar{0}$ and is closed under my successor operation $\bar{S}$, our two systems are isomorphic copies of one another.

The essence of the previous argument lies in the principle of *recursive definition*, which Dedekind proved for his arithmetic. Namely, in order to define an operation on the natural numbers, it suffices to describe merely how it starts at 0 and how it proceeds under successors, defining the next value at Sn in terms of the previous value at n. Every such recursive definition, Dedekind proved, yields a unique operation defined on the whole of the natural numbers $\mathbb{N}$. Namely, by induction, every number has a recursive solution proceeding at least that far, since any solution up to n can be extended to Sn; and further, any two solutions agree on their common domain, since there can be no least point of disagreement. The common answers of all such finite partial solutions, therefore, is a total solution defined on the whole set of natural numbers, and this solution is unique because no two solutions can have a least point of disagreement.

The familiar arithmetic operations can be defined recursively—like this for addition:

$$x + 0 = x \qquad \text{and} \qquad x + Sy = S(x + y),$$

and like this for multiplication:

$$x \cdot 0 = 0 \qquad \text{and} \qquad x \cdot Sy = x \cdot y + x.$$

One may proceed to verify all the usual arithmetic properties for these operations, such as associativity, commutativity, and distributivity, proved by induction.

1.9 Mathematical induction

The principle of mathematical induction, appearing in Dedekind arithmetic, in Peano's arithmetic, and implicitly in Frege's number conception, is a core arithmetic idea used to prove essentially all the fundamental facts of arithmetic. Let us take a brief mathematical detour, exploring some uses of the principle of mathematical induction.

For example, we may use induction to prove that every fraction can be placed into lowest terms. The claim is that every fraction has a representation p/q where the numerator and denominator have no nontrivial common factor. To prove this, consider any fraction. Let p be the smallest natural number arising as the numerator of a representation p/q of the given fraction. If p and q had a nontrivial common factor, we could factor it out, thereby obtaining a representation with a smaller numerator. But p was the smallest, and so this is impossible. Therefore, p/q must already be in lowest terms.

In elementary accounts, one sometimes sees a rigid $n \rightarrow n+1$ approach to induction, the *common induction principle*, asserting that if 0 has a certain property, and if that property is transferred from every number n to $n+1$, then every natural number has that property. That is, if $P0 \land \forall n(Pn \implies Pn+1)$, then $\forall n\, Pn$. The *strong induction principle*, in contrast, asserts that if we have a property P of natural numbers, with the feature that a number has that property whenever all smaller numbers have it, then indeed every number has that property. That is, if $\forall n\,[(\forall k < n\, Pk) \implies Pn]$, then $\forall n\, Pn$. The previous argument about lowest terms, however, used induction in the form of the *least-number principle*, asserting that every nonempty set of natural numbers has a least element. We had chosen p to be least with the property that it could be used as the numerator to represent the given fraction.

In fact, all these induction principles are equivalent to one another; one can prove the strong induction principle and the least number principle from the common induction principle and vice versa. Induction is robust; there are many equivalent ways to express it.

Fundamental theorem of arithmetic

Probably the reader is familiar (perhaps from elementary school) with the idea of factoring numbers into primes. For example, we may factor the number 315 first as $3 \cdot 105$, and then as $3 \cdot 5 \cdot 21$, and then ultimately into primes as $3 \cdot 5 \cdot 3 \cdot 7$. Do we always get the same factorization? I mean "the same" here in the sense that $3^2 \cdot 5 \cdot 7$ is the same as $5 \cdot 3 \cdot 7 \cdot 3$; we are allowed to rearrange the prime factors. How do we know that the factorization of a number is always unique in this sense? We call it *the* prime factorization, but perhaps we should have been saying *a* prime factorization.

We can factor the number 2678 into primes as $2 \cdot 13 \cdot 103$, for example, and perhaps in a comparatively small case like this, we might hope to try all the candidate factors and thereby become convinced that the factorization is unique (although actually to do this exhaustively takes a lot of work). That method is infeasible, however, for larger numbers, such as 345263846453524227283884746353537. How do we know that this number has a unique prime factorization? I would like to convince you that this is a much deeper and harder question than it may seem at first. The fact of the matter is that the usual elementary-school approach to prime factorization rarely touches on the question of uniqueness. We have become familiar with the uniqueness of prime factorizations simply by observing many instances of prime factorization without ever encountering a number with more than one factorization. But this is not proof.

Meanwhile, the fact that every number has a unique prime factorization is perhaps the first deep theorem of number theory, and it is considered so important and foundational that mathematicians have given it a name:

Theorem 3 (The fundamental theorem of arithmetic). *Every positive integer can be expressed uniquely as a product of primes.*

The fundamental theorem is making two claims, an existence claim and a uniqueness claim. The existence claim is that every natural number has a prime factorization, that is, at least one. The uniqueness claim is that these factorization are unique in the sense that any two factorizations of a number have simply rearranged the prime factors. Both claims can be proved using induction.

The existence claim is a bit easier. Suppose inductively that every number less than n has a prime factorization. If n itself is prime, then this is its prime factorization. If it is not prime, then it is a product ab of smaller numbers, which by the inductive assumption have their own prime factorizations $a = p_1 p_2 \cdots p_n$ and $b = q_1 q_2 \cdots q_m$. By juxtaposing these, we achieve a prime factorization of n:

$$n = ab = p_1 p_2 \cdots p_n \cdot q_1 q_2 \cdots q_m.$$

So every number has a prime factorization.

The uniqueness claim of the fundamental theorem of arithmetic is more delicate. We use the fact that a prime number p divides a product ab exactly when it divides one of the factors a or b. One may extend this property inductively to larger products. If we now have two prime factorizations of a number $p_1 p_2 \cdots p_k = q_1 q_2 \cdots q_r$, then we can use this property to see that p_1 must divide one of the factors on the right side, and since those are prime, p_1 must be one of the factors on the right. By canceling this common factor, we reduce to a smaller instance of the theorem. By the induction assumption, therefore, the primes that remain are rearrangements of each other, and so the two original products were also rearrangements of each other.

This theorem says that every positive integer is a product of primes; but what about the number 7? It is prime, to be sure, but is it a *product* of primes? The mathematicians will say, yes, it is a product consisting of just one factor, the number 7 itself. But you might reply, that is not a product at all! Meanwhile, mathematicians have learned that our mathematical theories generally become more robust and smoother to analyze if we include such trivial or degenerate instances in our main definitions. Every square is also a rectangle, and every equilateral triangle is also isosceles. One can find arguments online about whether squares and parallelograms should count as trapezoids, but the tendency in higher mathematics is to incorporate the degenerate cases into a concept.

Some mathematicians have emphasized that the process of getting a concept right in the degenerate case often leads one to discover and formulate the right collection of ideas for handling the general case in a robust manner. When a theory needs explicitly to exclude the empty set or some other trivial case—the disease known as "emptysetitis"—it is a sign that one has not yet found the right formulation. For example, is the empty topological space connected? Is the empty graph a finite connected planar graph? If so, it would contradict Euler's theorem that $v - e + f = 2$ for finite connected planar graphs.

One sometimes hears that 7^n means that we multiply 7 by itself n times. But this is inaccurate—an instance of the fence-post error[1]—since 7^2 means 7×7, which has only one instance of multiplication, not two. The exponent n refers to the number of factors, rather than to the number of multiplications. Thus, 7^1 is just 7 itself, with no multiplication at all, but it is regarded as a product with just one factor, and $7^0 = 1$ is a product with no factors, the empty product. This is the sense in which 1 is a product of prime numbers.

It is precisely these conventions that enable us to state the fundamental theorem with elegant simplicity: every positive integer can be expressed uniquely as a product of primes. This includes the number 1 as the empty product and includes the primes themselves, each as a product with only one factor.

Infinitude of primes

Let us next consider another classical theorem, often attributed to Euclid, the claim that there are infinitely many prime numbers. The first few primes are probably familiar:

$$2 \quad 3 \quad 5 \quad 7 \quad 11 \quad 13 \quad 17 \quad 19 \quad 23 \quad 29 \quad \cdots$$

But as one gets to larger numbers, they gradually become less common. Do they eventually run out? Or are there infinitely many prime numbers?

Indeed, there are infinitely many prime numbers. Following Euclid, let us prove that every finite list of prime numbers $p_1, p_2, \ldots, p_n$ can be extended. Let N be the result of multiplying them together, and adding one:

$$N = p_1 p_2 \cdots p_n + 1.$$

Since every natural number has a prime factorization, there must be some prime number q that is a divisor of N. But none of the primes p_i is a divisor of N because each of them leaves a remainder of 1 when dividing into N. Therefore, q is a new prime number, not on the previous list. So we can always find another prime number, and therefore there must be infinitely many.

One can prove this alternatively by contradiction. Namely, supposing toward contradiction that one has a finite list of *all* the primes $p_1, \ldots, p_n$, one then multiplies them together and adds one $N = p_1 p_2 \cdots p_n + 1$. This new number is not a multiple of any p_i, and so its prime factorization must involve new primes, not on the list. This contradiction our assumption that we had all the prime numbers on the list.

Does it matter whether we give a proof by contradiction or directly? That proof seems perfectly fine. Meanwhile, some mathematicians are very careful to point out that Euclid did not give his proof by contradiction. Rather, he proved, as we did here, that every finite list of primes can be extended. With good reason, we often prefer direct proofs

---

[1] A *fence-post error* is an off-by-one error that can arise when counting things delimited by other things. A fence with two segments has three posts; if you are holding pages 12 through 15 of a manuscript, you have four pages, not three; and if you spend five days in Paris, then you sleep there only four nights.

over proofs by contradiction. Direct proofs often carry information about how to construct the mathematical objects whose existence is being asserted. But more importantly, direct proofs often paint a fuller picture of mathematical reality. When one proves an implication $p \to q$ directly, one assumes p and then derives various further consequences p_1, p_2, and so on, before ultimately concluding q. Thus, one has derived a whole context about what it is like in the p worlds. Similarly, with a proof by contraposition, one assumes $\neg q$ and then derives further implications about what it is like in the worlds without q, before finally concluding $\neg p$. But in a proof by contradiction, however, one assumes both p and $\neg q$, something that is ultimately shown not to hold in any world; this tells us nothing about any mathematical world beyond the brute fact of the implication $p \to q$ itself.

1.10 Structuralism

Now that we have established Dedekind's categoricity result for arithmetic, let us discuss the philosophical position known as *structuralism*, in one form perhaps the most widely held philosophical position amongst mathematicians today. Contemporary structuralist ideas in mathematics tend to find their roots in Dedekind's categoricity result and the other classical categoricity results characterizing the central structures of mathematics, placing enormous importance on the role of isomorphism-invariance in mathematics. Much of the philosophical treatment of structuralism, meanwhile, grows instead out of Benacerraf's influential papers (1965, 1973).

The main idea of structuralism is that it just does not matter what numbers or other mathematical objects are, taken as individuals; what matters is the structures they inhabit, taken as a whole. Numbers each play their structural roles within a number system, and other mathematical objects play structural roles in their systems. The slogan of structuralism, according to Shapiro (1996, 1997), is that "mathematics is the science of structure."

A defining structural role played by the number zero in any copy of the ring of integers $\mathbb{Z}$ is that it is the additive identity. It also happens to be the unique additive idempotent, the only number z for which $z + z = z$; it is the only additively self-inverse number $z = -z$; and it is the smaller of the two multiplicative idempotents. So in general, there can be many ways to characterize the role played by a mathematical object. In the rational numbers $\mathbb{Q}$, the number $\frac{1}{2}$ is the only number whose sum with itself is the multiplicative identity $\frac{1}{2} + \frac{1}{2} = 1$. In the real field $\mathbb{R}$, the defining structural role played by $\sqrt{2}$ is that it is positive and its square is equal to 2, which is $1 + 1$, where 1 is the multiplicative identity.

Yet, one should not confuse structural roles with definability. Tarski's theorem on real-closed fields, after all, implies that the number π, being transcendental, is not definable in the real field $\mathbb{R}$ by any property expressible in the language of ordered fields. Yet it still plays a unique structural role, determined, for example, by how it cuts the rational numbers into those below and those above; only it makes exactly that same cut.

Definability versus Leibnizian structure

Let me elaborate. An object *a* in a structure *M* is *definable* in that structure if it has a property $\varphi(a)$ in *M* that it alone has—a property expressible in terms of the structural relations of *M*, which picks out this object *a* uniquely. This is relevant for structuralism, because the definition φ specifies explicitly the structural role played by the object in that structure. A structure is *pointwise definable* if every object in it is definable in this way.

In the directed graph pictured here, for example, node 2 is the unique node that is pointed at by a node that is not pointed at by any node (namely, 2 is pointed at by 1, which is not pointed at at all); and node 4 is the unique node that is pointed at by a node, yet does not itself point at any node. In fact, every node in this graph is characterized by a property expressible in terms of the pointing-at relation, so this graph is pointwise definable. A mathematical structure is *Leibnizian*, in contrast, if any two distinct objects in the structure can be distinguished by some property. In other words, a Leibnizian structure is one that fulfills Leibniz's principle on the *identity of indiscernibles* with respect to properties expressible in the language of that structure.

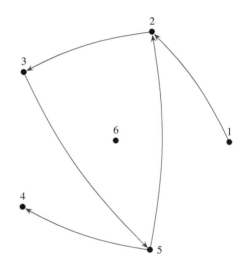

Every pointwise definable structure is Leibnizian, since the defining properties of two different objects will distinguish them. But the notions are distinct. For example, the real ordered field $\langle \mathbb{R}, +, \cdot, <, 0, 1 \rangle$ is Leibnizian, since for any two distinct real numbers $x < y$, there is a rational number $\frac{p}{q}$ between them, and *x* has the property that $x < \frac{p}{q}$, while *y* does not, and this property is expressible in the language of ordered fields. But this structure is not pointwise definable because there are only countably many possible definitions to use in this structure, but uncountably many real numbers, so they cannot all be definable.

Every Leibnizian structure must be *rigid*, meaning that it admits no nontrivial automorphism, because automorphisms are truth-preserving—any statement true of an individual in a structure will also be true of its image under any automorphism of the structure. If all individuals are discernible, therefore, then no individual can be moved to another. Because of this, we should look upon the Leibnizian property as a strong form of rigidity. These two concepts are not identical, however, because there can be rigid structures that are not Leibnizian. Every well-order structure, for example, is necessarily rigid, but when an order is sufficiently large—larger than the continuum is enough—then not every point can be characterized by its properties, simply because there aren't enough sets of formulas in the language to distinguish all the points, and so it will not be Leibnizian. Indeed, for any language $\mathcal{L}$, every sufficiently large $\mathcal{L}$-structure will fail to be Leibnizian for the same reason.

The *rigid relation principle*, introduced and investigated by Hamkins and Palumbo (2012), is the mathematical principle asserting that every set carries a rigid binary relation. This is a consequence of the well-order principle, because well-orders are rigid, but it turns out to be strictly weaker; it is an intermediate weak form of the axiom of choice, neither equivalent to the axiom of choice nor provable in ZF set theory without the axiom of choice.

Role of identity in the formal language

The nature of Leibnizian structures is often sensitive to the question of whether one has included the equality or identity relation $x = y$ in the formal language. In contemporary approaches to model theory and first-order logic, it often goes without saying that equality is included as a logical relation in every language and interpreted in every model as actual equality. This is ultimately a convention, of course, and one can easily and sensibly undertake a version of model theory without treating equality in this special manner.

When one omits equality from the language, then every model is elementarily equivalent to a model that violates the Leibnizian principle on the identity of indiscernibles. Specifically, for any model M in a language without equality, consider a new model M^* obtained by adding any number of duplicate elements for any or all of the elements of M, defining the atomic relations for the duplicates in the new structure M^* in accordance with the original structure. For example, in the rational order $\langle \mathbb{Q}, \leq \rangle$, we might consider the order $\langle \mathbb{Q}^*, \leq \rangle$ in which every rational number has two copies, each less-than-or-equal to the other and ordered with the other elements as one would expect, so that both copies of 0, for example, are less than any of the copies of positive elements, and so on. It follows inductively that any equality-free statement $\varphi(a_0, \ldots, a_n)$ true of individuals in M will also be true in M^* of any of their duplicates $\varphi(a_0^*, \ldots, a_n^*)$. In particular, the structure M^* will not be able to discern an individual from its copies, and so this structure will not be Leibnizian if indeed any nontrivial duplication occurred. Furthermore, the two models have exactly the same equality-free truth assertions; they are elementarily equivalent in that language. (Meanwhile, with equality we can distinguish the models, since $\leq$ is anti-symmetric in $\mathbb{Q}$, meaning that $\forall x, y\, (x \leq y \land y \leq x \rightarrow x = y)$, but this is not true in $\mathbb{Q}^*$.)

The philosophical point to make about this is that one can never expect a theory to give rise to the Leibnizian principle of identity of indiscernibles, unless the language includes the equality relation explicitly. In particular, nothing you say about a nonempty structure can possibly ensure that it is Leibnizian, unless equality is explicitly mentioned, since the structure has all the same equality-free assertions as the corresponding structure in which every individual has been duplicated.

Isomorphism orbit

While definability and even discernibility are sufficient for capturing the structural roles played by an object, they are not necessary, and a fuller account will arise from the notion of an isomorphism orbit. Specifically, two mathematical structures A and B are *isomorphic* if they are copies of one another, or more precisely, if there is an *isomorphism* $\pi : A \rightarrow B$ between them, a one-to-one correspondence or bijective map between the respective domains of the structures that respects the salient structural relations and operations. For example, an order isomorphism of linear orders $\langle L_1, \leq_1 \rangle$ and $\langle L_2, \leq_2 \rangle$ is a bijection $\pi : L_1 \rightarrow L_2$ between the domains of the orders that preserves the order structure from one to the other, meaning that $x \leq_1 y$ if and only if $\pi(x) \leq_2 \pi(y)$. An isomorphism of arithmetic structures $\langle Y, +, \cdot \rangle$ and $\langle Z, \oplus, \otimes \rangle$ is a bijection $\tau : Y \rightarrow Z$, for which $\tau(a+b) = \tau(a) \oplus \tau(b)$ and $\tau(a \cdot b) = \tau(a) \odot \tau(b)$, translating the structure from Y to Z. Every mathematical structural conception is accompanied by a corresponding isomorphism concept.

The isomorphism concept is intricately linked with that of formal language, which is a way of making precise exactly which mathematical structure one is considering. Whether a given one-to-one correspondence is an isomorphism depends crucially, after all, on which structural features are deemed salient. Is one considering the rational numbers only as an order, or as an ordered field? A given bijection may preserve only part of the structure.

Structural roles are respected by isomorphism, and indeed, they are *exactly* what is respected by isomorphism. An object a in structure A plays the same *structural role* as object b in structure B exactly when there is an isomorphism of A with B carrying a to b.

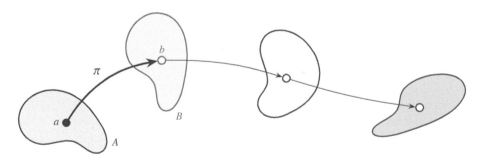

So let us consider the *isomorphism orbit* of an object in a structure, the equivalence class of the object/structure pair (a, A) with respect to the same-structural-role-as relation.

$$(a, A) \equiv (b, B) \qquad \Longleftrightarrow \qquad \exists \pi : A \cong B \quad \pi(a) = b.$$

This orbit tracks how a is copied to all its various isomorphic images in all the various structures isomorphic to A. And whether or not these objects are definable or discernible in their structures, it is precisely the objects appearing in the isomorphism orbit that play the same structural role in those structures that a plays in A.

Categoricity

A theory is *categorical* if all models of it are isomorphic. In such a case, the theory completely captures the structural essence of what it is trying to describe, characterizing that structure up to isomorphism. Dedekind, for example, had isolated fundamental principles of arithmetic and proved that they characterized the natural numbers up to isomorphism; any two models are isomorphic. In other words, he proved that his theory is categorical.

The analogous feat has long been performed for essentially all the familiar mathematical structures, and we have categorical characterizations not only of the natural numbers $\mathbb{N}$, but of the ring of integers $\mathbb{Z}$, the field of rational numbers $\mathbb{Q}$, the field of real numbers $\mathbb{R}$, the field of complex numbers $\mathbb{C}$, and many more (see sections 1.11 and 1.13). Daniel Isaacson (2011) emphasizes the role that categoricity plays in identifying particular mathematical structures. Namely, we become familiar with a structure; we find the essential features of that structure; and then we prove that those features axiomatically characterize the structure up to isomorphism. For Isaacson, this is what it means to identify a particular mathematical structure, such as the natural numbers, the integers, the real numbers, or indeed, even the set-theoretic universe.

Categoricity is central to structuralism because it shows that the essence of our familiar mathematical domains, including $\mathbb{N}$, $\mathbb{Z}$, $\mathbb{Q}$, $\mathbb{R}$, $\mathbb{C}$, and so on, are determined by structural features that we can identify and express. Indeed, how else could we ever pick out a definite mathematical structure, except by identifying a categorical theory that is true in it? Because of categoricity, we need not set up a standard canonical copy of the natural numbers, like the iron rod kept in Paris that defined the standard meter; rather, we can investigate independently whether any given structure exhibits the right structural features by investigating whether it fulfills the categorical characterization.

Invariably, for deep reasons, these categorical characterizations use second-order logic, meaning that their fundamental axioms involve quantification not only over the individuals of the domain, but also over arbitrary sets of individuals or relations on the domain. Dedekind's arithmetic, for example, asserts the induction axiom for arbitrary sets of natural numbers, and we shall see in section 1.11 that Dedekind's completeness axiom for the real numbers involves quantifying over arbitrary bounded sets of rational numbers.

The deep reasons are that a purely first-order theory, one whose axioms involve quantification only over the domain of individuals in a structure rather than over arbitrary sets of individuals, can never provide a categorical characterization of an infinite structure. This is a consequence of the Löwenheim-Skolem theorem, which shows that every first-order theory that is true in an infinite model is also true in models of diverse infinite cardinalities, which therefore cannot be isomorphic. Meanwhile, the Löwenheim-Skolem theorem does not apply to second-order theories, and it should be no surprise to find second-order axioms in the categorical theories characterizing our familiar mathematical structures.

Some philosophers object that we cannot identify or secure the definiteness of our fundamental mathematical structures by means of second-order categoricity characterizations. Rather, we only do so relative to a set-theoretic background, and these backgrounds are not absolute. The proposal is that we know what we mean by the structure of the natural numbers—it is a definite structure—because Dedekind arithmetic characterizes this structure uniquely up to isomorphism. The objection is that Dedekind arithmetic relies fundamentally on the concept of arbitrary collections of numbers, a concept that is itself less secure and definite than the natural-number concept with which we are concerned. If we had doubts about the definiteness of the natural numbers, how can we assuaged by an argument relying on the comparatively indefinite concept of "arbitrary collection"? Which collections are there? The categoricity argument takes place in a set-theoretic realm, whose own definite nature would need to be established in order to use it to establish definiteness for the natural numbers.

Structuralism in mathematical practice

I should like to contrast several forms of structuralism, distinguishing first a form of structuralism that is widespread amongst mathematicians—a form which I call *structuralism in practice*. Structuralism in practice involves an imperative about how to undertake mathematics, a view about which kinds of mathematical investigations will be fruitful. According to the structuralist-in-practice, mathematics is about mathematical structure, and mathematicians should state and prove only structuralist theorems in a structuralist manner. All of one's mathematical concepts should be invariant under isomorphisms.

The structuralist imperative. *For mathematical insight, investigate mathematical structure, the relations among entities in a mathematical system, and consider mathematical concepts only as being invariant under isomorphism. Therefore, do not concern yourself with the substance of individual mathematical objects, for this is mathematically fruitless as structure arises with any kind of object.*

According to the structuralist imperative, it would be mathematically misguided to state theorems about particular instantiations of mathematical structure; a theorem involving the real numbers, for example, should do so in a way that it becomes invariant under isomorphism; one should be able to replace the real numbers with any other complete ordered field, while preserving the truth of the theorem. The structuralist-in-practice dismisses questions about the "true nature" of mathematical objects—about what numbers "actually" are—as mathematically irrelevant.

It would accord with the structuralist imperative, for example, to prove a theorem about the countable random graph if one takes this term to refer to any countable graph with the finite pattern property, a feature that characterizes these graphs up to isomorphism. One might prove, for example, that the countable random graph is homogeneous or that it has diameter two or an infinite chromatic number; what this really means is that all such countable graphs with the finite pattern property have these features. Because the hypothesis is invariant under isomorphism, we do not care which particular copy of the countable random graph we are using, and this is the heart of structuralism.

It would be antistructuralist, in contrast, to state those theorems specifically only about the Rado graph if this is taken to refer to the specific graph relation on the natural numbers with an edge between n and m, where $n < m$, if the nth binary digit of m is 1; this graph happens to exhibit the finite pattern property, and therefore it is a specific instance of the countable random graph (once one has fixed a copy of the natural numbers).

Notice that it would be fine, logically, to prove something about the countable random graph by proving it specifically about the Rado graph, even using specific features of the Rado graph, provided that the theorem itself was invariant under graph isomorphisms, for then it would transfer from the Rado graph to all copies of the countable random graph. In this sense, it might seem that structuralism in practice requires only that one's theorems are structuralist, that is, that they are properly invariant under isomorphism.

Yet, the structuralist imperative recommends against that style of proof, against using nonstructural details of one's specific interpretation instances, even when they might seem convenient. The reason is that those details never lie at the core of the mathematical phenomenon—they are always a distraction—precisely because they cannot matter for the structuralist conclusion of the theorem. If you have a proof of an isomorphism-invariant theorem that uses incidental details of a specific instantiation, then the mathematical structuralist will say that you have a poor argument; you have missed the essential point; and your argument will not produce mathematical insight. In this sense, the structuralist imperative is a recommendation about mathematical efficacy. Namely, by undertaking structural arguments, we will stay closer on the trail of mathematical truth.

Meanwhile, it would be structuralist to prove theorems about the Rado graph if one was concerned with some of the extra structure inherent in that particular presentation of this graph. For example, the edge relation of the Rado graph is a computably decidable relation on the natural numbers, but not every copy of the countable random graph on the natural numbers is computably decidable. In this case, one is not really studying the countable random graph, with only its graph structure, but rather one is studying computable model theory, looking at the computational complexity of presentations of this graph. This is again structural, but with different additional structure beyond pure graph theory.

Consider a structuralist analogy with computer programming. A structuralist approach to programming treats its data objects only with respect to the structural features explicitly in the defining data types; this way of programming is often portable to other operating systems and implementations of the programming language. It would be antistructuralist for a program to use details of how a particular piece of data is represented on a particular system, to peek into the internal coding of data in the machine; this sneaky way of getting at the data might work fine at first, but it often causes portability issues because the methods can fail when one changes to a different machine, which might represent the data differently "under the hood," so to speak.

The structuralist imperative tends to lead one away from the junk-theorem phenomenon, for junk is often particularly objected to, specifically because it is antistructuralist (but consider question 1.19 as a counterpoint). The structuralist-in-practice dismisses the Julius Caesar objection as misguided, for it does not matter what the cardinal numbers are, so long as they obey the Cantor-Hume principle, and so we do not care if any of them are Julius Caesar or not. Indeed, we can easily define an interpretation of number in which Julius Caesar is the number 17, or not, and everything in our theory will work fine either way. There is nothing mathematical at stake in it.

Some mathematicians have emphasized that in some of the category-theoretic foundations, such as in ETCS, the formal languages provided for these systems are *necessarily* invariant under isomorphisms. When working in those languages, therefore, one cannot help but follow the structuralist imperative.

Eliminative structuralism

What I am calling *structuralism in practice* is closely related to *eliminative structuralism* (also called *post-rem structuralism*), defended by Benacerraf (1965)—namely, the view that mathematical structure is simply that which is instantiated in particular structures. Eliminative structuralism includes the nominalist claim that there is no abstract object that is the mathematical structure itself, beyond representations in particular instantiations. There is no abstract thing that is "the number 3"—any object can play that role in a suitable structure—and so talk of numbers and other particular mathematical objects is merely instrumental. Shapiro (1996) says, "Accordingly, numerals are not genuine singular terms, but are disguised bound variables." A reference to the number 3 really means: *in the model of Dedekind arithmetic at hand, the successor of the successor of the successor of zero.*

One difference between structuralism in practice and eliminative structuralism, however, is that the structuralist-in-practice drops the elimination claim, the nominalist ontological claim that abstract structural objects do not exist; rather, the structuralist-in-practice simply follows the structuralist imperative to pursue isomorphism-invariant mathematics, whether abstract structural objects exist or not. And since elimination is not part of the view, it would seem wrong to call it eliminative structuralism.

Another form of eliminative structuralism is the view of *modal structuralism*, also called *in-re structuralism* and defended by Geoffrey Hellman, according to which assertions about mathematical objects are to be understood modally as necessary claims about their possible instantiations. According to this view, mathematical structures are ontologically dependent on the systems that exemplify them. In extreme form, one might hope to reduce mathematical structure ultimately to concrete physical systems. And there is also a relative form of eliminative structuralism, which reduces structure to sets. Namely, according to *set-theoretic reductionism*, an extreme form of set-theoretic foundationalism, structure does not exist apart from its set-theoretic realizations, such as by means of the isomorphism orbit. This is different from merely using set theory as a foundation of mathematics, since one can propose set theory as a foundation simply by interpreting mathematical structure within set theory, without insisting that there is no structure outside of set theory.

Abstract structuralism

There is something a little puzzling about the structuralist mathematician, who follows the structuralist imperative, yet happily refers to *the* natural numbers and *the* number 17 and *the* real number π. If we only care about the natural numbers up to isomorphism, after all, then there is not any longer a unique mathematical object or structure corresponding to these terms, and so what is meant by "the" here? It would seem that the structuralist-in-practice should be referring instead to *a* natural numbers or *a* number 17. But mathematicians do not generally talk that way, even when they are structuralist. Structuralism seems to face a serious problem with singular reference.

To be sure, most mathematicians, when pressed about their singular references, articulate the structuralist-in-practice view. They say that it does not matter to them which copy of the natural numbers we use, and that by 17, they mean to refer to the object playing that role in whichever version we currently have. Thus, they have dutifully inserted Shapiro's disguised quantifiers.

But some philosophers aim for a more robust solution to the problem of singular reference. According to *abstract structuralism*, also known as *ante-rem structuralism*, defended by Stewart Shapiro (1997), Michael Resnik (1988), and others, the objects of mathematics, including numbers, functions, and sets, are inherently structural; they exist as purely structural abstract objects, positions within a structure, locations in a pattern of arrangement that might be realized in diverse instantiations. The quarterback is a position on an American football team, the role played by the person who calls the play, receives the hike and makes the passes. Each individual quarterback is a person rather than a position—a person who plays the role of quarterback on a particular team. Similarly, the natural number 3 is the third successor "location" in the natural number structure—the position that any particular copy of 3 fills in any particular instantiation of the natural numbers. On this view, mathematical structure exists independently of the particular systems instantiating that structure.

Abstract structuralism provides a direct account of the reference of singular terms in mathematics, explaining how *the* number 3 and *the* natural numbers can sensibly refer, even from a structuralist perspective, to the purely structural object or the abstract structural role played by these entities. One undertakes the Fregean process of abstraction from the same-structural-role relation, whose equivalence classes are precisely the isomorphism orbits. Every isomorphism orbit leads one by abstraction to a corresponding abstract structural role. Shapiro also argues, much like Maddy in the egg carton argument (mentioned on page 21), that abstract structuralism offers a solution to Benacerraf's epistemological problem: we gain access to finite instances of abstract structures and then proceed by abstraction to the structure itself.

The abstract structuralist is providing a structuralist account of mathematics by realizing mathematical objects as purely structural. Yet, the structuralist-in-practice will say that this form of abstract structuralism is not structuralist at all—it violates the structuralist imperative—precisely because it is concerned with what the mathematical objects are, even if the answer it provides is that they are purely structural. According to the structuralist-in-practice, such concerns are misguided; they never elucidate a mathematical phenomenon and are irrelevant for mathematical insight. The structuralist-in-practice will happily consider any classification invariant of the isomorphism orbit relation, such as the orbit itself (akin to Frege taking numbers as equinumerosity classes), without the need for a purely structural abstract object representing the structural role.

Yet, the abstract structuralist may reply, "Fine, we do not pursue abstract structuralism for mathematical insight, but rather as a philosophical investigation in mathematical ontology, aiming to understand what mathematical structure really is." The abstract structuralist seeks to identify and elucidate the essential nature of mathematical objects, a philosophical effort rather than a mathematical one. The abstract structuralist seeks to give an account of *what structure is*—the thing that the mathematicians take to be so fundamental.

1.11   What is a real number?

Let us consider the real continuum. The classical discovery of irrational numbers reveals gaps in the rational number line: the place where $\sqrt{2}$ would be, if it were rational, is a hole in the rational line. Thus, the rational numbers are seen to be incomplete. One seeks to complete them, to fill these holes, forming the real number line $\mathbb{R}$.

Dedekind cuts

Dedekind (1901, I.§3) observed how every real number cuts the line in two and found in that idea a principle expressing the essence of continuity:

> If all points of the straight line fall into two classes such that every point of the first class lies to the left of every point of the second class, then there exists one and only one point which produces this division of all points into two classes, this severing of the straight line into two portions.

For Dedekind, the real numbers are what we now call *Dedekind complete*: every cut is filled. In the rational line, some cuts, determined by a rational number, are already filled; but other cuts correspond to holes in the rational line, not yet filled. For any such unfilled cut, Dedekind proposes that we may imagine or "create" an irrational number in thought precisely to fill it. In this way, we shall realize the real number line as the Dedekind-completion of the rational number line.

> And if we knew for certain that space was discontinuous there would be nothing to prevent us, in case we so desired, from filling up its gaps, in thought, and thus making it continuous; this filling up would consist in a creation of new point-individuals and would have to be effected in accordance with the above principle. (Dedekind, 1901, I.§3)

Theft and honest toil

Russell explains how one may undertake this creation process explicitly, building the real numbers as a mathematical structure that fulfills Dedekind's completeness property. In a truly elegant construction, he forms the Dedekind-completion of the rational line from the set of all Dedekind cuts themselves, viewing each cut as constituting a single new point. A *Dedekind cut* in the rational line is a bounded nonempty initial segment of the rationals with no largest element. The no-largest-element requirement ensures that rational numbers are represented uniquely, since otherwise we could place the rational limit point on either side, forming two distinct cuts where only one is wanted.

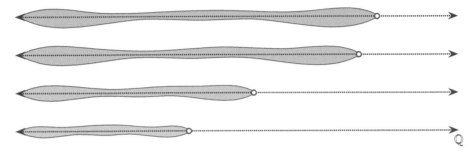

Toiling under Russell's direction, we form the set of all Dedekind cuts, viewing each as a single new point; we define the natural order upon them (it is just the inclusion order $\subseteq$ on the cuts); we prove easily that this new order is Dedekind complete (the union of any bounded set of cuts is itself a cut that is the least upper bound); we extend the field operations from the rational numbers to the set of cuts, defining what it means to add any two cuts or multiply them; and we prove that these operations and order make the set of cuts into an ordered field. Thus, we construct the real numbers as Dedekind cuts, forming a Dedekind-complete ordered field.

Although one can imagine the Dedekind cuts as arising from the real numbers, to do so is precisely the inverse of the intended logic. Rather, we seek to use the cuts to define *what the real numbers are*, or at least what they could be. According to this account, a real number is a Dedekind cut in the rational numbers. Indeed, Russell (1919, p. 71) makes a withering criticism of Dedekind's axiomatic approach, by which one postulates that the real numbers are Dedekind complete.

> The method of "postulating" what we want has many advantages; they are the same as the advantages of theft over honest toil. Let us leave them to others and proceed with our honest toil.

Russell's "honest toil" was to construct the real numbers via Dedekind cuts as described here, proving that the resulting structure is Dedekind complete, rather than merely postulating that the real numbers are already Dedekind complete.

Cauchy real numbers

An alternative continuity concept is provided by Augustin-Louis Cauchy, who was inspired by the idea that every real number is the limit of the various rational sequences converging to it. A sequence of real numbers is a *Cauchy sequence* if the points in the sequence become eventually as close as desired to one another. The continuity of the real numbers is expressed by *Cauchy completeness*, the property that every Cauchy sequence converges to a limiting real number.

The rational line, of course, is not Cauchy complete, for there are Cauchy sequences converging to where $\sqrt{2}$ would be, but there is no rational number there as the limit of this sequence. And it is similar for the other irrational numbers. But one may form the Cauchy completion of the rational numbers by considering all possible Cauchy sequences on them. Two such sequences are equivalent if their members eventually become as close to each other as desired, and we may form the real numbers as the collection of equivalence classes of Cauchy sequences. This admits a natural ordered field structure; it is Archimedean, which means that the finite sums $1 + 1 + \cdots + 1$ are unbounded; and it is Cauchy complete. According to this account, a real number is an equivalence class of Cauchy sequences.

Real numbers as geometric continuum

The ancient Greek conception of the continuum, in contrast, persisting through the ages, was inherently geometric: a real quantity is a length, area, or volume. According to the classical *number line* conception of number, advanced by René Descartes and taught in primary schools everywhere, a real number is a point on the number line, specified by an origin and a unit length.

One problem with this conception is that if a real number x is a length, a product xy is an area and xyz is a volume, then how are we to conceptualize expressions such as $x + xy + xyz$, which mix quantities of different dimensions? Can we add a length to an area or a volume? Quadratic expressions $ax^2 + bx + c$ become problematic. We all agree that $2 \times 3 = 6$, but if 2×3 is an area and 6 is a length, what does that mean? One can solve this, of course, by considering $6 = 1 \times 6$ also as an area, and similarly in higher dimensions.

Another problem is that one wants to express the idea that the geometric continuum itself is continuous. Dedekind does this by means of his cuts, asserting that every cut is filled.

Categoricity for the real numbers

David Hilbert identified the essential natural properties that we want to be true of the real numbers, which, it turns out, characterize the field of real numbers up to isomorphism. He specified that the real numbers are a maximal Archimedean ordered field—maximal in the sense that they cannot be extended to a larger Archimedean ordered field. This is a form of completeness precisely because the Dedekind completion of any Archimedean ordered field remains Archimedean. In modern terminology, the definition amounts to saying that the real numbers are a complete ordered field, using the Dedekind formulation of completeness, since the least-upper-bound property implies the Archimedean property, as I shall argue in the proof of theorem 4. Indeed, one can prove that the real numbers construed as Dedekind cuts or as equivalence classes of Cauchy sequence are complete ordered fields and thereby fulfill Hilbert's axioms.

What is a real number? What is the number π, for example, as a mathematical object? Is it a certain Dedekind cut? Is it an equivalence class of Cauchy sequences? A geometric length? Something else? The structuralist answers these questions by pointing to the categoricity result, asserting that there is only one complete ordered field up to isomorphism. The real numbers are a complete ordered field, and all such fields are isomorphic.

Theorem 4 (Huntington, 1903). *All complete ordered fields are isomorphic.*

Proof sketch. I claim first that every complete ordered field R is Archimedean, which means that there is no number in R that is larger than every finite sum $1 + 1 + \cdots + 1$. If there were such a number, then by completeness, there would have to be a least such upper bound b to these sums; but $b - 1$ would also be an upper bound, which is a contradiction. So every complete ordered field is Archimedean.

Suppose now that we have two complete ordered fields, $\mathbb{R}_0$ and $\mathbb{R}_1$. We form their respective prime subfields, that is, their copies of the rational numbers $\mathbb{Q}_0$ and $\mathbb{Q}_1$, by computing inside them all the finite quotients $\pm(1 + 1 + \cdots + 1)/(1 + \cdots + 1)$. This fractional representation itself provides an isomorphism of $\mathbb{Q}_0$ with $\mathbb{Q}_1$, indicated below with blue dots and arrows:

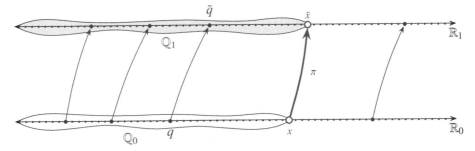

Next, by the Archimedean property, every number $x \in \mathbb{R}_0$ determines a cut in $\mathbb{Q}_0$, indicated in yellow, and since $\mathbb{R}_1$ is complete, there is a counterpart $\bar{x} \in \mathbb{R}_1$ filling the corresponding cut in $\mathbb{Q}_1$, indicated in violet. Thus, we have defined a map $\pi : x \mapsto \bar{x}$ from $\mathbb{R}_0$ to $\mathbb{R}_1$. This map is surjective, since every $y \in \mathbb{R}_1$ determines a cut in $\mathbb{Q}_1$, and by the completeness of $\mathbb{R}_0$, there is an $x \in \mathbb{R}_0$ filling the corresponding cut. Finally, the map π is a field isomorphism since it is the continuous extension to $\mathbb{R}_0$ of the isomorphism of $\mathbb{Q}_0$ with $\mathbb{Q}_1$. $\square$

This result characterizes the structure of the real numbers in the same way that Dedekind's arithmetic axioms characterize the structure of the natural numbers. We found the fundamental principles for the real continuum and proved that they determine that structure up to isomorphism. Thus, we have identified the real numbers $\mathbb{R}$ as a mathematical structure.

According to structuralism, it is not necessary to pick out a particular complete ordered field, an official copy, since the only mathematically relevant property of the real numbers—the only property that should be used in a mathematical argument—is that they constitute a complete ordered field. Individual real numbers are comprehended by their roles within such a structure. As we noted earlier, $\sqrt{2}$ is the unique object in whichever complete ordered field you have selected, that happens to be positive and to square to the number 2 in that field, where 2 is the number $1 + 1$ in that field, where 1 is the unique multiplicative identity in that field. This is the structural role played by $\sqrt{2}$. In any complete ordered field, every rational number is algebraically definable, and every real number is characterized by the cut that it makes in the rational numbers. It follows that the real field $\mathbb{R}$ is a Leibnizian structure: any two real numbers are discernible in the language of fields.

Kevin Buzzard (2019) highlights the question of structuralism by inquiring: How do we know that a theorem proved using the Dedekind-cut real numbers is also true of Cauchy-completion real numbers? Why is it that a mathematical assertion involving the real numbers, even if only incidentally, when true for the Dedekind real numbers, must also be true when one uses the Cauchy real numbers? There would seem to be an enormous pile of mathematical material that would have to be proved isomorphism-invariant in order to make such sweeping general conclusions, and has this work actually been done?

As a community, mathematicians in current practice are highly structuralist, often insistently so. It would be considered very strange to prove a theorem involving the real numbers by insisting that one is using the Dedekind real numbers as opposed to the Cauchy real numbers, for example, unless one were specifically concerned with the additional structural features that those formulations of the real numbers involved. Because of this widespread practice, the vast bulk of mathematical development is indeed structuralist and follows the structuralist imperative with regard to the central mathematical structures, including the natural numbers, the integers, the real numbers, and so on. Therefore, the enormous pile of isomorphism-invariant material that Buzzard claims must be undertaken has in fact already been undertaken—this is the standard practice of normal mathematics—and this is why we may deduce that mathematical statements involving the real numbers do not depend on which particular copy of the real numbers we are using.

Categoricity for the real continuum

We characterized the real numbers above by the fact that they form a complete ordered field and all such fields are isomorphic. This categoricity argument, therefore, uses the algebraic properties of the real numbers—the fact that they form an ordered field—as an essential part of the characterization. It turns out, however, that we may also characterize the real number line purely by its order properties rather than its algebraic properties as an ordered field.

Specifically, let us consider the real number line $\langle \mathbb{R}, < \rangle$ with only the order structure. Viewed as a topological space under the order topology, this is known as the *real continuum*. What can we say about it? Well, this is a linear order, of course, since any two real numbers are comparable; and it is endless, meaning that there is neither a largest nor a smallest real number; and it is densely ordered, meaning that between any two real numbers, there is another; and it is Dedekind complete, meaning that every cut in the real number line is filled. Thus, the real number line is an endless, complete, dense linear order. This is not yet enough to characterize the real number line, however, for there are other such orders, not isomorphic to the real number line, such as the endless long line, for those who are familiar with it.

One additional property, however, will enable a characterization to go through. It suffices to add that the real number line has a countable dense subset. That is, there is a subset $\mathbb{Q} \subseteq \mathbb{R}$, the set of rational numbers, which is (1) dense in the real number line in the sense that every nontrivial interval (a, b) of real numbers contains elements of the subset; and (2) the subset $\mathbb{Q}$ is countable, as discussed in chapter 3. All these properties together now determine the real numbers order up to isomorphism.

Theorem 5. *Any two complete endless dense linear orders with countable dense sets are order isomorphic.*

The essence of the proof is Cantor's back-and-forth method, which shows that the two countable dense suborders are isomorphic, for indeed, Cantor shows that any two countable, endless, dense linear orders are isomorphic. One can then lift this isomorphism from the suborders to the whole order using the completeness of the orders, just as we did in the case of the complete ordered fields.

I mention this categoricity result in part because a fascinating foundational issue arises when one considers a small variation of it, weakening the countable-dense-set requirement to what is called the *countable chain condition*, which asserts that every family of nonoverlapping intervals is countable. The real number line has the countable chain condition, since if we have a family of nonoverlapping intervals in the real number line, then inside each one we may pick a rational number, and we will never pick the same rational number twice since they do not overlap; so the family must have been countable.

Question 6. Are all complete endless dense linear orders with the countable chain condition isomorphic?

In other words, do these properties characterize the real number line? The answer is subtle and fascinating. A positive answer is known as *Suslin's hypothesis*, while a counterexample order, a complete endless dense linear order with the countable chain condition, but which is not isomorphic to the real number line, is called a *Suslin line*. The extremely interesting situation is that this question cannot be settled using the standard axioms of set theory; Suslin's hypothesis is an *independent* statement, neither provable nor refutable from the axioms of set theory; it is consistent either way. In particular, the question of whether the real numbers are categorically characterized by the property of Suslin's hypothesis is itself independent, neither provable nor refutable from the axioms of set theory. We shall discuss the independence phenomenon at length in chapter 8.

1.12   Transcendental numbers

The *real algebraic* numbers are the real numbers that solve a nontrivial polynomial equation over the integers, such as $x^2 = 2$, which is solved by $x = \pm\sqrt{2}$, or the equation $x^2 - x - 1 = 0$, which is solved by the golden ratio number, $\frac{1+\sqrt{5}}{2}$, and its conjugate, $\frac{1-\sqrt{5}}{2}$. One can form algebraic numbers with radical expressions such as $\sqrt[5]{5 + \sqrt{10}}$, but it is a deep theorem that not all algebraic numbers can be described in such a manner by radicals. A real number is *transcendental* if is not algebraic. The proof that transcendental numbers exist, due to Joseph Liouville, can be seen as an higher analogue of the Pythagorean proof that irrational numbers exist, a continuation of a thread of reasoning picked up again after two thousand years.

Liouville defined a class of *lacunic* numbers, whose expansions had increasingly large lakes of zeros. He considered, for example, the following specific number:

$$0.110001000000000000000001000\ldots$$

now known as the *Liouville constant*, which has the digit 1 in decimal places given by the factorials 1, 2, 6, 24, and so on. For any lacunic number, if the lakes grow sufficiently in size, then the isolated 1s between the lakes eventually do not interfere with one another in rational polynomial expressions. In particular, these parts of the number cannot conspire to cancel each other out in algebraic combinations. Precisely for this reason, such numbers cannot solve a nontrivial algebraic equation over the rational numbers, and therefore they are transcendental. The lakes of zeros in the Liouville constant have a very interesting consequence, investigated by the number theorists—namely, that this number has rational approximations with surprising accuracy: 0.11 is correct to 5 decimal places, and 0.110001 is correct to 23 places. The rational approximation up to the closer shore of each lake extends in accuracy across it to the opposite shore.

We now know that e and π and many other famous real numbers are transcendental, and Cantor's arguments, discussed in chapter 3, show in a precise sense that most real numbers are transcendental.

The transcendence game

Consider the following mathematical game. You start with a number x, either real or complex, and you are allowed to form a next number from the current number either by adding 1, subtracting 1, multiplying by a nonzero integer, or multiplying by x. You win if you can produce the number zero by this process. For example, if we start with $x = \sqrt{3} + 2$, then we could subtract 1 four times to form $\sqrt{3} - 2$, and then multiply by x to form $(\sqrt{3} + 2)(\sqrt{3} - 2) = 3 - 4 = -1$, and then add 1 to form 0. We won!

The amazing fact is that the numbers x for which you can win in this game are exactly the algebraic numbers. If a person ever wins the game with a number x, then by unwrapping the winning sequence of moves and expressing them as an equation, one finds an integer polynomial of which x is a root, thereby showing that x is algebraic. In the example given here, we in effect computed $(x - 4)x + 1 = 0$. Conversely, if x is algebraic, then it is the solution of a polynomial equation over the integers, such as $2x^4 + 5x^3 - 7x^2 + 3x + 1 = 0$. In this case, we may simply factor x out of successive pairs of terms like this:

$$
\begin{aligned}
2x^4 + 5x^3 - 7x^2 + 3x + 1 &= (2x^3 + 5x^2 - 7x + 3)x + 1 \\
&= ((2x^2 + 5x - 7)x + 3)x + 1 \\
&= (((2x + 5)x - 7)x + 3)x + 1.
\end{aligned}
$$

From the resulting expression, starting at the x of the inner $2x + 5$, we can successively read off how to form 0 from x by the following process: we multiply x by 2, add 5 (which means add 1 five times), multiply by x, subtract 7, multiply by x, add 3, multiply by x, and add 1. This forms the original polynomial, which is 0, and so we have won the game. This factoring process will show that one can win the transcendence with any algebraic number, and so the winning numbers are exactly the algebraic numbers.

In particular, one will not be able to win the game with e or π, as these numbers are transcendental. Meanwhile, it is a fun exercise to prove that a number is rational if and only if you can win the game without using the multiply-by-x rule.

1.13   Complex numbers

Given the real numbers, one proceeds to the complex numbers $\mathbb{C}$, motivated by the enticing, yet perhaps terrifying, possibility that the imaginary unit $i = \sqrt{-1}$ exists as an actual number. One wants to consider complex numbers of the form $a + bi$, where a and b are real. What is a complex number?

We can easily construct a natural presentation of the complex field by means of the complex plane. Specifically, since complex numbers are to have the form $a + bi$ for real numbers a and b, let us think of the number $a + bi$ as represented by the pair (a, b), a point in the plane. We may define the usual coordinate-wise addition operation $(a, b) + (c, d) = (a + c, b + d)$, but we use a certain strange multiplication operation, defined by $(a, b) \cdot (c, d) = (ac - bd, ad + bc)$. This definition exactly implements the distributive consequences of $i^2 = -1$ in the product

$$
\begin{aligned}
(a + bi)(c + di) &= ac + adi + bci + bdi^2 \\
&= (ac - bd) + (ad + bc)i.
\end{aligned}
$$

By identifying $a + bi$ with the point (a, b), we have realized the complex numbers as points in the plane, now called the *complex plane*, and we understand the complex arithmetic simply as certain elementary operations defined on those points. So there is no terrifying mystery after all in the complex numbers. We can construct them from the real numbers.

Platonism for complex numbers

We formed the complex numbers by extending the real numbers with a solution of the equation $z^2 + 1 = 0$, using the solution $z = i$. It is a remarkable fact, known as the *fundamental theorem of algebra*, that by adding this one solution, the complex numbers thereby become algebraically closed: every nontrivial polynomial equation over the complex numbers has a full solution there. The complex numbers are the algebraic closure of the real numbers.

What is a complex number, actually? Imagine that at your death, you are astonished to meet God in Heaven, who informs you, "Yes, you were completely right about platonism for the real numbers—there they are!" He points across the way, and there you see them— the real numbers, each of them a perfect platonic ideal of its kind. You find the numbers π, e, $\sqrt{2}$, each where you expect them. "But," God continues, "you were wrong about platonism for the complex numbers; you have to construct them from the real numbers as pairs (a, b), with the parentheses and comma and everything."

The situation is absurd because we expect that our mathematical ontology should treat similar kinds of mathematical objects similarly; if the real numbers are real, then the complex numbers should be as well. Is this a slippery slope for platonism? Once one admits a real existence for one kind of mathematical object or structure, why not more? Soon, we shall find ourselves in plenitudinous platonism. But what of mathematical structures that might differ in their level of abstraction? Some philosophers propose that the natural numbers have a more definite existence than real numbers, and that while platonism is correct for the natural numbers, it is not for the real numbers. Is the allegory relevant for them? Perhaps not; perhaps the difference in abstraction makes natural numbers and real numbers fundamentally different in kind, unlike the real and complex numbers of the allegory.

Categoricity for the complex field

Like the real numbers, the complex field $\mathbb{C}$ admits a categorical characterization. Namely, the complex field is uniquely characterized up to isomorphism as being the algebraic closure of a complete ordered subfield, the real numbers. Any two fields like that are isomorphic, since their real subfields will be isomorphic and this isomorphism will extend to the algebraic closure. The complex field is also characterized up to isomorphism as the unique algebraically closed field of characteristic 0 having size continuum. One can express the concept of having size continuum in second-order logic by asserting that there is a bijection with a subset that is a real continuum.

Thus, each of our familiar number systems—the natural numbers $\mathbb{N}$, the integer ring $\mathbb{Z}$, the rational field $\mathbb{Q}$, the real field $\mathbb{R}$ and the complex field $\mathbb{C}$—admit categorical characterizations. Precisely because of these categorical accounts, we are able to pick out and refer to these structures simply by describing what is true in them, rather than by having to exhibit sample instances of the structures. We don't need to present a particular constructed copy of $\mathbb{C}$ to refer to the complex field, because we can just say that we are referring to the algebraically closed field of characteristic 0 having size continuum. To my way of thinking, this ability to refer to structures without needing to exhibit particular instances is a core part of the deep connection between categoricity results in mathematics and the philosophy of structuralism.

A complex challenge for structuralism?

Although one conventionally describes i as "the square root of negative one," nevertheless one might reply to this, "Which one?" in light of the fact that $-i$ also is such a root:

$$(-i)^2 = (-1 \cdot i)^2 = (-1)^2 i^2 = i^2 = -1.$$

Indeed, the complex numbers admit an automorphism, an isomorphism of themselves with themselves, induced by swapping i with $-i$—namely, complex conjugation:

$$z = a + bi \qquad \mapsto \qquad \bar{z} = a - bi.$$

The conjugation map preserves the field structure, since $\overline{y+z} = \bar{y} + \bar{z}$ and $\overline{y \cdot z} = \bar{y} \cdot \bar{z}$, and therefore the complex field is not a rigid mathematical structure. Since conjugation swaps i and $-i$, it follows that i can have no structural property in the complex numbers that $-i$ does not also have. So there can be no principled, structuralist reason to pick one of them over the other. Is this a problem for structuralism? It does seem to be a problem for singular terms, since how do we know that the i appearing in my calculations this week is the same number as what will appear in your calculations next week? Perhaps my i is your $-i$, and we do not even realize it.

If one wants to understand mathematical objects as abstract positions within a structure, as in abstract structuralism, then one must grapple with the fact that in light of the conjugation automorphism, the numbers i and $-i$ play exactly the same roles in this structure (see Shapiro, 2012). The numbers i and $-i$ have the same isomorphism orbit with respect to the complex field, and so in this sense, although distinct, they each play exactly the same structural role in $\mathbb{C}$. This would seem to undermine the idea that mathematical objects *are* abstract positions in a structure, since we want to regard these as distinct complex numbers.

Furthermore, there is nothing special about the numbers i and $-i$ in this argument. For example, the numbers $\sqrt{2}$ and $-\sqrt{2}$ also happen to play the same structural role in the complex field $\mathbb{C}$, because there is an automorphism of $\mathbb{C}$ that swaps them (although one uses the axiom of choice to prove this). Contrast this with the real field $\mathbb{R}$, where $\sqrt{2}$ and $-\sqrt{2}$ are of course discernible, since one is positive and the other is negative, and the order is definable from the field operations in $\mathbb{R}$ via $x \leq y \leftrightarrow \exists u \; x + u^2 = y$. It follows that the real number field is not definable in the complex field by any assertion in the language of fields. In fact, there is an enormous diversity of automorphisms of the complex field; one may move $\sqrt[3]{2}$, for example, to one of the nonreal cube roots of 2, such as $\sqrt[3]{2}(\sqrt{3}i - 1)/2$.

Therefore, the numbers $\sqrt[3]{2}$ and $\sqrt[3]{2}(\sqrt{3}i - 1)/2$ are indiscernible in the complex field—there is no property expressible in the language of fields that will distinguish them. Indeed, except for the rational numbers, every single complex number is part of a nontrivial orbit of automorphic copies, from which it cannot be distinguished in the field structure. So the same issue as with i and $-i$ occurs with every irrational complex number. For this reason, it is problematic to try to identify complex numbers with the abstract positions or roles that the numbers play in the complex field.

Meanwhile, one recovers the uniqueness of the structural roles simply by augmenting the complex numbers with additional natural structure. Specifically, once we augment the complex field $\mathbb{C}$ with the standard operators for the real and imaginary parts:

$$\text{Re}(a + bi) = a \qquad \text{Im}(a + bi) = b,$$

then the expanded structure $\langle \mathbb{C}, +, \cdot, \text{Re}, \text{Im} \rangle$ becomes *rigid*, meaning that it has no nontrivial automorphisms. Thus, every complex number plays a unique structural role in this new structure, which is Leibnizian. This additional structure is implicit in the complex plane conception of the complex numbers, which is part of why the number i appears fine as a singular term—it refers to the point $(0, 1)$ in the complex plane—whereas $-i$ refers to $(0, -1)$. The complex plane is not merely a field, for it carries along its coordinate information by means of the real-part and imaginary-part operators, making it rigid. In the complex plane, every complex number plays a different role.

Structure as reduct of rigid structure

This situation, where a natural nonrigid structure is made rigid by natural additional structure, is extremely common in mathematics. Examples abound. The additive group of integers $\langle \mathbb{Z}, + \rangle$ admits an automorphism by negation, but is made rigid with the multiplicative structure $\langle \mathbb{Z}, +, \cdot \rangle$ or the order structure $\langle \mathbb{Z}, +, < \rangle$. The rational order $\langle \mathbb{Q}, < \rangle$ is a countable endless dense linear order and therefore highly nonrigid—every point looks the same, and indeed any two finite sets of the same size are order-automorphic—but becomes rigid with the field structure $\langle \mathbb{Q}, +, \cdot, < \rangle$. The complex field $\langle \mathbb{C}, +, \cdot \rangle$ has $2^{2^{\aleph_0}}$ many automorphisms, but is made rigid by incorporating the coordinate structure. Every group G with at least three elements is nonrigid, but elements are distinguished when the group is given a particular presentation, such as by means of generators and relations or as permutations of a particular set.

The pattern is that a particular nonrigid structure is realized as a reduct substructure of another structure that is rigid, thereby resolving the problem of reference, since we may refer to the objects of the nonrigid structure by reference to their roles in the expanded structure. I claim that this pattern is inherent in mathematical practice. The reason is that precisely because of the reference problem, it is difficult for us ever actually to present or specify a nonrigid structure, except by presenting it as a reduct substructure of a structure in which the objects are individuated. How else are we coherently to specify the structure on those objects in the first place? We don't start with a naked copy of $\mathbb{C}$ and then seek to impose an orientation on it that will enable us to resolve i from $-i$. Rather, we proceed oppositely: instances of mathematical structures are obtained from richer contexts where the objects were already individuated. We might build a copy of $\mathbb{C}$ from ordered pairs of real numbers, for example, where we can discern $(0, 1)$ from $(0, -1)$ and therefore i from $-i$ in this particular copy of $\mathbb{C}$. Every particular copy of $\mathbb{C}$ and indeed every particular mathematical structure of any kind arises similarly from a context in which the objects are individuated.

When using ZFC set theory as a foundation of mathematics, this philosophical observation becomes a mathematical theorem: every set is a reduct substructure of a rigid structure, a structure in which every individual plays a distinct structural role. The reason is that every set is a subset of a transitive set, and every transitive set is rigid with respect to the $\in$ membership relation. Indeed, the set-theoretic universe $\langle V, \in \rangle$ as a whole is rigid—any two objects in the set-theoretic universe are therefore distinguishable as sets and play different set-theoretic roles (see the argument on page 286). Therefore, every mathematical structure that can be realized in set theory at all can be realized as a reduct substructure of a rigid structure. We can refer to distinct individuals in the original structure by the distinct structural roles they play in the larger context.

1.14 Contemporary type theory

Mathematics overflows with mathematical objects of many types: we have natural numbers, integers, real numbers, sets of numbers, lines in the plane, functions on the real numbers, topological spaces, sequences and series, and on and on and on. All these different types of mathematical objects have different natures and features, and each type follows certain rules. It makes sense to apply a function on the real numbers only to input objects that are themselves real numbers. Type theory is concerned with keeping track of all these different types and their interaction rules, and also with the formal methods by which we might form new types. At bottom, type theory recognizes the strongly typed nature of mathematics and takes the idea of a "type" seriously. The most general forms of type theory have the capability to represent essentially arbitrary mathematical structure within the types, and they can be used as a foundational theory, much like set theory. One forms new types from old types in a vast, transfinite hierarchy, just as in the cumulative hierarchy of set theory.

Similar type-changing issues are often made completely explicit in programming languages, where one sometimes must change the type of a number from `int` to `float`, for example, to apply operations that are available in floating-point arithmetic, but not in integer arithmetic. The strongly typed nature of some programming languages allows one to test for typing errors in a program, simply by checking that functional and relational expressions are used in accordance with their types.

Meanwhile, elementary mathematical usage is often relaxed about typing. Every natural number is commonly also taken as an integer; every integer is also rational; every rational number is also real, and every real number is also a complex number:

$$\mathbb{N} \ \subseteq \ \mathbb{Z} \ \subseteq \ \mathbb{Q} \ \subseteq \ \mathbb{R} \ \subseteq \ \mathbb{C}$$

From this view, we may think of the number 57 either as a natural number or as a real number, or even as a complex number, and it is always the same number.

The type-theoretic account of mathematics, however, encourages us to take more care with these type distinctions. According to the type theorists, the natural number 57 is not the same thing as the integer 57 or the real number 57, although we have canonical embeddings or translations from each of these number systems to the next. We can translate an integer to the real numbers by applying the type-changing translation $\tau_{\mathbb{Z},\mathbb{R}}$. For example, if 57 is an integer, then $\tau_{\mathbb{Z},\mathbb{Q}}(57)$ would be the corresponding rational number 57, and $\tau_{\mathbb{Z},\mathbb{R}}(57)$ would be the real number 57.

Should we regard our conventional arithmetic calculations as ridden with invisible type-adjusting morphisms? According to type theory, we will find more robust mathematical theories and insight if we respect mathematical typing. Meanwhile, ordinary mathematical usage commonly ignores these types and takes the natural number 57 to be identical to the integer 57. The question is whether there would be a mathematical or philosophical advantage in making a robust distinction.

1.15   More numbers

The complex numbers are extended by the *quaternions*, a system with a noncommutative multiplication, where numbers have the form $a + bi + cj + dk$, where i, j, and k are the quaternion units. The quaternions are in turn extended by the *octonians*, which have a nonassociative multiplication and eight dimensions; they are mentioned in the popular science press for some connections with dark matter and theoretical physics. *Infinitesimal* numbers will be discussed in chapter 2, along with the associated *hyperreal* numbers, and the transfinite *ordinal* numbers will be covered in chapter 3. Meanwhile, the fascinating *surreal* numbers, introduced by John Conway, combine features of the real numbers, the hyperreal numbers, and the ordinals in one elegant system, a vast universal number system that unifies all numbers, great and small.

1.16   What is a philosophy for?

What is the purpose of a philosophy of mathematics? What is at stake in these philosophical disputes? Some philosophers might answer by saying that we are striving to get at the essential nature of mathematics, to understand what mathematics is about and why it works as it does. We hope to give an account of the nature of mathematical objects and mathematical claims, and how we come to know them. We have so many fundamental philosophical questions about the nature of mathematics, and we seek to answer them, or at least clarify the philosophical issues and lay out the terrain of possible answers. Other philosophers, in contrast, object that there often seems to be little mathematically at stake in the philosophical debates, and yet one might seek to harness philosophical insight for mathematical ends. Different philosophies of mathematics might guide one toward more meaningful mathematical questions or more fruitful avenues of mathematical investigation. A successful philosophy of mathematics, therefore, might be one with a greater mathematical payoff—one that leads to more valuable mathematical insights.

To describe one instance, consider the dispute over set-theoretic pluralism, currently raging in the philosophy of set theory (discussed in chapter 8). At issue is the nature of set-theoretic existence, and in particular the question of whether there is just one set-theoretic universe, an intended model that the theory is about, or whether there are multiple parallel conceptions of set theory giving rise to a plural multiverse of set theory.

In the appendix of my initial contribution to this debate, I wrote:

> The mathematician's measure of a philosophical position may be the value of the mathematics to which it leads. Thus, the philosophical debate here may be a proxy for: where should set theory go? Which mathematical questions should it consider? Hamkins (2012)

The multiverse view might lead one to consider how the various models of set theory interact, whereas the universe view might lead one to try to discover fundamental features of the one true set-theoretic universe. These are quite different mathematical projects, and so one's philosophy of set theory will direct one to different mathematical efforts.

1.17   Finally, what is a number?

I am truly very sorry, but we do not know, fully, what numbers are. The problem of mathematical ontology is not solved. Although we have an abundance of philosophical perspectives on the matter, to my way of thinking, none of them is entirely satisfactory. We do not really know ultimately what it means to make existence assertions in mathematics. Our work is cut out for us in the philosophy of mathematics.

> When you finish a PhD in mathematics, they take you to a special room and explain that i isn't the only imaginary number—turns out that ALL the numbers are imaginary, even the ones that are real. Kate Owens (2020)

Questions for further thought

1.1   What is the difference, if any, between a number and a numeral?

1.2   The numerator of $\frac{4}{6}$ is not the same as the numerator of $\frac{2}{3}$, yet we usually say that these numbers are identical. How can one thing have a property that another does not, and yet they are the same thing?

1.3   Explain why the set $\{\varnothing\}$ is a number in all three of the early twentieth-century number interpretations—the Frege conception, the Zermelo conception, and the von Neumann conception. All three of them agree that $\{\varnothing\}$ is a number, but do they agree on which number it is? Are there any other sets that are numbers with respect to all three number conceptions?

1.4   Provide formal logical expressions asserting that "there is exactly one P" and "there are exactly three Ps" and "there are no Ps." Can you express "there are infinitely many Ps"?

1.5   According to the Fregean equinumerosity-class conception of number, how big is the number 2? How big is 1? How big is zero? How does the answer change if we use the Zermelo numbers or the von Neumann numbers?

1.6   Provide several classification invariants for the has-read-all-the-same-books-as relation and for the has-attended-exactly-the-same-theater-performances-as relation, considered as equivalence relations on the set of all people.

1.7   Argue that for any equivalence relation, the equivalence classes form a classification invariant for that relation. Does this provide a Fregean abstraction solution for any concept of equivalence?

1.8   What happens with Frege's conception of the cardinal successor when it is applied to infinite sets?

1.9   How well does Frege's development of arithmetic and his number concept accord with the philosophy of structuralism?

1.10   Verify the development of $\leq$ on the natural numbers as described in this chapter, proving from first principles in Dedekind's axiomatization that $\leq$ is a discrete linear order.

1.11   Prove that $\sqrt{3}$ is irrational. What is the most general version of this theorem that you can prove?

1.12   At the heart of structuralism seems to be the idea that when a structure A is isomorphic to a structure B, then A and B share all the same mathematical features. To what extent and why is this true?

1.13   How many characterizations of 0 can you find in the ring of integers? And 1?

1.14   Does every integer play a different role in the ring of integers? What if one considers only the additive structure $\langle \mathbb{Z}, + \rangle$?

1.15   Can a structuralist mathematician coherently refer to *the* natural numbers or *the* real numbers?

1.16   To what extent, if any, does structuralism require mathematical objects to be definable? Can a mathematical object play a unique structural role in a structure without being formally definable in that structure?

1.17   How does structuralism handle nonrigid mathematical structures, that is, structures with a nontrivial automorphism? Is the particular kind of structuralism relevant here?

1.18   Consider the football/baseball sports metaphor for structuralism, by which the structural role of a mathematical object is analogous to a position on a sports team, such as the quarterback in

(American) football or the first baseman in baseball. Which common sports exhibit the same challenge in terms of player roles that the complex numbers pose for the form of structuralism by which mathematical objects *are* the structural roles that the object play in the structure?

1.19 Are structuralists afflicted by the junk theorem phenomenon? For example, if one interprets arithmetic in set theory but follows the structuralist imperative, then cannot one still prove set-theoretic facts about the numbers, however they might be interpreted in set theory? For example, the numbers will all have power sets and distinct ones will have different elements, and so on, since these are facts true of any set. Is this junk?

1.20 Show that the number $\sqrt[5]{7 + \sqrt[3]{10}}$ is real algebraic by finding a polynomial, with integer coefficients, having it as a root. (Hint: Think about $x^5 - 7$, and then do something more.)

1.21 Can you win the transcendence game starting with $x = \sqrt{5} + \sqrt{7}$?

1.22 Summarize Dedekind's categoricity argument for the natural numbers. How is it relevant for structuralism? What role do categoricity arguments play in mathematics for a structuralist?

1.23 Explain the meaning of Leopold Kronecker's assertion, "*Die ganzen Zahlen hat der liebe Gott gemacht, alles andere ist Menschenwerk*" ("God made the integers, all the rest is the work of man").

1.24 It is a model-theoretic fact that different models of set theory can have different nonisomorphic instantiations of the natural numbers. That is, there can be models of set theory M_1 and M_2, such that $\mathbb{N}^{M_1}$ and $\mathbb{N}^{M_2}$, the respective finite von Neumann ordinals of these models, are not isomorphic. What is the significance of this, if any, for Dedekind's categoricity characterization of the natural numbers?

1.25 In light of the fact that both i and $-i$ are square roots of -1, how are we to know that what you call i is the same as what I call i? If someone says, "the square root of minus one," suppose that I respond, which one? Does this cause a problem of reference for uses of the complex numbers?

1.26 Discuss the issue of categoricity with respect to the natural numbers, the integers, the rational numbers, the real numbers, and the complex numbers. Does the structuralist have a uniform treatment of these cases?

Further reading

Paul Benacerraf (1965, 1973). The influential articles in which Benacerraf makes his fundamental objection. The title of the first, "What numbers could not be," is of course a play on the title of Dedekind's (1888) article, "Was sind und was sollen die Zahlen? (What are numbers and what should they be?)."

W. D. Hart (1991). Hart presents Benacerraf's argument in the form of a dilemma: the abstract nature of the mathematical objects that seem necessary in order to make the claims of mathematics true is precisely what interferes with our having knowledge of mathematical truths.

Joanne E. Snow (2003). An enlightening account of the views of and exchanges between Dedekind, Cantor, Heine, and Weierstrass on the concept of the real numbers and the continuum.

Stewart Shapiro (2012). A delightful discussion of the imaginary unit i as a singular term.

Dirk Greimann (2003). An elucidating account of Frege's Julius Caesar problem, emphasizing a subtle structuralist aspect of Frege's views and arguing that he took the Julius Caesar problem beyond the antistructuralist metaphysical issue.

Gideon Rosen (2018). A Stanford Encyclopedia of Philosophy (SEP) entry providing an excellent summary of views on the nature of abstract objects.

Kevin C. Klement (2019). An informative account of Bertrand Russell's brand of logicism, placing it in the broad historical development of logicism.

Button and Walsh (2018). A sweeping account of philosophically relevant issues and aspects of model theory, such as the informative discussion (p. 157) of the possible philosophical attitudes and responses to Dedekind's arithmetic categoricity result.

D. E. Knuth (1974). An unusual text, a rare piece of mathematical fiction, telling the tale of how one might undertake the imaginative development of the theory of surreal numbers from their mysterious first principles through the dialogue of a young couple marooned on an island. Reviewer John W. Dawson, Jr. (MR0472278) says, "it is entertaining, and its plot does incorporate that sine qua non of contemporary literary success, an episode of illicit sex."

Doron Zeilberger (2007). A talk on nothing, entitled "   " (but without the quotation marks), which also describes the beginnings of Conway's game account of numbers.

Credits

The terminology of Hume's principle was introduced by George Boolos. The concept of a Leibnizian model was introduced by Ali Enayat (2004). Theorem 4 appears essentially in Eward V. Huntington (1903), following up on an unpublished 1902 axiomatization of the positive real numbers. Oswald Veblen (1904) defined the notion of categoricity for a system of axioms, citing John Dewey for the terminology and Huntington as a principal example (although Dedekind's categoricity result for arithmetic would have been an earlier example). Thanks to Philip Ehrlich (2020) and Umberto Cavasinni (2020) for helping to reveal this historical information in answer to the question I had posed about it on MathOverflow, Hamkins (2020d). I heard a version of the God-in-Heaven platonism allegory from Peter Koellner, who related it to me in my office at New York University during a visit he made to New York. The view of the surreal numbers under the slogan "All numbers great and small" is due to Philip Ehrlich (2012), in his monograph of the same title. I learned a version of the transcendence game from Mark Dominus (2016).

2 Rigor

Abstract. Let us consider the problem of mathematical rigor in the development of the calculus. Informal continuity concepts and the use of infinitesimals ultimately gave way to the epsilon-delta limit concept, which secured a more rigorous foundation while also enlarging our conceptual vocabulary, enabling us to express more refined notions, such as uniform continuity, equicontinuity, and uniform convergence. Nonstandard analysis resurrected the infinitesimals on a more secure foundation, providing a parallel development of the subject. Meanwhile, increasing abstraction emerged in the function concept, which we shall illustrate with the Devil's staircase, space-filling curves, and the Conway base 13 function. Finally, does the indispensability of mathematics for science ground mathematical truth? Fictionalism puts this in question.

The subject of calculus, developed independently by Newton and Leibniz—accompanied by a century of raging dispute over the proportion of credit due to each of them—is concerned fundamentally with the idea of instantaneous rates of change, particularly for functions on the real numbers. The class of *continuous* functions becomes centrally important. But what does it mean, precisely, for a function to be continuous?

2.1 Continuity

In natural language, one distinguishes between a continuous process, which is one that proceeds in an unbroken manner, without interruption, and a process performed continually, which means without ending. For example, you might hope that your salary payments arrive continually in the coming decades, but it is not necessary that they do so continuously, since it will be fine to receive a separate payment each month.

Informal account of continuity

In mathematics, a continuous function is one whose graph is unbroken in a sense. What is this sense? Perhaps an informal continuity concept suffices at first. In my junior high school days, my teachers would say:

A function is continuous if you can draw it without lifting your pencil.

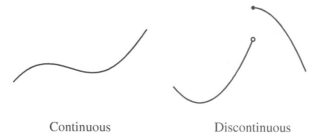

Continuous   Discontinuous

This statement conveys the idea that a jump discontinuity, as occurs in the middle of the red function, should disqualify a function from being continuous, because you would have to lift your pencil to jump across the gap. But surely it is inadequate to support precise mathematical argument; and it is inaccurate in the fine detail, if one considers that the lead of a pencil has a certain nonzero width, and furthermore, the material coming off the pencil consists of discrete atoms of carbon. So let us take it as a suggestive metaphor rather than as a mathematical definition.

In introductory calculus classes, one often hears a slightly better statement:

> A function f is continuous at c if the closer and closer x gets to c, the closer and closer $f(x)$ gets to $f(c)$.

This is an improvement, by suggesting that one can obtain increasingly good approximations to the value of a continuous function at a point by applying the function to increasingly good approximations to the input; we view $f(x)$ as an approximation of $f(c)$ when x is an approximation of c.

But the definition is still much too vague. Worse, it is not quite right. Suppose you were to walk through Central Park in New York, proceeding uptown from Central Park South. As you walk north, you would be getting closer and closer (if only slightly) to the North Pole. But you would not be getting *close* to the North Pole, since you would remain thousands of miles away from it. The problem with the definition above is that it does not distinguish between the idea of getting closer and closer to a quantity and the idea of getting close to it. How close does it get? How close is close enough? The definition does not tell us.

To make the same point differently, consider the elevation function of a hiker as she descends a gently sloped plateau toward its edge, where a dangerous cliff abruptly drops. As she approaches the cliff's edge, she is getting closer and closer to the edge, and her elevation gets closer and closer to the elevation of the valley floor (since she is descending, even if only slightly), but the elevation function is not continuous, since there is an abrupt vertical drop at the cliff's edge, a jump discontinuity, if she were to proceed that far.

The definition of continuity

A more correct definition should therefore not speak of "closer and closer," but should rather concern itself with exactly how close x is to c and how close $f(x)$ is to $f(c)$, and how these degrees of closeness are related. This is precisely what the epsilon-delta definition of continuity achieves.

Definition 7. A function f on the real numbers is *continuous* at the point c if for every positive $\epsilon > 0$, there is $\delta > 0$ such that whenever x is within δ of c, then $f(x)$ is within ϵ of $f(c)$. The function overall is said to be continuous if it is continuous at every point.

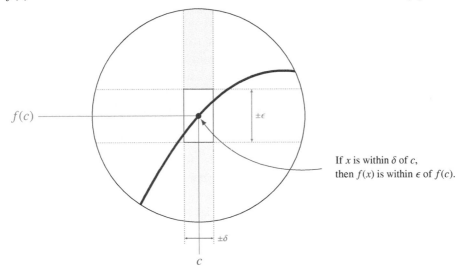

If x is within δ of c,
then $f(x)$ is within ϵ of $f(c)$.

In the figure, the y values within ϵ of $f(c)$ are precisely those within the horizontal green band, while the x values within δ of c are those within the vertical red band. The diagram therefore illustrates a successful choice of δ, a situation (and you will explain precisely why in question 2.3) where every x within δ of c has $f(x)$ within ϵ of $f(c)$.

We may express the continuity of f at c succinctly in symbols as follows:

$$\forall \epsilon > 0 \; \exists \delta > 0 \; \forall x \, [x \text{ within } \delta \text{ of } c \implies f(x) \text{ within } \epsilon \text{ of } f(c)].$$

The quantifier symbol $\forall$ is to be read as "for all" and the symbol $\exists$ as "there exists." So a function f is continuous at a point c, according to what this says, if for any desired degree of accuracy ϵ, there is a degree of closeness δ, such that any x that is that close to c will have $f(x)$ within the desired accuracy of $f(c)$. In short, you can ensure that $f(x)$ is as close to $f(c)$ as you want by insisting that x is sufficiently close to c.

The continuity game

Consider the continuity game. In this game, your role is to defend the continuity of the function f. The challenger presents you with a value c and an $\epsilon > 0$, and you must reply with a $\delta > 0$. The challenger can then pick any x within δ of c, and you win, provided that $f(x)$ is indeed within ϵ of $f(c)$. In question 2.1, you will show that the function f is continuous if and only if you have a winning strategy in this game.

Many assertions in mathematics have such alternating $\forall \exists$ quantifiers, and these can always be given the *strategic* reading for the game, in which the challenger plays instances of the universal $\forall$ quantifier and the defender answers with witnesses for $\exists$. Mathematically complex assertions often have many alternations of quantifiers, and these correspond to longer games. Perhaps because human evolution took place in a challenging environment of essentially game-theoretic human choices, with consequences for strategic failures, we seem to have an innate capacity for the strategic reasoning underlying these complex, alternating-quantifier mathematical assertions. I find it remarkable how we can leverage our human experience in this way for mathematical insight.

Estimation in analysis

Let us illustrate the epsilon-delta definition in application by proving that the sum of two continuous functions is continuous. Suppose that f and g are both continuous at a point c, and consider the function $f + g$, whose value at c is $f(c) + g(c)$. To see that this function is continuous at c, we shall make what is known as an $\epsilon/2$ *argument*. Consider any $\epsilon > 0$. Thus also, $\epsilon/2 > 0$. Since f is continuous, there is $\delta_1 > 0$ such that any x within δ_1 of c has $f(x)$ within $\epsilon/2$ of $f(c)$. Similarly, since g is continuous, there is $\delta_2 > 0$ such that any x within δ_2 of c has $g(x)$ within $\epsilon/2$ of $g(c)$. Let δ be the smaller of δ_1 and δ_2. If x is within δ of c, therefore, then it is both within δ_1 of c and within δ_2 of c. Consequently, $f(x)$ is within $\epsilon/2$ of $f(c)$ and $g(x)$ is within $\epsilon/2$ of $g(c)$. It follows that $f(x) + g(x)$ is within ϵ of $f(c) + g(c)$, since each term has error less than $\epsilon/2$, and thus we have won this instance of the continuity game. So $f + g$ is continuous, as desired.

This argument illustrates the method of "estimation," so central to the subject of real analysis, by which one delimits the total error of a quantity by breaking it into pieces that are analyzed separately. One finds not only $\epsilon/2$ arguments, but also $\epsilon/3$ arguments, breaking the quantity into three pieces, and $\epsilon/2^n$ arguments, splitting into infinitely many pieces, with the error in the nth piece at most $\epsilon/2^n$. The point is that because

$$\sum_{n=1}^{\infty} \frac{\epsilon}{2^n} = \epsilon,$$

one thereby bounds the total error by ϵ, as desired. Let me emphasize that this use of the word *estimate* does not mean that one is somehow guessing how much the difference can be, but rather one is proving absolute bounds on how large the error could possibly be.

The analyst's attitude can be expressed by the slogan:

> In algebra, it is *equal, equal, equal.*
> But in analysis, it is *less-than-or-equal, less-than-or-equal, less-than-or-equal.*

In algebra, one often proceeds in a sequence of equations, aiming to solve them exactly, while in analysis, one proceeds in a sequence of inequalities, mounting an error estimate by showing that the error is less than one thing, which is less than another, and so on, until ultimately, it is shown to be less than the target ϵ, as desired. The exact value of the error is irrelevant; the point, rather, is that it can be made as small as desired.

Limits

The epsilon-delta idea enables a general formalization of the limit concept. Namely, one defines that the limit of $f(x)$ as x approaches c is the quantity L, written like this:

$$\lim_{x \to c} f(x) = L,$$

if for any $\epsilon > 0$, there is $\delta > 0$ such that any x within δ of c (but ignoring $x = c$) has $f(x)$ within ϵ of L.

But why all the fuss? Do limits and continuity require such an overly precise and detailed treatment? Why can't we get by with a more natural, intuitive account? Indeed, mathematicians proceeded with an informal, intuitive account for a century and half after Newton and Leibniz. The epsilon-delta conception of limits and continuity was a long time coming, achieving its modern form with Weierstrass and with earlier use by Cauchy and Bolzano, after earlier informal notions involving infinitesimals, which are infinitely small quantities. Let us compare that usage with our modern method.

2.2 Instantaneous change

In calculus, we seek to understand the idea of an *instantaneous* rate of change. Drop a steel ball from a great tower; the ball begins to fall, with increasing rapidity as gravity pulls it downward, until it strikes the pavement—watch out! If the height is great, then the ball might reach terminal velocity, occurring when the force of gravity is balanced by the force of air friction. But until that time, the ball was *accelerating*, with its velocity constantly increasing. The situation is fundamentally different from the case of a train traveling along a track at a constant speed, a speed we can calculate by solving the equation:

$$\text{distance} = \text{rate} \times \text{time}.$$

For the steel ball, however, if we measure the total elapsed time of the fall and the total distance, the resulting rate will be merely an average velocity. The average rate over an interval, even a very small one, does not quite seem fully to capture the idea of an instantaneous rate of change.

Infinitesimals

Early practitioners of calculus solved this issue with infinitesimals. Consider the function $f(x) = x^2$. What is the instantaneous rate of change of f at a point x? To find out, we consider how things change on an infinitesimally small interval—the interval from x to $x + \delta$ for some infinitesimal quantity δ. The function accordingly changes from $f(x)$ to $f(x + \delta)$, and so the average rate of change over this tiny interval is

$$\frac{f(x + \delta) - f(x)}{\delta} = \frac{(x + \delta)^2 - x^2}{\delta}$$
$$= (x^2 + 2x\delta + \delta^2 - x^2)/\delta$$
$$= (2x\delta + \delta^2)/\delta$$
$$= 2x + \delta.$$

Since δ is infinitesimal, this result $2x + \delta$ is infinitely close to $2x$, and so we conclude that the instantaneous change in the function is $2x$. In other words, the *derivative* of x^2 is $2x$.

Do you see what we did there? Like Newton and Leibniz, we introduced the infinitesimal quantity δ, and it appeared in the final result $2x + \delta$, but in that final step, just like them, we said that δ did not matter anymore and could be treated as zero. But we could not have treated it as zero initially, since then our rate calculation would have been $\frac{0}{0}$, which makes no sense.

What exactly is an infinitesimal number? If an infinitesimal number is just a very tiny but nonzero number, then we would be wrong to cast it out of the calculation at the end, and also we would not be getting the *instantaneous* rate of change in f, but rather only the *average* rate of change over an interval, even if it was a very tiny interval. If, in contrast, an infinitesimal number is not just a very tiny number, but rather infinitely tiny, then this would be a totally new kind of mathematical quantity, and we would seem to need a much more thorough account of its mathematical properties and how the infinitesimals interact with the real numbers in calculation. In the previous calculation, for example, we were multiplying these infinitesimal numbers by real numbers, and in other contexts, we would be applying exponential and trigonometric functions to such expressions. To have a coherent theory, we would seem to need an account of why this is sensible.

Bishop Berkeley (1734) makes a withering criticism of the foundations of calculus.

> And what are these same evanescent Increments? They are neither finite Quantities nor Quantities infinitely small, nor yet nothing. May we not call them the ghosts of departed quantities?

Berkeley's mocking point is that essentially similar-seeming reasoning can be used to establish nonsensical mathematical assertions, which we know are wrong. For example, if δ is vanishingly small, then 2δ and 3δ differ by a vanishingly small quantity. If we now treat that difference as zero, then $2\delta = 3\delta$, from which we may conclude $2 = 3$, which is absurd. Why should we consider the earlier treatment of infinitesimals as valid if we are not also willing to accept this conclusion? It seems not to be clear enough when we may legitimately treat an infinitesimal quantity as zero and when we may not, and the early foundations of calculus begin to seem problematic, even if practitioners were able to avoid erroneous conclusions in practice. The foundations of calculus become lawless.

Modern definition of the derivative

The epsilon-delta limit conception addresses these objections and establishes a new, sound foundation for calculus, paving the way for the mature theory of real analysis. The modern definition of the derivative of a function f is given by

$$f'(x) = \lim_{h \to 0} \frac{f(x+h) - f(x)}{h},$$

provided that this limit exists, using the epsilon-delta notion of limit we mentioned earlier. Thus, one does not use just a single infinitesimal quantity δ, but rather one in effect uses many various increments h and takes a limit as h goes to zero. This precise manner of treating limits avoids all the paradoxical issues with infinitesimals, while retaining the essential intuition underlying them—that the continuous functions are those for which small changes in input cause only small changes in the output, and the derivative of a function at a point is obtained from the average rate of change of the function over increasingly tiny intervals surrounding that point.

2.3   An enlarged vocabulary of concepts

The enlarged mathematical vocabulary provided by the epsilon-delta approach to limits expands our capacity to express new, subtler mathematical concepts, greatly enriching the subject. Let us get a taste of these further refined possibilities.

Strengthening the continuity concept, for example, a function f on the real numbers is said to be *uniformly* continuous if for every $\epsilon > 0$, there is $\delta > 0$ such that whenever x and y are within δ, then $f(x)$ and $f(y)$ are within ϵ. But wait a minute—how does this differ from ordinary continuity? The difference is that ordinary continuity is a separate assertion made at each point c, with separate ϵ and δ for each number c. In particular, with ordinary continuity, the value δ chosen for continuity at c can depend not only on ϵ, but also on c. With uniform continuity, in contrast, the quantity δ may depend only on ϵ. The same δ must work uniformly with every x and y (the number y in effect plays the role of c here).

Consider the function $f(x) = x^2$, a simple parabola, on the domain of all real numbers. This function is continuous, to be sure, but it is not uniformly continuous on this domain, because it becomes as steep as one likes as one moves to large values of x. Namely, for any $\delta > 0$, if one moves far enough away from the origin, then the parabola becomes sufficiently steep so that one may find numbers x and y very close together, differing by less than δ, while x^2 and y^2 have changed by a huge amount. For this reason, there can be no single δ that works for all x and y, even when one takes a very coarse value of

ϵ. Meanwhile, using what is known as the *compactness* property of closed intervals in the real number line (expressed by the Heine-Borel theorem), one can prove that every continuous function f defined on a closed interval in the real numbers $[a, b]$ is uniformly continuous on that interval.

The uniform continuity concept arises from a simple change in quantifier order in the continuity statement, which one can see by comparing:

A function f on the real numbers is *continuous* when

$$\forall y \; \forall \epsilon > 0 \; \exists \delta > 0 \; \forall x \; [x, y \text{ within } \delta \implies f(x), f(y) \text{ within } \epsilon],$$

whereas f is *uniformly continuous* when

$$\forall \epsilon > 0 \; \exists \delta > 0 \; \forall x, y \; [x, y \text{ within } \delta \implies f(x), f(y) \text{ within } \epsilon].$$

Let us explore a few other such variations—which concepts result this way? The reader is asked to provide the meaning of these three statements in question 2.5 and to identify in each case exactly which functions exhibit the property:

$$\exists \delta > 0 \; \forall \epsilon > 0 \; \forall x, y \; [x, y \text{ within } \delta \implies f(x), f(y) \text{ within } \epsilon].$$
$$\forall \epsilon > 0 \; \forall x, y \; \exists \delta > 0 \; [x, y \text{ within } \delta \implies f(x), f(y) \text{ within } \epsilon].$$
$$\forall \epsilon > 0 \; \forall \delta > 0 \; \forall x \; \exists y \; [x, y \text{ within } \delta \text{ and } f(x), f(y) \text{ within } \epsilon].$$

Also requiring $x \neq y$ in the last example makes for an interesting, subtle property.

Errett Bishop (1977) describes the epsilon-delta notion of continuity as "common sense," and I am inclined to agree. To my way of thinking, however, the fact that minor syntactic variations in the continuity statement lead to vastly different concepts seems to underline a fundamental subtlety and fragility of the epsilon-delta account of continuity, and the delicate nature of the concept may indicate how greatly refined it is.

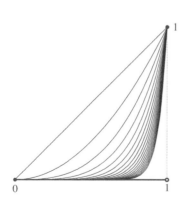

Suppose we have a sequence of continuous functions f_0, f_1, f_2, ..., and they happen to converge pointwise to a limit function $f_n(x) \to f(x)$. Must the limit function also be continuous? Cauchy made a mistake about this, claiming that a convergent series of continuous functions is continuous. But this turns out to be incorrect. For a counterexample, consider the functions x, x^2, x^3, ... on the unit interval, as pictured here in blue. These functions are each continuous, individually, but as the exponent grows, they become increasingly flat on most of the interval, spiking to 1 at the right. The limit function, shown in red, accordingly has constant value 0, except at $x = 1$, where it has a jump discontinuity up to 1. So the convergent limit of continuous functions is not necessarily continuous. In Cauchy's defense, he had a convergent series $\sum_n f_n(x)$ rather than a point-wise convergent limit $\lim_n f_n(x)$, which obscures the counterexamples, although one may translate between sequences and series via successive differences, making the two formulations equally wrong. Meanwhile, Imre Lakatos (2015 [1976]) advances a more forgiving view of Cauchy's argument in its historical context.

One finds a correct version of the implication by strengthening pointwise convergence to uniform convergence $f_n \rightrightarrows f$, which means that for every $\epsilon > 0$, there is N such that every function f_n for $n \geq N$ is contained within an ϵ tube about f, meaning that $f_n(x)$ is within ϵ of $f(x)$ for every x. The uniform limit of continuous functions is indeed continuous by an $\epsilon/3$ argument: if x is close to c, then $f_n(x)$ is eventually close to $f(x)$ and to $f_n(c)$, which is close to $f(c)$. More generally, if a merely pointwise convergent sequence of functions forms an *equicontinuous family*, which means that at every point c and for every $\epsilon > 0$, there is a $\delta > 0$ that works for every f_n at c, then the limit function is continuous.

What I am arguing is that the epsilon-delta methods do not serve merely to repair a broken foundation, leaving the rest of the structure intact. We do not merely carry out the same old modes of reasoning on a shiny new (and more secure) foundation. Rather, the new methods introduce new modes of reasoning, opening doors to new concepts and subtle distinctions. With the new methods, we can make fine gradations in our previous understanding, now seen as coarse; we can distinguish between continuity and uniform continuity or among pointwise convergence, uniform convergence, and convergence for an equicontinuous family. This has been enormously clarifying, and our mathematical understanding of the subject is vastly improved.

2.4 The least-upper-bound principle

The subject of real analysis can be founded upon the *least-upper-bound principle*, a version of Dedekind completeness. Taking this as a core principle, one proceeds to prove all the familiar foundational theorems, such as the intermediate value theorem, the Heine-Borel theorem and many others. In a sense, the least-upper-bound principle is to real analysis what the induction principle is to number theory.

Least-upper-bound principle. *Every nonempty set of real numbers with an upper bound has a least upper bound.*

An upper bound of a set $A \subseteq \mathbb{R}$ is simply a real number r, such that every element of A is less than or equal to r. The number r is the *least* upper bound of A, also called the *supremum* of A and denoted $\sup(A)$, if r is an upper bound of A and $r \leq s$ whenever s is an upper bound of A.

Consequences of completeness

Let us illustrate how this fundamental principle is used by sketching proofs of a few familiar elementary results in real analysis. Consider first the *intermediate-value theorem*, which asserts that if f is a continuous function on the real numbers and d is an intermediate value between $f(a)$ and $f(b)$, then there is a real number c between a and b, with $f(c) = d$. In short, it asserts that for continuous functions, every intermediate value is realized.

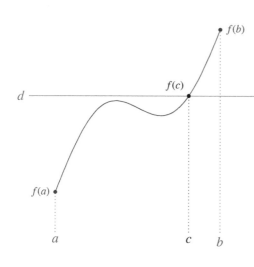

To prove this from the least-upper-bound principle, assume that $a < b$ and $f(a) < d < f(b)$, and consider the set A of all real numbers x in the interval $[a, b]$ for which $f(x) < d$. This set is nonempty because a is in A, and it is bounded by b. By the least-upper-bound principle, therefore, it has a least upper bound $c = \sup(A)$. Consider the value of $f(c)$. If $f(c)$ is too small, meaning that $f(c) < d$, then by the continuity of f, the values of $f(x)$ for x near c will all also be less than d. But this would mean that A has some elements above c, contrary to $c = \sup(A)$. Therefore, $f(c) \geq d$. Similarly, if $f(c)$ is too large, meaning that $f(c) > d$, then again by the continuity of f, the values of $f(x)$ for x in a small neighborhood of c will all be above d, contradicting $c = \sup(A)$, since there must be elements of A as close as desired to c, and those elements x must have $f(x) < d$ by the definition of A. Therefore, it must be that $f(c) = d$, and we have found the desired point c realizing the intermediate value d.

Consider next the *Heine-Borel theorem*, which asserts that the unit interval $[0, 1]$ is *compact*. What this means is that whenever $\mathcal{U}$ is a set of open intervals covering the closed unit interval in the real numbers $[0, 1]$, then there are finitely many open intervals of $\mathcal{U}$ that already cover $[0, 1]$. Huh? For someone new to real analysis, the importance of this open-cover conclusion may not be readily apparent; perhaps it may even seem a little bizarre. Why would it be important that every open cover of a closed interval admits a finite subcover? The answer is that this property is vitally important, and indeed, it is difficult to overstate the importance of the compactness concept, which is the key to thousands of mathematical arguments. The underlying ideas with which it engages grow ultimately into the subject of topology. Let us have a small taste of it.

To prove the Heine-Borel theorem, fix the open cover $\mathcal{U}$, consisting of open intervals in the real numbers, such that every number $x \in [0, 1]$ is in some $U \in \mathcal{U}$. Consider the set B consisting of all $x \in [0, 1]$, such that the interval $[0, x]$ is covered by finitely many elements of $\mathcal{U}$. Thus, $0 \in B$, since $[0,0]$ has just the point 0, which is covered by some open set in $\mathcal{U}$. Let b be the least upper bound of B in $[0, 1]$. If $b = 1$, then we can cover $[0, 1]$ with finitely many elements of $\mathcal{U}$, and we're done. So assume that $b < 1$. Since $\mathcal{U}$ covers the entire interval, there is some open interval $U \in \mathcal{U}$ with $b \in U$. Since U contains a neighborhood of b, which is the supremum of B, there must be a point $x \in B$ with $x \in U$. Therefore, we can cover $[0, x]$ with finitely many open sets from $\mathcal{U}$, and the set U itself covers the rest of the interval $[x, b]$, and so we have covered the interval $[0, b]$ with finitely many open intervals from $\mathcal{U}$. But because the open interval must spill strictly beyond b, there must be elements of B strictly larger than b, which contradicts the assumption that b is the least upper bound of B. So we have proved the theorem.

Another application of the least-upper-bound principle is the fact that every nested descending sequence of closed intervals has a nonempty intersection. That is, if

$$[a_0, b_0] \supseteq [a_1, b_1] \supseteq [a_2, b_2] \supseteq \cdots,$$

then there is a real number z inside every interval $z \in [a_n, b_n]$. One can simply let $z = \sup_n a_n$. More generally, using the Heine-Borel theorem, one can prove that every nested descending sequence of nonempty closed sets in a real number interval has a nonempty intersection. For if there were no such real number z, then the complements of those closed sets would form an open cover of the original interval, with no finite subcover, contrary to the Heine-Borel theorem.

Continuous induction

Earlier I made an analogy between the least-upper-bound principle in real analysis and the principle of mathematical induction in number theory. This analogy is quite strong in light of the principle of continuous induction, which is an inductionlike formulation of the least-upper-bound principle that can be used to derive many fundamental results in real analysis.

Principle of continuous induction. *If A is a set of nonnegative real numbers such that*

1. *$0 \in A$;*
2. *If $x \in A$, then there is $\delta > 0$ with $[x, x + \delta) \subseteq A$;*
3. *If x is a nonnegative real number and $[0, x) \subseteq A$, then $x \in A$.*

Then A is the set of all nonnegative real numbers.

In plain language, the principle has the anchoring assumption that 0 is in the set A, and the inductive properties that whenever all the numbers up to a number x are in A, then x itself is in A; and whenever a number x is in A, then you can push it a little higher, and all the numbers between x and some $x + \delta$ are in A. The conclusion, just as in the principle of induction, is that any such set A must contain every number (in this case, every nonnegative real number).

Yuen Ren Chao (1919), describes a similar principle like this:

> The theorem is a mathematical formulation of the familiar argument from "the thin end of the wedge," or again, the argument from "the camel's nose."
>
> Hyp. 1. Let it be granted that the drinking of half a glass of beer be allowable.
>
> Hyp. 2. If any quantity, x, of beer is allowable, there is no reason why $x + \delta$ is not allowable, so long as δ does not exceed an imperceptible amount Δ.
>
> Therefore any quantity is allowable.

The principle of continuous induction can be proved from the least-upper-bound principle, by considering the supremum r of the set of numbers a for which $[0, a] \subseteq A$. This supremum must be in A because $[0, r) \subseteq A$; but if $r \in A$, then it should be at least a little larger than it is, which is a contradiction. The converse implication also holds, as the reader will prove in question 2.16, and so we have three equivalent principles: continuous induction, least upper bounds, and Dedekind completeness.

Mathematicians are sometimes surprised by the principle of continuous induction—by the idea that there is a principle of induction on the real numbers using the real-number order—because there is a widely held view, or even an entrenched expectation, that induction is fundamentally discrete and sensible only with well-orders. Yet here we are with an induction principle on the real numbers based on the continuous order.

The principle of continuous induction is quite practical and can be used to establish much of the elementary theory of real analysis. Let us illustrate this by using the principle to give an alternative inductive proof of the Heine-Borel theorem. Suppose that $\mathcal{U}$ is an open cover of $[0, 1]$. We shall prove by continuous induction that every subinterval $[0, x]$ with $0 \le x \le 1$ is covered by finitely many elements of $\mathcal{U}$. This is true of $x = 0$, since $[0, 0]$ has just one point. If $[0, x]$ is covered by finitely many sets from $\mathcal{U}$, then whichever open set contains x must also stick a bit beyond x, and so the same finite collection covers a nontrivial extension $[0, x + \delta]$. And if every smaller interval $[0, r]$ for $r < x$ is finitely covered, then take an open set containing x, which must contain $(r, x]$ for some $r < x$; by combining that one open set with a finite cover of $[0, r]$, we achieve a finite cover of $[0, x]$. So by continuous induction, every $[0, x]$ has a finite subcover from $\mathcal{U}$. In particular, $[0, 1]$ itself is finitely covered, as desired.

2.5 Indispensability of mathematics

What philosophical conclusion can we make from the fact that mathematical tools and vocabulary seem to lie at the very core of nearly every contemporary scientific theory? How remarkable that at every physical scale, from the microscopic to the cosmic, our best scientific theories are thoroughly mathematical. Why should this be? The laws of Newtonian physics are expressed in universal differential equations that explain the interaction of forces and motion, unifying our understanding of diverse physical phenomena, from the harmonic oscillations of a mass on a spring to the planetary motions of the heavenly bodies; our best theory of electromagnetism posits unseen electrical and magnetic fields that surround us all, enveloping the Earth; relativity theory explains the nature of space and time with exotic mathematical geometries; quantum mechanics uses Hilbert spaces; string theory uses still more abstract mathematical objects; and all the experimental sciences, including the social sciences, make fundamental use of mathematical statistics. Physicist Paul Dirac (1963) describes the situation like this:

> It seems to be one of the fundamental features of nature that fundamental physical laws are described in terms of a mathematical theory of great beauty and power, needing quite a high standard of mathematics for one to understand it. You may wonder: Why is nature constructed along these lines? One can only answer that our present knowledge seems to show that nature is so constructed. We simply have to accept it. One could perhaps describe the situation by saying that God is a mathematician of a very high order, and He used very advanced mathematics in constructing the universe. Our feeble attempts at mathematics enable us to understand a bit of the universe, and as we proceed to develop higher and higher mathematics we can hope to understand the universe better.[2]

Thus, mathematics appears to be indispensable for physics and other sciences.

On the basis of this, Hilary Putnam and Willard Van Orman Quine mount the *indispensability* argument for mathematical realism, arguing that we ought to have an ontological commitment to the objects that are indispensably part of our best scientific theories (see Putnam (1971) for a classic presentation). Just as a scientist finds grounds for the existence of unseen microscopic organisms on the basis of the well supported germ theory of disease, even in the absence of direct observations of those organisms, and just as a scientist might commit to the atomic theory of matter or the existence of electrons or a molten iron core in the Earth, or black holes, or wave functions, even when the evidence for them in our well supported theories is indirect, then similarly, according to the indispensability argument, we should find grounds for the existence of the abstract mathematical objects that appear in our best theories. Quine emphasizes a view of *confirmational wholism*, by which theories are confirmed only as a whole. If mathematical claims are an indispensable part of the theory, then they are part of what is confirmed.

Science without numbers

Attacking the indispensability argument at its heart, Hartry H. Field (1980) argues that the truth of mathematics is not actually indispensable for science. Defending a nominalist approach to mathematics, he argues that we do not require the actual existence of these abstract mathematical objects in order to undertake a successful scientific analysis.

---

[2] One should not misunderstand Dirac's views on God, however, for he also said, "If we are honest—and scientists have to be—we must admit that religion is a jumble of false assertions, with no basis in reality. The very idea of God is a product of the human imagination." Dirac, quoted by Werner Heisenberg in (Heisenberg, 1971, pp. 85–86).

Rather, Field points out that there is a kind of logical error underlying the indispensability argument. Namely, even if the mathematical theories are indispensable to the scientific analysis, this is not a reason to suppose that the mathematical theories are actually true. It would be sufficient, for example, if the mathematical claims formed merely a *conservative extension* of the scientific theory. Specifically, as mentioned in chapter 1, a theory S is *conservative* over another theory T that it extends with respect to a class of assertions $\mathcal{L}$ if, whenever S proves an $\mathcal{L}$ assertion, this assertion is already provable in T. In other words, the stronger theory S tells us no new $\mathcal{L}$ facts that we could not already know on the basis of T. This does not mean that the theory S is useless or unhelpful, however, for perhaps S unifies our knowledge somehow or is more explanatory or makes reasoning easier, even if ultimately, no new $\mathcal{L}$ assertions will be proved in S.

In the case of the indispensability argument, our scientific theory S describes the nature of the physical world, but this theory includes mathematical claims making existence assertions about various mathematical objects. Let N be the nominalist fragment of the scientific theory, omitting the mathematical claims. If the full theory S were conservative over N concerning assertions about the physical world, then we could safely use the full theory S to make deductions about the physical world, whether or not the mathematical claims it makes are true. In the mathematically augmented theory, the scientist may safely reason as if the mathematical part of the theory were true, without needing to commit to the truth of those additional mathematical assertions.

Instances of this same pattern have arisen entirely in mathematics. Consider the early use of the complex numbers in mathematics, before the nature of imaginary numbers was well understood. In those early days, skeptical mathematicians would sometimes use the so-called imaginary numbers, using expressions like $1 + \sqrt{-5}$, even when they looked upon these expressions as meaningless, because in the end, the imaginary parts of their calculations would sometimes cancel out and they would find the desired real number solution to their equation. It must have been mystifying to see calculations proceeding through the land of nonsense, manipulating those imaginary numbers with ordinary algebra, and yet somehow working out in the end to a real number solution that could be verified independently of the complex numbers. My point here is that even if a mathematician did not commit to the actual existence of the complex numbers, the theory of complex numbers was conservative over their theory of real numbers, so far as assertions about the real numbers are concerned. So even a skeptical mathematician could safely reason as if imaginary numbers actually existed.

Field is arguing similarly for applications of mathematics in general. Ultimately, it does not matter, according to Field, whether the mathematical claims made as part of a scientific theory are actually true or false, provided that the theory is conservative over the nonmathematical part of the theory, so far as physical assertions are concerned, for in this case, the scientist can safely reason as if the mathematical claims were true.

Impressively, Field attempts to show how one can cast various scientific theories without any reference to mathematical objects, replacing the usual theories with nominalized versions, which lack a commitment to the existence of mathematical objects. He provides a nominalist account of Newtonian spacetime and of the Newtonian gravitational theory. One basic idea, for example, is to use physical arrangements as stand-ins for mathematical quantities. One may represent an arbitrary real number, for example, by the possible separations of two particles in space, and then refer to that number in effect by referring to the possible locations of those particles. Thus, one avoids the need for abstract objects.

Critics of Field point out that although he has strived to eliminate numbers and other abstract mathematical objects, nevertheless his ontology is rich, filled with spatiotemporal regions and other objects that can be seen as abstract. Shall we take physics to be committed to these abstract objects? Also, the nominalized theory is cumbersome and therefore less useful for explanation and insight in physics—isn't this relevant for indispensability?

Fictionalism

Fictionalism is the position in the philosophy of mathematics, according to which mathematical existence assertions are not literally true, but rather are a convenient fiction, useful for a purpose, such as the applications of mathematics in science. According to fictionalism, statements in mathematics are similar in status to statements about fictional events. An arithmetic assertion p, for example, can be interpreted as the statement, "According to the theory of arithmetic, p." It is just as one might say, "According to the story by Beatrix Potter (1906), Jeremy Fisher enjoys punting." In his nominalization program for science, Field is essentially defending a fictionalist account of mathematics in science. Even if the mathematical claims are not literally true, the scientist can reason as if they were true.

I find it interesting to notice how the fictionalist position might seem to lead one to nonclassical logic in mathematics. Let us suppose that in the story of *Jeremy Fisher*, the cost of his punt is not discussed; and now consider the statement, "Jeremy Fisher paid more than two shillings for his punt." It would be wrong to say, "According to the story by Beatrix Potter, Jeremy Fisher paid more than two shillings for his punt." But it would also be wrong to say, "According to the story by Beatrix Potter, Jeremy Fisher did *not* pay more than two shillings for his punt." The story simply has nothing to say on the matter. So the story asserts neither p nor $\neg p$. Is this a violation of the law of excluded middle?

As I see it, no, this is not what it means to deny the law of excluded middle. While one asserts neither p nor $\neg p$, still, one asserts $p \vee \neg p$, since according to the story by Beatrix Potter, we may reasonably suppose that either Jeremy Fisher did pay more than two shillings for his punt or he did not, since Jeremy Fisher's world is presented as obeying such ordinary logic. Another way to see this point is to consider an incomplete theory T in classical logic. Since the theory is incomplete, there is a statement p not settled by the theory, and so in the theory, we do not assert p and we do not assert $\neg p$, and yet we do assert $p \vee \neg p$, and we have not denied the law of excluded middle. Fictional accounts are essentially like incomplete theories, which do not require one to abandon classical logic.

The theory/metatheory distinction

In a robust sense, fictionalism is a retreat from the object theory into the metatheory. Let us make the theory/metatheory distinction. In mathematics, the *object theory* is the theory describing the mathematical subject matter that the theory is about. The *metatheory*, in contrast, places the object theory itself under mathematical analysis and looks into metatheoretic issues concerning it such as provability and consistency. In the object theory—take ZFC set theory, for instance—one asserts that there are sets of all kinds, including well-orders of the real numbers and diverse uncountable sets of vast cardinality, while in the metatheory, typically with only comparatively weak arithmetic resources suitable for managing the theory, one might make none of those existence claims outright, and instead one asserts like the fictionalist merely that "According to the theory ZFC, there are well-orders of the real numbers and uncountable sets of such-and-such vast cardinality." In this way, fictionalism amounts exactly to the metamathematical move. For this reason, there are affinities between fictionalism and formalism (see chapter 7), for the formalist also retreats into the metatheory, taking himself or herself not to have asserted the existence of infinite sets, for example, but asserting instead merely that, according to the theory at hand, there are infinite sets.

2.6   Abstraction in the function concept

The increasing rigor in mathematical analysis was also a time of increasing abstraction in the function concept. What is a function? In the naive account, one specifies a function by providing a rule or formula for how to compute the output value y in terms of the input value x. We all know many examples, such as $y = x^2 + x + 1$ or $y = \sqrt{1 - x^2}$ or $y = e^x$ or $y = \sin(x)$. But do you notice that already these latter two functions are not expressible directly in terms of the algebraic field operations? Rather, they are *transcendental* functions, already a step up in abstraction for the function concept.

The development of proper tools for dealing with power series led mathematicians to consider other more general function representations, such as power series and Fourier series, as given here for the exponential function e^x and the sawtooth function $s(x)$:

$$e^x = \sum_{n=0}^{\infty} \frac{x^n}{n!} \qquad\qquad s(x) = \frac{2}{\pi} \sum_{n=0}^{\infty} \frac{(-1)^{n+1}}{n} \sin(nx).$$

The more general function concept provided by these kinds of representations enabled mathematicians to solve mathematical problems that were previously troubling. One can look for a solution of a differential equation, for example, by assuming that it will have a certain series form and then solve for the particular coefficients, ultimately finding that indeed the assumption was correct and there is a solution of that form. Fourier used his Fourier series to find solutions of the heat equation, for example, and these series are now pervasive in science and engineering.

The Devil's staircase

Let us explore a few examples that might tend to stretch one's function concept. Consider the *Devil's staircase*, pictured below.

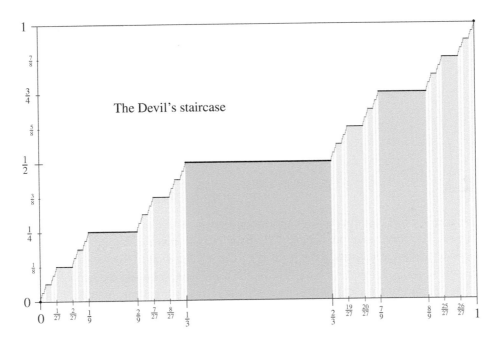

The Devil's staircase

This function was defined by Cantor using his middle-thirds set, now known as the *Cantor set*, and serves as a counterexample to what might have been a certain natural extension of the fundamental theorem of calculus, since it is a continuous function from the unit interval to itself, which has a zero derivative at almost every point with respect to the Lebesgue measure, yet it rises from 0 to 1 during the interval. This is how the Devil ascends from 0 to 1 while remaining almost always motionless.

To construct the function, one starts with value 0 at the left and 1 at the right of the unit interval. Interpolating between these, one places constant value $\frac{1}{2}$ on the entire middle-third interval; this leaves two intervals remaining, one on each side. Next, one places interpolating values $\frac{1}{4}$ and $\frac{3}{4}$ on the middle-thirds of those intervals, and so on, continuing successively to subdivide. This defines the function on the union of all the resulting middle-thirds sections, indicated in orange in the figure, and because of the interpolation, the function continuously extends to the entire interval. Since the function is locally constant on each of the middle-thirds intervals, the derivative is zero there, and those intervals add up to full measure one. The set of points that remain after omitting all the middle-thirds intervals is the Cantor set.

Space-filling curves

Next, consider the phenomenon of space-filling curves. A *curve* is a continuous function from the one-dimensional unit interval into a space, such as the plane, a continuous function $c : [0, 1] \rightarrow \mathbb{R}^2$. We can easily draw many such curves and analyze their mathematical properties.

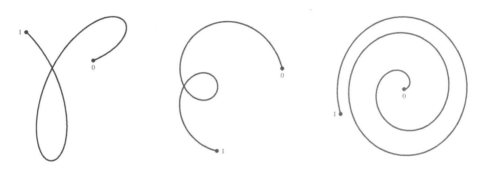

A curve in effect describes a way of traveling along a certain path: one is at position $c(t)$ at time t. The curve therefore begins at the point $c(0)$; it travels along the curve as time t progresses; and it terminates at the point $c(1)$. Curves are allowed to change speed, cross themselves, and even to move backward on the same path, retracing their route. For this reason, one should not identify the curve with its image in the plane, but rather the curve is a way of tracing out that path.

Since the familiar curves that are easily drawn and grasped seem to exhibit an essentially one-dimensional character, it was a shocking discovery that there are *space-filling curves*, which are curves that completely fill a space. Peano produced such a space-filling curve, a continuous function from the unit interval to the unit square, with the property that every point in the square is visited at some time by the curve. Let us consider Hilbert's space-filling curve, shown here, which is a simplification.

The Hilbert curve can be defined by a limit process. One starts with a crude approximation, a curve consisting of three line segments, as in the figure at the upper left of the next page; each approximation is then successively replaced by one with finer detail, and the final Hilbert curve itself is simply the limit of these approximations as they get finer and finer. The first six iterations of the process are shown here. Each approximation is a curve that starts at the lower left and then wriggles about, ultimately ending at the lower right. Each approximation curve consists of finitely many line segments, which we should imagine are traversed at uniform speed. What needs to be proved is that as the approximations get finer, the limiting values converge, thereby defining the limit curve.

One can begin to see this in the approximations to the Hilbert curve above. If we travel along each approximation at constant speed, then the halfway point always occurs just as the path crosses the center vertical line, crossing a little bridge from west to east (look for this bridge in each of the six iterations of the figure—can you find it?). Thus, the halfway point of the final limit curve h is the limit of these bridges, and one can see that they are converging to the very center of the square. So $h(\frac{1}{2}) = (\frac{1}{2}, \frac{1}{2})$. Because the wriggling is increasingly local, the limit curve will be continuous; and because the wriggling gets finer and finer and ultimately enters every region of the square, it follows that every point in the square will occur on the limit path. So the Hilbert curve is a space-filling curve. To my way of thinking, this example begins to stretch, or even break, our naive intuitions about what a curve is or what a function (or even a continuous function) can be.

Conway base-13 function

Let us look at another such example, which might further stretch our intuitions about the function concept. The Conway base 13 function $C(x)$ is defined for every real number x by inspecting the tridecimal (base 13) representation of x and determining whether it encodes a certain secret number, the Conway value of x. Specifically, we represent x in tridecimal notation, using the usual ten digits $0, \ldots, 9$, plus three extra numerals $\oplus$, $\ominus$ and $\odot$, having values 10, 11 and 12, respectively, in the tridecimal system. If it should happen that the tridecimal representation of x has a final segment of the form

$$\oplus a_0 a_1 \cdots a_k \odot b_0 b_1 b_2 b_3 \cdots,$$

where the a_i and b_j use only the digits 0 through 9, then the Conway value $C(x)$ is simply the number $a_0 a_1 \cdots a_k . b_0 b_1 b_2 b_3 \ldots$, understood now in decimal notation. And if the tridecimal representation of x has a final segment of the form

$$\ominus a_0 a_1 \cdots a_k \odot b_0 b_1 b_2 b_3 \cdots,$$

where again the a_i and b_j use only the digits 0 through 9, then the Conway value $C(x)$ is the number $-a_0 a_1 \cdots a_k . b_0 b_1 b_2 b_3 \ldots$, in effect taking $\oplus$ or $\ominus$ as a sign indicator for $C(x)$. Finally, if the tridecimal representation of x does not have a final segment of one of those forms, for example, if it uses the $\oplus$ numeral infinitely many times or if it does not use the $\odot$ numeral at all, then we assign the default value $C(x) = 0$. Some examples may help illustrate the idea:

x	$\mapsto$	$C(x)$
$12.35432 \oplus 3 \odot 14159265 \cdots$	$\mapsto$	$3.14159265 \cdots$
$1.231 \ominus 2 \odot 718281828 \cdots$	$\mapsto$	$-2.718281828 \cdots$
$-1342 \ominus \oplus 12 \oplus 52.123 \odot 7686767 \cdots$	$\mapsto$	$52123.7686767 \cdots$
$1.2 \oplus 3 \ominus 4 \oplus 5 \ominus 6 \oplus 7 \ominus 8 \oplus \cdots$	$\mapsto$	0

The point is that we can easily read off the decimal value of $C(x)$ from the tridecimal representation of x. On the left, we have the input number x, given in tridecimal representation, and on the right is the encoded Conway value $C(x)$, in decimal form, with a default value of 0 when no number is encoded. Note that we ignore the tridecimal point of x in the decoding process.

Let us look a little more closely at the Conway function to discover its fascinating features. The key thing to notice is that every real number y is the Conway value of some real number x, and furthermore, you can encode y into the final segment of x after having already specified an arbitrary finite initial segment of the tridecimal representation of x. If you want the tridecimal representation of x to start in a certain way, go ahead and do that, and then simply add the numeral $\oplus$ or $\ominus$ after this, depending on whether y is positive or negative, and then list off the decimal digits of y to complete the representation of x, using the numeral $\odot$ to indicate where the decimal point of y should be. It follows that $C(x) = y$. Because we can specify the initial digits of x arbitrarily, before the encoding of y, it follows that every interval in the real numbers, no matter how tiny, will have a real number x, with $C(x) = y$. In other words, the restriction of the Conway function to any tiny interval results in a function that is still onto the entire set of real numbers.

Consequently, if we were to make a graph of the Conway function, it would look totally unlike the graphs of other, more ordinary functions. Here is my suggestive attempt to represent the graph:

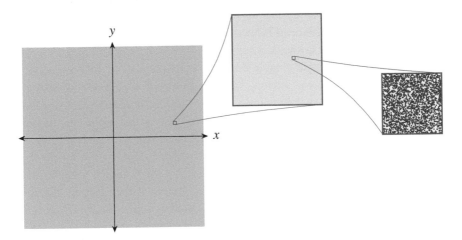

The values of the function are plotted in blue, you see, but since these dots appear so densely in the plane, we cannot easily distinguish the individual points, and the graph appears as a blue mass. When I draw the graph on a chalkboard, I simply lay the chalk sideways and fill the board with dust.

But the Conway function is indeed a function, with every vertical line having exactly one point, and one should imagine the graph as consisting of individual points $(x, C(x))$. If we were to zoom in on the graph, as suggested in the figure, we might hope to begin to discern these points individually. But actually, one should take the figure as merely suggestive, since after any finite amount of magnification, the graph would of course remain dense, with no empty regions at all, and so those empty spaces between the dots in the magnified image are not accurate. Ultimately, we cannot seem to draw the graph accurately in any fully satisfactory manner.

Another subtle point is that the default value of 0 actually occurs quite a lot in the Conway function, since the set of real numbers x that fail to encode a Conway value has full Lebesgue measure. In this precise sense, almost every real number is in the default case of the Conway function, and so the Conway function is almost always zero. We could perhaps have represented this aspect of the function in the graph with a somewhat denser hue of blue lying on the x-axis, since in the sense of the Lebesgue measure, most of the function lies on that axis. Meanwhile, the truly interesting part of the function occurs on a set of measure zero, the real numbers x that do encode a Conway value.

Notice that the Conway function, though highly discontinuous, nevertheless satisfies the conclusion of the intermediate value theorem: for any $a < b$, every y between $f(a)$ and $f(b)$ is realized as $f(c)$ for some c between a and b.

We have seen three examples of functions that tend to stretch the classical conception of what a function can be. Mathematicians ultimately landed with a very general function concept, abandoning all requirements functions must be defined by a formula of some kind; rather, a function is simply a certain kind of relation between input and output: a function is any relation for which an input gives at most one output, and always the same output.

In some mathematical subjects, the function concept has evolved further, growing horns in a sense. Namely, for many mathematicians, particularly in those subjects using category theory, a function f is not determined merely by specifying its domain X, and the function values $f(x)$ for each point in that domain $x \in X$. Rather, one must also specify what is called the *codomain* of the function, the space Y of intended target values for the function $f : X \to Y$. The codomain is not the same as the range of the function, because not every $y \in Y$ needs to be realized as a value $f(x)$. On this concept of the function, the squaring function $s(x) = x^2$ on the real numbers can be considered as a function from $\mathbb{R}$ to $\mathbb{R}$ or as a function from $\mathbb{R}$ to $[0, \infty)$, and the point would be that these would count as two different functions, which happen to have the same domain $\mathbb{R}$ and the same value x^2 at every point x in that domain; but because the codomains differ, they are different functions. Such a concern with the codomain is central to the category-theoretic conception of morphisms, where with the composition of functions $f \circ g$, for example, one wants the codomain of g to align with the domain of f.

2.7 Infinitesimals revisited

For the final theme of this chapter, let us return to the infinitesimals. Despite the problematic foundations and Berkeley's criticisms, it will be good to keep in mind that the infinitesimal conception was actually extremely fruitful and led to many robust mathematical insights, including all the foundational results of calculus. Mathematicians today routinely approach problems in calculus and differential equations essentially by considering the effects of infinitesimal changes in the input to a function or system.

For example, to compute the volume of a solid of revolution $y = f(x)$ about the x-axis, it is routine to imagine slicing the volume into infinitesimally thin disks. The disk at x has radius $f(x)$ and infinitesimal thickness dx (hence volume $\pi f(x)^2 \, dx$), and so the total volume between a and b, therefore, is

$$V = \pi \int_a^b f(x)^2 \, dx.$$

Another example arises when one seeks to compute the length of the curve traced by a smooth function $y = f(x)$. One typically imagines cutting it into infinitesimal pieces and observing that each tiny piece is the hypotenuse ds of a triangle with infinitesimal legs *dx* and *dy*. So by an infinitesimal instance of the Pythagorean theorem, we see $ds = \sqrt{dx^2 + dy^2}$, from which one "factors out" *dx*, obtaining $ds = \sqrt{1 + (dy/dx)^2}\, dx$, and therefore the total length of the curve is given by

$$s = \int_a^b \sqrt{1 + \left(\frac{dy}{dx}\right)^2}\, dx.$$

Thus, one should not have a cartoon understanding of developments in early calculus, imagining that it was all bumbling nonsense working with the ghosts of departed quantities. On the contrary, it was a time of enormous mathematical progress and deep insights of enduring strength. Perhaps this situation gives a philosopher pause when contemplating the significance of foundational issues for mathematical progress—must one have sound foundations in order to advance mathematical knowledge? Apparently not. Nevertheless, the resolution of the problematic foundations with epsilon-delta methods did enable a far more sophisticated mathematical analysis, leading to further huge mathematical developments and progress. It was definitely valuable to have finally fixed the foundations.

Nonstandard analysis and the hyperreal numbers

What an astounding development it must have been in 1961, when Abraham Robinson introduced his theory of nonstandard analysis. This theory, arising from ideas in mathematical logic and based on what are now called the hyperreal numbers $\mathbb{R}^*$, provides a rigorous method of handling infinitesimals, having many parallels to the early work in calculus. I look upon this development as a kind of joke that mathematical reality has played on both the history and philosophy of mathematics.

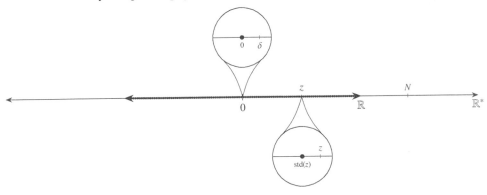

The way it works is as follows. The hyperreal number system $\mathbb{R}^*$ is an ordered field extending the real numbers, but having new infinite and infinitesimal numbers. In the figure, the real number line—meaning all of it, the entire real number line—is indicated in blue. The hyperreal number line, indicated in red, thus extends strictly beyond the end of the real number line. It may be difficult to imagine at first, but indeed this is what the construction of the hyperreal numbers produces. The hyperreal number *N* indicated in the figure, for example, is larger than every real number, and the number δ is positive, being strictly larger than 0 but smaller than every positive real number. One could imagine that $\delta = 1/N$, since the reciprocal of an infinite number will be infinitesimal and conversely.

Every real number is surrounded by a neighborhood of those hyperreal numbers infinitesimally close to it, and so if one were to zoom in on a real number in the hyperreal numbers, one would ultimately find a window containing only that real number and the hyperreal numbers infinitely close to it. Every hyperreal z that is bounded by standard real numbers is infinitesimally close to a unique standard real number, called the standard part of z and denoted as $\text{std}(z)$, as indicated in the figure.

Part of what makes the hyperreal numbers attractive for nonstandard analysis is their further remarkable property that any statement in the formal language of analysis that is true for the real numbers $\mathbb{R}$ is also true for the hyperreal numbers $\mathbb{R}^*$. The *transfer principle* asserts that every real number a and function f on the real numbers have nonstandard counterparts a^* and f^* in the hyperreal numbers, such that any assertion $\varphi(a, f)$ about a and f that is true in the real numbers $\mathbb{R}$ is also true of the counterparts $\varphi(a^*, f^*)$ in the hyperreal numbers $\mathbb{R}^*$. Thus, the real numbers have an elementary embedding $a \mapsto a^*$ into the hyperreal numbers $\mathbb{R} \preccurlyeq \mathbb{R}^*$.

Because of the transfer principle, we know exactly what kinds of arithmetic and algebraic operations are allowed in the hyperreal numbers—they are the same as those allowed in the real number system. Because of the transfer principle, we will have concepts of infinite integers, infinite prime numbers, infinite powers of 2, and so on, and we can divide an interval in the real numbers into infinitely many subintervals of equal infinitesimal length. These are exactly the kinds of things that one wants to do in calculus, and nonstandard analysis provides a rigorous foundation for it.

Calculus in nonstandard analysis

Let us illustrate this by explaining how the derivative is treated in nonstandard analysis. One computes the derivative of a function f by using an infinitesimal hyperreal number, without any limit process or epsilon-delta arguments. To find the derivative of $f(x) = x^2$, for example, let δ be a positive infinitesimal hyperreal. We compute the rate-of-change quotient over the interval from x to $x + \delta$,

$$\frac{f(x + \delta) - f(x)}{\delta} = \frac{(x + \delta)^2 - x^2}{\delta} = 2x + \delta,$$

with the same algebraic steps as performed earlier, and then simply take the standard part of the result, arriving at

$$\text{std}(2x + d) = 2x.$$

Thus, the derivative of x^2 is $2x$. The use of the standard-part operation in effect formalizes how one can correctly treat infinitesimals—it explains exactly how the ghosts depart! Thus, nonstandard analysis enables a fully rigorous use of infinitesimals.

Nonstandard analysis has now grown into a mature theory, providing an alternative conception parallel to the classical theory. There are several philosophical perspectives one might naturally adopt when undertaking work in nonstandard analysis, which I would like now to explain. Some of this material is technical, however, and so readers not familiar with ultrapowers, for example, might simply skip over it.

Classical model-construction perspective

In this approach, one thinks of the nonstandard universe as the result of an explicit construction, such as an ultrapower construction. In the most basic instance, one has the standard real number field structure $\langle \mathbb{R}, +, \cdot, 0, 1, \mathbb{Z} \rangle$, and you perform the ultrapower construction with respect to a fixed nonprincipal ultrafilter U on the natural numbers (or on some other set if this were desirable). The ultrapower structure $\mathbb{R}^* = \mathbb{R}^{\mathbb{N}}/U$ is then taken as a conception of the hyperreal numbers, an ordered non-Archimedean field, which therefore has infinitesimal elements.

In time, however, one is led to want more structure in the pre-ultrapower model, so as to be able to express more ideas, which will each have nonstandard counterparts. One will have constants for every real number, a predicate for the integers $\mathbb{Z}$, or indeed for every subset of $\mathbb{R}$, and a function symbol for every function on the real numbers, and so on. In this way, one gets the nonstandard analogue z^* of every real number z, the set of nonstandard integers $\mathbb{Z}^*$, and nonstandard analogues f^* for every function f on the real numbers, and so on. Before long, one also wants nonstandard analogues of the power set $P(\mathbb{R})$ and higher iterates. In the end, what one realizes is that one might as well take the ultrapower of the entire set-theoretic universe $V \to V^{\omega}/U$, which amounts to doing nonstandard analysis with second-order logic, third-order, and indeed α-order for every ordinal α. One then has the copy of the standard universe V inside the nonstandard realm V^*, which one analyzes and understands by means of the ultrapower construction itself.

Some applications of nonstandard analysis have required one to take not just a single ultrapower, but an iterated ultrapower construction along a linear order. Such an ultrapower construction gives rise to many levels of nonstandardness, and this is sometimes useful. Ultimately, as one adds additional construction methods, this amounts just to adopting all model theory as one's toolkit. One will want to employ advanced saturation properties, or embeddings, or the standard system, and so on. There is a well developed theory of models of arithmetic that uses quite advanced methods. To give a sample consequence of saturation, every infinite graph, no matter how large, arises as an induced subgraph of a nonstandard-finite graph in every sufficiently saturated model of nonstandard analysis. This sometimes can allow you to undertake finitary constructions with infinite graphs, with the cost being a move to the nonstandard context.

Axiomatic approach

Most applications of nonstandard analysis, however, do not rely on the details of the ultra-power or iterated ultrapower constructions, and so it is often thought worthwhile to isolate the general principles that make the nonstandard arguments succeed. Thus, one writes down the axioms of the situation. In the basic case, one has the standard structure $\mathbb{R}$, and so on, perhaps with constants for every real number (and for all subsets and functions in the higher-order cases), with a map to the nonstandard structure $\mathbb{R}^*$, so that every real number a has its nonstandard version a^* and every function f on the real numbers has its nonstandard version f^*. Typically, the main axioms would include the transfer principle, which asserts that any property expressible in the language of the original structure holds in the standard universe exactly when it holds of the nonstandard analogues of those objects in the non-standard realm. The transfer principle amounts precisely to the elementarity of the map $a \mapsto a^*$ from standard objects to their nonstandard analogues. One often also wants a *saturation principle*, expressing that any sufficiently realizable type is actually realized in the nonstandard model, and this just axiomatizes the saturation properties of the ultrapower. Sometimes one wants more saturation than one would get from an ultrapower on the natural numbers, but one can still achieve this by larger ultrapowers or other model-theoretic methods.

Essentially the same axiomatic approach works with the higher-order case, where one has a nonstandard version of every set-theoretic object, and a map $V \to V^*$, with nonstandard structures of any order. And similarly, one can axiomatize the features that one wants to use in the iterated ultrapower case, with various levels of standardness.

"The" hyperreal numbers?

As with most mathematical situations where one has both a construction and an axiomatic framework, it is usually thought better to argue from the axioms, when possible, than to use details of the construction. And most applications of nonstandard analysis that I have seen can be undertaken using only the usual nonstandardness axioms. A major drawback of the axiomatic approach to nonstandard analysis, however, is that the axiomatization is not categorical. There is no unique mathematical structure fulfilling the hyperreal axioms, and no structure that is "the" hyperreal numbers up to isomorphism. Rather, we have many candidate structures for the hyperreal numbers, of various cardinalities; some of them extend one another, and all of them satisfy the axioms, but they are not all isomorphic.

For this reason, and despite the common usage that one frequently sees, it seems in-correct, and ultimately not meaningful, to refer to "the" hyperreal numbers, even from a structuralist perspective. The situation is totally unlike that of the natural numbers, the integers, the real numbers, and the complex numbers, where we do have categorical char-acterizations. To be sure, the structuralist faces challenges with the use of singular terms even in those cases (challenges we discussed in chapter 1), but the situation is far worse with the hyperreal numbers. Even if we are entitled to use singular terms for the natural numbers and the real numbers, with perhaps some kind of structuralist explanation as to what this means, nevertheless those structuralist explanations will fail completely with the hyperreal numbers, simply because there is no unique structure that the axioms identify. We just do not seem entitled, in any robust sense, to make a singular reference to the hyperreal numbers.

And yet, one does find abundant references to "the" hyperreal numbers in the literature. My explanation for why this has not caused a fundamental problem is that for the purposes of nonstandard analysis, that is, for the goal of establishing truths in calculus about the real numbers by means of the nonstandard real numbers, the nonisomorphic differences between the various hyperreal structures simply happen not to be relevant. All these structures are non-Archimedean ordered fields with the transfer principle, and it happens that these properties alone suffice for the applications of these fields that are undertaken. In this sense, it is as though one fixes a particular nonstandard field $\mathbb{R}^*$—and for the applications considered, it does not matter much which one—and then reference to "the" hyperreal numbers simply refer to that particular $\mathbb{R}^*$ that has been fixed. It is as though the mathematicians are implementing Stewart Shapiro's disguised bound quantifiers, but without a categoricity result.

I wonder whether this lack of categoricity may explain the hesitancy of some mathematicians to study nonstandard analysis; it prevents one from adopting a straightforward structuralist attitude toward the hyperreal numbers, and it tends to push one back to the model-theoretic approach, which more accurately conveys the complexity of the situation.

Meanwhile, we do have a categoricity result for the surreal numbers, which form a nonstandard ordered field of proper class size, one that is saturated in the model-theoretic sense—as in, every set-sized cut is filled. In the standard second-order set theories such as Gödel-Bernays set theory with the principle of global choice (a class well-ordering of the universe) or Kelley-Morse set theory, one can prove that all such ordered fields are isomorphic. This is a sense in which "the hyperreal numbers" might reasonably be given meaning by taking it to refer to the surreal numbers, at the cost of dealing with a proper class structure.

Radical nonstandardness perspective

This perspective on nonstandard analysis involves an enormous foundational change in one's mathematical ontology. Namely, from this perspective, rather than thinking of the standard structures as having analogues inside a nonstandard world, one instead thinks of the nonstandard world as the "real" world, with a "standardness" predicate picking out parts of it. On this approach, one views the real numbers as including both infinite and infinitesimal real numbers, and one can say when and whether two finite real numbers have the same standard part, and so on. With this perspective, we think of the "real" real numbers as what from the other perspective would be the nonstandard real numbers, and then we have a predicate on that, which amounts to the range of the star map in the other approach. So some real numbers are standard, some functions are standard, and so on.

In an argument involving finite combinatorics, for example, someone with this perspective might casually say, "Let N be an infinite integer" or "Consider an infinitesimal rational number." (One of my New York colleagues sometimes talks this way.) That way of speaking may seem alien for someone not used to this perspective, but for those that adopt it, it is productive. These practitioners have drunk deeply of the nonstandardness wine; they have moved wholly into the nonstandard realm—a new plane of existence.

Extreme versions of this idea adopt many levels of standardness and nonstandardness, extended to all orders. Karel Hrbacek (1979, 2009) has a well developed theory like this for nonstandard set theory, with an infinitely deep hierarchy of levels of standardness. There is no fully "standard" realm according to this perspective. In Hrbacek's system, one does not start with a standard universe and go up to the nonstandard universe, but rather, one starts with the full universe (which is fundamentally nonstandard) and goes down to deeper and deeper levels of standardness. Every model of ZFC, he proved, is the standard universe inside another model of the nonstandard set theories he considers.

Translating between nonstandard and classical perspectives

Ultimately, my view is that the choice between the three perspectives I have described is a matter of taste, and any argument that can be formulated in one of the perspectives has analogues in the others. In this sense, there seems to be little at stake, mathematically, between the perspectives. And yet, as I argued in section 1.16, divergent philosophical views can lead one toward different mathematical questions and different mathematical research efforts.

One can usually translate arguments not only amongst the perspectives of nonstandard analysis, but also between the nonstandard realm and the classical epsilon-delta methods. Terence Tao (2007) has described the methods of nonstandard analysis as providing a smooth way to manage one's ϵ arguments. One might say, "This δ is smaller than anything defined using that ϵ." It is a convenient way to undertake error estimation. Tao similarly points out how ultrafilters can be utilized in an argument as a simple way to manage one's estimates; if one knows separately that for each of the objects a, b, and c, there is a large measure (with respect to the ultrafilter) of associated witnesses, then one can also find a large measure of witnesses working with all three of them.

For some real analysts, however, it is precisely the lack of familiarity with ultrafilters or other concepts from mathematical logic that prevents them from appreciating the nonstandard methods. In this sense, the preference for or against nonstandard analysis appears to be in part a matter of cultural upbringing.

H. Jerome Keisler wrote an introductory calculus textbook, *Elementary Calculus: An Infinitesimal Approach* (1976), intended for first-year university students, based on the ideas of nonstandard analysis, but otherwise achieving what this genre of calculus textbook achieves. It is a typical thick volume, with worked examples on definite integrals and derivatives and optimization problems and the chain rule, and so on, all with suitable exercises for an undergraduate calculus student. It looks superficially like any of the other standard calculus textbooks used in such a calculus class. But if you peer inside Keisler's book, in the front cover alongside the usual trigonometric identities and integral formulas you will find a list of axioms concerning the interaction of infinite and infinitesimal numbers, the transfer principle, the algebra of standard parts, and so on. It is all based on nonstandard analysis and is fundamentally unlike the other calculus textbooks. The book was used for a time, successfully, in the calculus classes at the University of Wisconsin in Madison.

There is an interesting companion tale to relate concerning the politics of book reviews. Paul Halmos, editor of the *Bulletin of the American Mathematical Society*, requested a review of Keisler's book from Errett Bishop, who decades earlier had been his student but who was also prominently known for his constructivist views in the philosophy of mathematics—views that are deeply incompatible with the main tools of nonstandard analysis. The review was predictably negative, and it was widely criticized, notably by Martin Davis (1977), later in the same journal. In response to the review, Keisler (1977) remarked (see also Davis and Hausner, 1978), that the choice of Bishop to review the book was like "asking a teetotaler to sample wine."

Criticism of nonstandard analysis

Alan Connes mounted a fundamental criticism of nonstandard analysis, remarking in an interview

> At that time, I had been working on nonstandard analysis, but after a while I had found a catch in the theory.... The point is that as soon as you have a nonstandard number, you get a nonmeasurable set. And in Choquet's circle, having well studied the Polish school, we knew that every set you can name is measurable; so it seemed utterly doomed to failure to try to use nonstandard analysis to do physics. Goldstein and Skandalis (2007)

What does he mean? To what is he referring? Let me explain.

> "as soon as you have a nonstandard number, you get a nonmeasurable set."

Every nonstandard natural number N gives rise to a certain notion of largeness for sets of natural numbers: a set $X \subset \mathbb{N}$ is large exactly if $N \in X^*$. In other words, a set X is large if it expresses a property that the nonstandard number N has. No standard finite set is large, and furthermore, the intersection of any two large sets is large and any superset of a large set is large. Thus, the collection $\mathcal{U}$ of all these large sets X forms what is called a nonprincipal ultrafilter on the natural numbers. We may identify the large sets with elements of Cantor space $2^{\mathbb{N}}$, which carries a natural probability measure, the coin-flipping measure, and so $\mathcal{U}$ is a subset of Cantor space.

But the point to be made now is that a nonprincipal ultrafilter cannot be measurable in Cantor space, since the full bit-flipping operation, which is measure-preserving, carries $\mathcal{U}$ exactly to its complement, so $\mathcal{U}$ would have to have measure $\frac{1}{2}$, but $\mathcal{U}$ is a tail event, invariant by the operation of flipping any finite number of bits, and so by Kolmogorov's zero-one law, it must have measure 0 or 1.

> "in the Polish school, ... every set you can name is measurable."

Another way of saying that a set is easily described is to say that it lies low in the descriptive set-theoretic hierarchy, famously developed by the Polish logicians, and the lowest such sets are necessarily measurable. For example, every set in the Borel hierarchy is measurable, and the Borel context is often described as the domain of *explicit* mathematics.

Under stronger set-theoretic axioms, such as large cardinals or projective determinacy, the phenomenon rises to higher levels of complexity, for under these hypotheses, it follows that all sets in the projective hierarchy are Lebesgue measurable. This would include any set that you can define by quantifying over the real numbers and the integers and using any of the basic mathematical operations. Thus, any set you can name is measurable.

The essence of the Connes criticism is that one cannot construct a model of the hyperreal numbers in any concrete or explicit manner, because one would thereby be constructing explicitly a nonmeasurable set, which is impossible. Thus, nonstandard analysis is intimately wrapped up with ultrafilters and weak forms of the axiom of choice. For this reason, it seems useless so far as any real-world application in science and physics is concerned.

Questions for further thought

2.1 Argue that f is continuous at c if and only if the defender has a winning strategy in the continuity game. Does your argument use the axiom of choice in order to prove the forward implication—that is, in order that the strategy would play a particular δ when faced with a particular ϵ? Can you eliminate the need for AC in this argument?

2.2 Argue from the epsilon-delta definitions of limit and continuity that f is continuous at c if and only if $\lim_{x \to c} f(x) = f(c)$.

2.3 Which precise detail about the relationship of the blue graph of f to the yellow-shaded box in the figure used in definition 7 on page 55 shows that this δ is a suitable choice for that ϵ? For example, would this δ work if ϵ were half as large?

2.4 Design a *uniform continuity game*, between defender and challenger, and argue that f is uniformly continuous if and only if the defender has a winning strategy in this game.

2.5 Explain in plain language the meaning of the various quantifier variations of the continuity statement made in section 2.3. For each of the three properties, identify the class of functions exhibiting it.

2.6 Compute the derivative of the function $f(x) = x^3$ in three different ways: (1) using infinitesimals in the classical style; (2) using the epsilon-delta limit approach and the definition of the derivative; and (3) using nonstandard analysis. Compare the three approaches. Can you identify exactly where Berkeley's ghost of a departed quantity appears in your calculation?

2.7 Provide an infinitesimals-based proof of the first fundamental theorem of calculus, which asserts that $\frac{d}{dx}\int_a^x f(t)dt = f(x)$ for any continuous real-valued function f defined on the real numbers. Can you provide an infinitesimals-style account of the definite integral $\int_a^b f(t)\,dt$?

2.8 Explain the difference between continuity and uniform continuity of a function, including the precise definitions and illustrative examples. Next, express the distinction as well as you can using only the old-style infinitesimal conceptions, rather than the modern epsilon-delta formulations.

2.9 How are the indispensability arguments affected by the fact that mathematical theories are often interpretable in one another? Does one dispense with the need for the rational numbers by interpreting them within the integers, or indeed within the natural numbers? And how about the real numbers? Do we dispense with the need for arithmetic and real analysis by using set theory instead? Or by using category theory or some other mathematical foundation? Does this affect the success of Field's nominalization program, if he has interpreted mathematical objects in terms of other abstract objects, such as spatio-temporal locations?

2.10 Can one use an indispensability argument to ground mathematical truth by arguing that one area of mathematics is indispensable for its applications in another part of mathematics? For example, perhaps linear algebra is indispensable for its applications in the theory of differential equations. And further, if the latter theory is indispensable for science, does this mean that linear algebra also is indispensable for science?

2.11 To what extent does Field's nominalization argument succeed for those who wish to remove not only the existence claims of abstract mathematical objects from our scientific theories, but also claims about concrete physical phenomena not directly observed. For example, can we mount the same style of argument in favor of skeptical scientific theories claiming that there are no germs or pathogens, except those in microscope plates, or that the center of the Earth does not exist?

2.12 Introduce a modal operator $\Box p$ to mean, "according to the story *Jeremy Fisher* by Beatrix Potter, p." How would you translate the meaning of the dual operator $\Diamond p = \neg\Box\neg p$? Which modal principles do these operators obey? For example, do we have $\Box p \rightarrow p$ or $\Box(p \rightarrow q) \rightarrow (\Box p \rightarrow \Box q)$? What about assertions with nested modalities, such as $\Diamond\Box p \rightarrow \Box\Diamond p$ or $\Box p \rightarrow \Box\Box p$? In light of the fact that the story of Jeremy Fisher makes no claims about any person named Beatrix Potter or any stories such a person may or may not have written, including a story called *Jeremy Fisher*, is this grounds to accept $\neg\Box\Box p$ for every p? Meanwhile, she does refer to herself in the first person, including an illustration, in *The Tale of Samuel Whiskers or, The Roly-Poly Pudding* (1908, p. 61).

2.13 In what way does fictionalism amount to a retreat from the object theory into the metatheory?

2.14 How well does fictionalism handle metamathematical claims? Suppose that T is our main object theory, such as set theory, and that we have undertaken a metamathematical argument to make claims about what is or is not provable in T. How does a fictionalist handle these metamathematical assertions? If these also are fictional, then is the theory T inside that meta-mathematical fiction the same theory T that the T fiction is about?

2.15   If Field is correct and our best scientific theories, including their mathematical components, are conservative over the nominalized versions of those theories, then is this a reason why we would no longer need those nominalized theories? After all, if the mathematical extensions truly are conservative, then we may safely use them for scientific reasoning. Does this undermine the nominalization program?

2.16   Argue that the principle of continuous induction is equivalent to the least-upper-bound principle in the real numbers, and so all three are forms of Dedekind completeness.

2.17   Using the epsilon-delta definition of continuity, explain why the Conway function is not continuous.

2.18   Concerning the nonstandard analysis approach to calculus, what is the significance, if any, of the fact that the hyperreal numbers lack a categoricity result? Does this fact undermine the nonstandard analysis approach to calculus?

2.19   In light of the absence of a categoricity result for the hyperreal numbers, how are we to understand truth assertions made about the hyperreal numbers?

2.20   Can one easily reconcile nonstandard analysis with a structuralist approach to the philosophy of mathematics?

2.21   Are sound foundations necessary for insightful advances in mathematical knowledge?

Further reading

H. Jerome Keisler (2000). This unusual calculus text carries out a development of elementary calculus using infinitesimals as they are founded in nonstandard analysis.

Pete Clark (2019). A delightful essay undertakes the development of real analysis upon the principle of continuous induction.

Mark Colyvan (2019). An excellent SEP entry on the indispensability arguments; a very good summary with many additional references.

Mark Balaguer (2018). Another excellent SEP entry, explaining fictionalism in mathematics.

Hilary Putnam (1971). A classic account of the indispensability argument.

Hartry H. Field (1980). Undertaking Field's nominalization program.

Credits

Some of the material in this chapter, including several of the figures, are adapted from my book, *Proof and the Art of Mathematics* (Hamkins, 2020c). The history of the principle of continuous induction is not fully clear, but Pete Clark (2019) provides many references and finds versions of it going back at least to Chao (1919) and Khintchin (1923). Chao's principle is actually not equivalent to the principle of continuous induction, however, because he insists on a uniform increase of at least Δ, and this is why he can also omit the analogue of condition (3). My remarks about the Connes interview are based upon the answer I had posted to MathOverflow (see Hamkins, 2011), in response to a question about it posted by Robert Haraway. See Blackwell and Diaconis (1996) for a detailed proof that nonprincipal ultrafilters form a nonmeasurable tail event.

3 Infinity

Abstract. We shall follow the allegory of Hilbert's hotel and the paradox of Galileo to the equinumerosity relation and the notion of countability. Cantor's diagonal arguments, meanwhile, reveal uncountability and a vast hierarchy of different orders of infinity; some arguments give rise to the distinction between constructive and nonconstructive proof. Zeno's paradox highlights classical ideas on potential versus actual infinity. Furthermore, we shall count into the transfinite ordinals.

3.1 Hilbert's Grand Hotel

Let us hear the parable of Hilbert's Grand Hotel. The hotel is truly grand, with infinitely many rooms, each a luxurious, full-floor suite, numbered like the natural numbers: 0, 1, 2, and so on endlessly. The hotel happens to be completely full—every room has a guest, infinitely many in all. But meanwhile, a new guest has arrived, wanting a room. He inquires at the reception desk. What is the manager to do? The manager says cooly, "No problem." He sends an announcement to all the current guests: everyone must change to the next higher room:

$$\text{Room } n \quad \mapsto \quad \text{Room } n + 1.$$

There was fine print in the check-in agreement, you see, stipulating that guests might have to change rooms under the management's direction. So everyone complies, with the result that room 0 has now become vacant, available for the new guest.

On the weekend, a crowd shows up all at once—1000 guests—and the manager organizes a great migration, instructing all the current guests move up 1000 rooms:

$$\text{Room } n \quad \mapsto \quad \text{Room } n + 1000.$$

This makes available all the rooms from 0 to 999, accommodating the crowd.

Hilbert's bus

The following week, a considerably larger crowd arrives: Hilbert's bus, which is completely full. The bus has infinitely many seats, numbered like the natural numbers, so there is seat 0, seat 1, seat 2, and so on; in every seat, there is a new guest. They all want to check in to the hotel.

Can the manager accommodate everyone? Well, he cannot, as before, ask all the guests to move up infinitely many rooms, since every individual room has a finite room number, and so he cannot simply move everyone up and make sufficient space at the bottom of the hotel. But can he somehow rearrange the current guests to make room for the bus passengers? Please try to figure out a method on your own before reading further.

Did you find a solution? The answer is yes, the guests can be accommodated. There are many ways to do it, but let me tell you how this particular manager proceeded. First, the manager directed that the guest currently in room n should move to room $2n$, which freed up all the odd-numbered rooms. Then he directed the passenger in seat s to take room $2s + 1$, which is certainly an odd number, and people in different seats got different odd-numbered rooms:

$$\text{Room } n \quad \mapsto \quad \text{Room } 2n$$
$$\text{Seat } s \quad \mapsto \quad \text{Room } 2s + 1$$

In this way, everyone was accommodated with a room of their own, with the previous guests in the even-numbered rooms and the bus passengers in the odd-numbered rooms.

Hilbert's train

Next, Hilbert's train arrives. The train has infinitely many train cars, each with infinitely many seats, and every seat is occupied. The passengers are each identified by two pieces of information: their car number c and their seat number s; and every passenger is eager to check into the hotel.

Can the manager accommodate them all? Yes, indeed. As a first step, the manager can direct that the current resident of room n should move to room $2n$, which again frees up all the odd-numbered rooms. Next, the manager directs that the passenger in car c, seat s

should check into room $3^c 5^s$:

$$\text{Room } n \quad \mapsto \quad \text{Room } 2n$$
$$\text{Car } c, \text{ Seat } s \quad \mapsto \quad \text{Room } 3^c 5^s$$

Since $3^c 5^s$ is certainly an odd number, it is available for a new guest, and different train passengers will take up different rooms because of the uniqueness of the prime factorization. So again, everyone is accommodated.

3.2 Countable sets

The main concept at play in this allegory is *countability*, for a set is countable when it fits into Hilbert's hotel, when it can be placed in a one-to-one correspondence with a set of natural numbers. Another way to think about it is that a nonempty set A is countable if and only if one can enumerate the elements of A by a list, possibly with repetition, indexed with the natural numbers:

$$a_0, \; a_1, \; a_2, \; a_3, \; \ldots$$

We allow repetition in order to include the finite sets, which we also regard as countable.

The arguments that we gave for Hilbert's hotel show that the union of two countable sets is countable, since you can correspond one of them with even numbers and the other with odd numbers. The integers $\mathbb{Z}$, for example, form a countable set, for they are the union of two countably infinite sets, the positive integers and the nonpositive integers. One can also see that $\mathbb{Z}$ is countable simply by enumerating it like this:

$$0, \; 1, \; -1, \; 2, \; -2, \; 3, \; -3, \; \ldots$$

A stronger and perhaps surprising result is that the union of countably many countable sets is still countable. One can see this with the argument in the treatment of Hilbert's train, since the train itself consisted of countably many countable sets—infinitely many train cars, each infinite.

But let us give another argument for this fundamental result. The claim is that we may place the natural numbers $\mathbb{N}$ in one-to-one correspondence with the set of *pairs* of natural numbers. If we consider the pairs (n, k) of natural numbers as forming the lattice points in a grid, as pictured here, then the claim is that the number of lattice points is the same as the number of natural numbers. Notice that the lattice points on the various columns of this grid form countably many countable sets. We want somehow to reorganize this two-dimensional grid of points into a one-dimensional list of points.

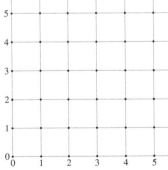

Imagine walking through the grid on the winding path pictured here. We start at the origin and then proceed to zig and zag progressively up and down the diagonals. As we walk, we shall eventually encounter any given lattice point. As we encounter the successive points on the path, let us associate them each with successive natural numbers—the first with 0, the next with 1, and then 2, and so on, and similarly associate every natural number n with the nth point (x_n, y_n) encountered on this path. Thus,

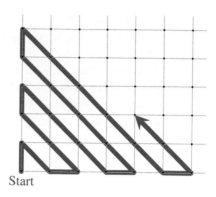

Start

the natural numbers are in correspondence with the pairs of natural numbers.

A variation of the argument enables us to provide a simple polynomial formula for the correspondence. Namely, instead of following the winding purple path up and down as before, let us instead follow each diagonal upward successively in turn, starting at the origin, as pictured here in red. The points encountered before you get to the point (x, y) consist of the upward diagonals, having lengths 1, 2, 3, and so on, plus the $y + 1$ points on the final diagonal. So the total number of points preceding (x, y) is given by the sum

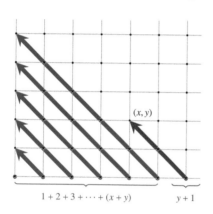

$$[1 + 2 + \cdots + (x + y)] \quad + \quad y.$$

This is equal to $\frac{1}{2}(x+y)(x+y+1)+y$, which thereby provides a polynomial pairing function for the natural numbers.

3.3 Equinumerosity

Countability is an instance of the more general underlying *equinumerosity* concept, by which two sets A and B are *equinumerous* if their elements can be placed in a one-to-one correspondence with each other. The set of people in this room, for example, is equinumerous with the set of noses, simply because we each have a nose and nobody has two (I presume nobody has an extra nose hidden in their pocket). We can verify this instance of equinumerosity without counting either the people or the noses, and this is a sense in which equinumerosity is conceptually prior to number and counting. Meanwhile, in terms of cognitive development for young children, the concepts evidently arise together; (Sarnecka and Wright, 2013, from the abstract) finds "that children either understand both cardinality [counting] and equinumerosity or they understand neither."

We had discussed equinumerosity in chapter 1 in connection with Frege's number concept and the Cantor-Hume principle, which asserts that our number concept is tracked by equinumerosity, in that two classes have the same number of elements exactly when they are equinumerous. That discussion was preoccupied with the finite numbers, for Frege was striving to reduce arithmetic to logic by means of this principle. Cantor, in contrast, focuses the equinumerosity relation particularly on the infinite sets.

Galileo wrote presciently on the confusing difficulties of infinitary equinumerosities in his final scientific work, *Dialogues Concerning Two New Sciences* (1638), centuries before Frege and Cantor, in a dialogue between the characters Simplicio and Salviati, who observe that the natural numbers can be placed in a one-to-one correspondence with the perfect squares by associating each number n with its square n^2:

$$0 \quad 1 \quad 2 \quad 3 \quad 4 \quad 5 \quad 6 \quad \cdots$$
$$0 \quad 1 \quad 4 \quad 9 \quad 16 \quad 25 \quad 36 \quad \cdots$$

The situation is paradoxical, since while there seem to be more numbers than squares, as not every number is a square, nevertheless the correspondence suggests that there are the same number of numbers as squares. The example, therefore, identifies tension between the idea that any two collections in one-to-one correspondence have the same size (later known as the Cantor-Hume principle), and *Euclid's principle*, which asserts that any proper part of an object is smaller than the original object. Galileo recognized that we must give one of them up.

Galileo's characters similarly observe that the points on line segments of different lengths can be associated in a one-to-one correspondence by a fan of similarity, like this:

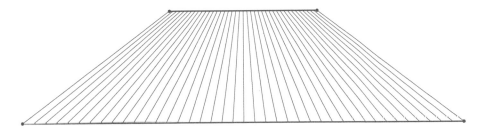

Because the points on the two segments are placed by this similarity into a one-to-one correspondence, it seems paradoxical that the smaller segment would have the same number of points as the larger segment, while at the same time, being shorter, it also seems to have fewer points.

Indeed, Galileo's characters observed that a finite open line segment can be placed in a one-to-one correspondence with an entire line! In the figure at the top of the next page, every point in the interior of the red segment at top drops down to the semicircle and is then radially projected out to the line; every point on the line is realized, and so this is a one-to-one correspondence between the line segment and the line:

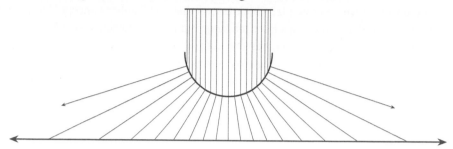

Galileo concludes, through his character Salviati, that comparisons of size simply do not apply in the realm of the infinite:

> So far as I see we can only infer that the totality of all numbers is infinite, that the number of squares is infinite, and that the number of their roots is infinite; neither is the number of squares less than the totality of all the numbers, nor the latter greater than the former; and finally the attributes "equal," "greater," and "less," are not applicable to infinite, but only to finite, quantities. When therefore Simplicio introduces several lines of different lengths and asks me how it is possible that the longer ones do not contain more points than the shorter, I answer him that one line does not contain more or less or just as many points as another, but that each line contains an infinite number. (Galileo Galilei, 1914 [1638], p. 32)

3.4 Hilbert's half-marathon

Returning to our allegory, everyone is in town for the big race, and here it comes: Hilbert's half-marathon—this crowd likes fractions—with the runners densely packed in amongst themselves. The runners each have a race number, one for every nonnegative rational number:

Can the hotel manager accommodate all the runners? Yes. As usual, he frees up the odd-numbered rooms by moving the current guests to double their current room number. Then, he places runner p/q in room $3^p 5^q$, where p/q is in lowest terms:

$$\text{Room } n \qquad \mapsto \qquad \text{Room } 2n$$
$$\text{Runner } \frac{p}{q} \qquad \mapsto \qquad \text{Room } 3^p 5^q$$

In this way, the entire half-marathon fits into Hilbert's hotel.

Cantor's cruise ship

After the race, a shadow falls upon the hotel, as Cantor's cruise ship pulls into the harbor. The ship carries a passenger for every real number. There is passenger $\sqrt{2}$, passenger e, passenger π, and so on, each excited finally to have arrived. Every ship passenger has a ticket, with a distinct ticket number printed upon it (a real number), and the cruise was fully booked: every real number arises on some ticket. The integers are in first class, and then the rational and algebraic numbers, followed by enormous crowds of passengers with transcendental ticket numbers. As the passengers disembark, they all want to check in to Hilbert's Grand Hotel. Can the hotel accommodate all of Cantor's passengers?

3.5 Uncountability

The fascinating answer, perhaps shocking or at least confusing, is no, the passengers on Cantor's cruise ship will not fit into Hilbert's hotel. Cantor's profound discovery was that the set of real numbers is *uncountable*. The real numbers cannot be placed in a one-to-one correspondence with the natural numbers. In particular, this shows that there are different sizes of infinity, for the infinity of the real numbers is a larger infinity than the infinity of the natural numbers.

The claim is that the set of real numbers $\mathbb{R}$ is uncountable, that it cannot be put in one-to-one correspondence with the natural numbers $\mathbb{N}$, and therefore the size of $\mathbb{R}$ is a higher-order infinity than that of $\mathbb{N}$. To prove this, let us suppose toward contradiction that the real numbers form a countable set. Thus, the real numbers can be enumerated in a list $r_1, r_2, r_3, \ldots$, in which every real number appears. Using this list, let us describe a particular real number z, by specifying its decimal digits:

$$z = 0.d_1 d_2 d_3 \cdots$$

We choose the digit d_n specifically to be different from the nth digit of the decimal representations of r_n. Let us take $d_n = 1$, unless the nth digit of a decimal representation of r_n is 1, in which case we use $d_n = 7$ instead. In the construction, we ensure that $z \neq r_n$ exactly at stage n, precisely because these two numbers differ in their nth decimal digits (and note that z, having all digits 1 or 7, has a unique decimal expansion). Since this is true for every n, it follows that the real number z does not appear on the list. But this contradicts our assumption that every real number appears on the list. And so the real numbers form an uncountable set, completing the proof.

To deepen our understanding of the argument, let me illustrate the construction with an explicit example. Perhaps the initial list of numbers happened to begin like this:

$$r_1 = \pi$$
$$r_2 = e$$
$$r_3 = \sqrt{2}$$
$$r_4 = \ln(2)$$
$$r_5 = \sqrt{29}$$
$$\vdots$$

We may line up the decimal expansions of these numbers after the decimal place, like this:

$$r_1 = 3.\,1\,4\,1\,5\,9\,2\,6\,5\,3\,5\,8\cdots$$
$$r_2 = 2.7\,1\,8\,2\,8\,1\,8\,2\,8\,4\,5\cdots$$
$$r_3 = 1.4\,1\,4\,2\,1\,3\,5\,6\,2\,3\,7\cdots$$
$$r_4 = 0.6\,9\,3\,1\,4\,7\,1\,8\,0\,5\,5\cdots$$
$$r_5 = 5.3\,8\,5\,1\,6\,4\,8\,0\,7\,1\,3\cdots$$
$$\vdots \qquad \ddots$$

By inspecting those red diagonal digits, we see that the resulting diagonal real number z defined in the proof will be

$$z = 0.\,7\,7\,1\,7\,1\,\cdots$$

since the construction said to use 1 as the nth digit, unless the nth digit of r_n is 1, in which case we are to use 7 as the digit. By consulting the red digits on the diagonal, this process results in the real number z above. The real number z is called the *diagonal real number* for the given enumeration, because the digits of z are defined by reference to the highlighted diagonal digits on the original list of numbers, the nth digit of the nth number. For this reason, Cantor's argument method is known as the *diagonal method*, a method that is extremely robust, appearing in perhaps thousands of mathematical arguments, many of them quite abstract. We shall see a few more diagonal arguments in this chapter.

Let me briefly address a subtle but ultimately minor issue caused by the fact that some real numbers have two representations, such as with $0.9999\cdots = 1.0000\cdots$, and for this reason, it is not correct in general to speak of "the" decimal expansion of a real number. Since nonuniqueness arises only in connection with such an eventually-all-nines/all-zeros situation, and since our diagonal real number z has only 1s and 7s, it will have a unique representation. Consequently, and indeed because of this property itself, we will achieve $z \neq r_n$ automatically if r_n should have two representations. So this problem is moot.

Cantor's original argument

As a historical matter, Cantor's original argument did not refer to the digits of the numbers, in the manner of argument we have given here. Rather, he argued as follows: If the real numbers form a countable set, then we may enumerate them in a list r_1, r_2, r_3, and so on. We may then construct a nested sequence of closed intervals, choosing the nth interval specifically to exclude the number r_n:

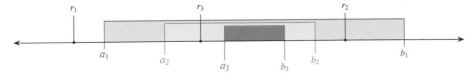

That is, we choose the first interval $[a_1, b_1]$ specifically to exclude the number r_1, we choose the next interval $[a_2, b_2]$ so as to exclude r_2, and so on, progressively through the natural numbers. Let $z = \sup_n a_n$ be the supremum of the left endpoints of the intervals. This is a real number that is therefore inside all the intervals. It is consequently different from every number r_n because those numbers were each systematically excluded from the intervals. So z is a real number that is not on the orginal list of numbers. But this contradicts the assumption that the list had included every real number. So the set of real numbers is uncountable.

Some authors have remarked on the difference between the two arguments, claiming that Cantor's original argument did not have the diagonalization idea, which, they say, came only later. My perspective is that this criticism is wrong, and that although the two proofs are superficially different, the first using diagonalization and the second using a nested sequence of closed intervals, nevertheless they are fundamentally the same construction.

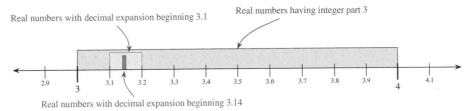

The reason is that to specify the first several digits of a real number in the decimal expansion is exactly to restrict to a certain interval of real numbers, the numbers whose expansion begins with those digits. And so to specify an additional digit of the diagonal real number so as to be different from r_n amounts to restricting to an interval of numbers excluding r_n. So the two arguments are describing the same underlying idea.

Mathematical cranks

Cantor's theorem and proof are sometimes misunderstood in certain characteristic ways, and there is the phenomenon of the mathematical "crank," an unfortunate person who takes himself or herself as an unrecognized genius who has refuted the established dogma of mathematics, but who actually has done nothing of the sort. I sometimes receive email or manuscripts from people claiming that Cantor's proof is all wrong, or that all of mathematics is wrong, a conspiracy of delusion. Sadly, the objections they raise are without force.

Let me try to give a taste of it. In some manuscripts, the proposal is essentially that after having constructed Cantor's diagonal real number z, if it does not appear on the original list $r_1, r_2, r_3, \ldots$, then one should simply *add it to the list*! Put the diagonal real number z in front, they say, and everything is fine. This proposal, of course, is without merit. It simply does not matter that the diagonal real number can appear on another list, different from the original list. The point of the diagonal real number is that because it is not on the original list, it refutes the supposition that the original list already contained all real numbers. This contradiction shows that it is impossible to make a list containing all real numbers, and that is exactly what it means to say that the set of real numbers $\mathbb{R}$ is uncountable.

Another strategy one sees is that a person will simply ignore Cantor's argument and instead try to mount his or her own direct proof that the set of real numbers is countable. For example, the person might start by proving that the set of real numbers with finite terminating decimal expansions is countable, and next, point out that every real number is a limit of such numbers; and so, they conclude, there are only countably many real numbers.

This argument starts out fine, since indeed, the numbers with a finite terminating decimal expansion are precisely the rational numbers expressible with a denominator that is a power of 10, and so there are only countably many such numbers. The second step is also fine, since indeed, every real number is a limit of a sequence of such rational numbers, which approximate it ever more closely. The last step of the argument, however, is not correct; and indeed, the claim that the set of limit points of a countable set is countable is usually left with no proof or with an erroneous proof, and the conclusion is simply incorrect. The set of real numbers does have a countable dense set, but this does not mean that the real numbers themselves are countable.

3.6 Cantor on transcendental numbers

Cantor used his theorem to provide a new proof of Liouville's 1844 theorem that transcendental numbers exist. Indeed, Cantor proved not only that transcendental real numbers exist, but furthermore, *most* real numbers are transcendental. Specifically, while there are only countably many algebraic real numbers, there are uncountably many transcendental real numbers—a strictly larger infinity.

To prove this, Cantor observed first that there are only countably many real algebraic numbers. A real number r is algebraic, you may recall from chapter 1, if it is a root of a nontrivial polynomial in one variable over the integers. In algebra, one proves that every such polynomial has only finitely many roots. For example, $\sqrt{2}$ is one of two roots of the polynomial $x^2 - 2$, the other being $-\sqrt{2}$. In this way, every algebraic real number is uniquely determined by specifying a finite list of integers: the coefficients of a polynomial of which it is a root and its place in order among the roots of that polynomial. Since the collection of all finite sequences of integers is countable, there are consequently only countably many algebraic real numbers.

Next, since Cantor had proved that there are uncountably many real numbers, it follows that there must be some real numbers that are not algebraic, which is to say that they are transcendental. Indeed, since the union of two countable sets is countable, there must be uncountably many transcendental real numbers, but only countably many algebraic numbers. In this sense, most real numbers are transcendental.

Constructive versus nonconstructive arguments

Let me highlight a certain subtle issue raised by this argument. The claim that there are transcendental numbers is an existential claim, a claim that a certain kind of mathematical object exists (a transcendental number). Existential claims in mathematics can often be proved with a *constructive* proof, a proof that explicitly exhibits the mathematical object in question. A nonconstructive proof, in contrast, might prove that there is such an object without providing any specific instance and without directing us to any specific example of that kind of object. A nonconstructive proof of an existential claim, for example, might proceed by deriving a contradiction from the assumption that there are no such objects.

Consider the following argument: there must be a cat in this warehouse, since we do not see any mice. The argument is nonconstructive, since it does not exhibit any specific cat; a constructive argument would show us the actual cat. Similarly, when we argue that there must be a transcendental real number because otherwise there would be only countably many real numbers, then we do not seem to learn of any specific transcendental number, while a constructive argument like Liouville's would exhibit an actual instance. Mathematicians often prefer constructive proofs, when possible, if the difficulty of the argument is comparable. Are constructive arguments more convincing? Perhaps, because when the object is explicitly provided, then it seems less likely that there has been a mistake in the argument; when you show me an actual cat in the warehouse, it is easy to believe that there is a cat in the warehouse. Nevertheless, constructive proofs can often be fussy and more complicated than the alternatives, precisely because they make a more detailed claim. Another reason to prefer constructive proofs is that they often seem to provide more information about the nature of the object asserted to exist than merely the raw statement of existence. When you show me the actual cat, then I might see its markings and learn whether it is friendly, and so forth.

One must be aware that there is a certain clash in the usage of the *constructive* terminology. Namely, on the one hand, mathematicians commonly use the term in the loose way I have just described, to refer to a situation where a mathematical argument provides a means of building—in some sense—the mathematical object that the argument is about, even in cases where those means might be infinitary or noncomputable. But meanwhile, on the other hand, mathematicians and philosophers following the philosophical position known as *constructivism* use the term with a more exact meaning, imposing much stricter requirements on what counts as constructive. For example, in this usage, constructive arguments are not allowed even to use the law of excluded middle, such as might arise in an argument by cases, even though such a use might still be considered constructive in the more permissive ordinary usage. We shall discuss this topic further in chapter 5. For now, let us proceed with the looser, common usage.

So, is Cantor's proof of the existence of transcendental numbers constructive? Liouville's earlier 1844 argument was certainly constructive; he provided a specific real number, the Liouville constant, by specifying its digits, and he proved that it was transcendental. Does Cantor's argument also provide specific transcendental numbers? Or is it instead only a pure existence proof, showing that there must be some transcendental numbers, but we know nothing about them, like the unseen cats in the warehouse.

One occasionally, or even often, finds mathematicians asserting that indeed, Cantor's argument is not constructive; I once saw an extremely prominent mathematician make this assertion to hundreds of people in a big public lecture. Mathematicians making this claim usually explain that Cantor's proof of the uncountability of the set of real numbers proceeds by contradiction—it is uncountable because any proposed enumeration is inadequate— and one then deduces, also by contradiction, that there must be transcendental numbers because otherwise there would be only countably many real numbers, which is a second contradiction. This way of arguing indeed appears nonconstructive, since it proves that some real numbers must be transcendental, but it does not seem to exhibit any particular transcendental number.

Nevertheless, I claim that Cantor's argument, undertaken properly, is actually constructive and does provide specific transcendental numbers. To see this, notice first that Cantor's proof that the algebraic numbers are countable provides an explicit method for enumerating them and second that Cantor's diagonalization method provides for any given enumeration of real numbers a method of constructing a specific number—the diagonal real—which does not appear on that enumeration. By combining these two steps, therefore, first enumerating the algebraic numbers and then producing the corresponding diagonal real number, we thereby construct a specific real number that is not algebraic. So Cantor's proof is constructive after all.

3.7   On the number of subsets of a set

Cantor generalized his diagonal argument to show that every set has strictly more subsets than elements. No set X is equinumerous with its *power set* $P(X)$, the set of all subsets of X. Let us give the following argument: We suppose toward contradiction that a set X is equinumerous with its power set $P(X)$. So there is a one-to-one correspondence between the elements of X and the subsets of X, associating each element $a \in X$ with a subset $X_a \subseteq X$ in such a way that every subset is realized. Given this association, let us define the diagonal set:

$$D = \{\, a \in X \mid a \notin X_a \,\}.$$

This is a perfectly good subset of X. The key thing to notice is that for every element a, we have $a \in D$ exactly when $a \notin X_a$. This immediately implies that $D \neq X_a$ for every $a \in X$. So D is a subset that is not associated to any element $a \in X$, contrary to our assumption that we had a one-to-one association of the elements of X with all the subsets of X. So $P(X)$ is not equinumerous with X. In fact, the argument shows that there can be no surjective map from a set X to its power set.

This argument follows an abstract form of Cantor's essential diagonal idea, since we have placed an element a into the diagonal subset D depending on whether a is in X_a. That is, at coordinate a, we work with set X_a, in effect working on the abstract diagonal of $X \times X$.

So far, we have proved that no set X is equinumerous with its power set $P(X)$. Meanwhile, it is easy to see that every set X has at least as many subsets as elements, for we can associate each element $x \in X$ with the singleton subset $\{x\}$. Therefore, the power set of a set is strictly larger than the set, $X < P(X)$.

On the number of infinities

The power set of the natural numbers $P(\mathbb{N})$, for example, is larger than $\mathbb{N}$, and so this is an uncountable set. By iterating the power set operation, we may therefore produce larger and larger sizes of infinity:

$$\mathbb{N} < P(\mathbb{N}) < P(P(\mathbb{N})) < P(P(P(\mathbb{N}))) < \cdots$$

Each successive power set is a strictly larger uncountable infinity, and so in particular, there are infinitely many different sizes of infinity.

How many infinities are there? Well, infinitely many, as we have just said. But which infinity? Suppose that we have κ many different infinite cardinals κ_i, indexed by elements i in some index set I of size κ. So each cardinal κ_i is the cardinality of some set X_i. Let us fix particular such sets X_i, and let $X = \bigcup_{i \in I} X_i$ be the union of all these sets, which is therefore at least as large as any of them. The power set $P(X)$, therefore, is strictly larger than every set X_i. So we have found a cardinality larger than every κ_i. What we have proved is that no set of cardinal numbers can have all the cardinal numbers.

The general conclusion, therefore, is:

There are more different infinities than any given infinity.

We shall pursue this further in chapter 8. While the argument here used the axiom of choice in picking the sets X_i, nevertheless a version of it can be undertaken in Zermelo-Fraenkel set theory without the axiom of choice. Meanwhile, the conclusion cannot be proved in the weaker Zermelo system, which lacks the replacement axiom; indeed, Zermelo's theory is consistent with the existence of only countably many different cardinalities.

Russell on the number of propositions

Bertrand Russell (1903) mounts a related argument, with credit to Cantor, that there are strictly more propositions than objects. First, he argues directly that there are at least as many propositions as objects:

> Or again, take the class of propositions. Every object can occur in some proposition, and it seems indubitable that there are at least as many propositions as there are objects. For, if u be a fixed class, "x is a u" will be a different proposition for every different value of x; ... we only have to vary u suitably in order to obtain propositions of this form for every possible x, and thus the number of propositions must be at least as great as that of objects. (Russell, 1903, §348)

Second, using Cantor's diagonalization method, he argues that there cannot be the same number of propositional functions as objects:

> Again, we can easily prove that there are more propositional functions than objects. For suppose a correlation of all objects and some propositional functions to have been affected, and let φ_x be the correlate of x. Then "not-$\varphi_x(x)$," i.e. "φ_x does not hold of x," is a propositional function not contained in the correlation; for it is true or false of x according as φ_x is false or true of x, and therefore it differs from φ_x for every value of x. (Russell, 1903, §348)

The formal diagonal logic of this argument is nearly identical to Cantor's proof that a set X has more subsets than elements.

On the number of possible committees

Let us recast the argument in an anthropomorphic form by arguing that the number of possible committees that may be formed from a given body of people is strictly larger than the number of people, whether there are finitely or infinitely many people. Let us presume that a committee may be formed from any set of people from that body. So we have the general committee, of which everyone is a member, and the empty committee, of which nobody is a member, and all the one-person committees and the various two-person and three-person committees, and so on. From the one-person committees, we see immediately that there are at least as many committees as people. This by itself does not show that there are more committees than people, however, for in the infinite case, perhaps there could still be a one-to-one correspondence between people and committees. So let us argue via

Cantor diagonalization that there is no such correspondence. Suppose toward contradiction that there is a one-to-one correspondence between people and committees. Let us interpret this correspondence as a committee-naming scheme, by which we name each committee after a different person. Let us now form the committee D, consisting of all and only those people that are not members of the committee that is named after them. This is a perfectly good committee, and so it must be named after someone, person d. Let us ask the question, is person d a member of committee D? If yes, then d is a member of the committee named after d, in which case d should *not* be a member of D; but if d is not a member of D, then d is not a member of the committee named after d, and so d *should* be a member of D. In either case, we have reached a contradiction, and so there can be no such naming system, and so the number of committees must therefore be strictly larger than the number of people.

This reasoning is essentially similar to that used by Cantor when he proved that the number of subsets of a set is greater in cardinality than the set itself, and it is also essentially similar to the reasoning we shall use in the case of the Russell paradox in chapter 8. In one sense, the reasoning does not extend to the class V of all sets, since in fact every sub*set* of V is a set and hence an element of V, and so there are just as many subsets of V as elements. But since V is a class and not a set, the analogous generalization would consider the subclasses of V rather than the subsets, and in this case, the argument does work. Namely, in all the standard class theories, such as Gödel-Bernays set theory and Kelley-Morse set theory, there can be no class enumeration of all the subclasses of V, as once again one can diagonalize against any such candidate.

The diary of Tristram Shandy

Let us turn now to some fun paradoxes of infinity. Consider first the paradox of Tristram Shandy, introduced by Russell in his *Principles of Mathematics*:

> Tristram Shandy, as we know, took two years writing the history of the first two days of his life, and lamented that, at this rate, material would accumulate faster than he could deal with it, so that he could never come to an end. Now I maintain that, if he had lived for ever, and not wearied of his task, then, even if his life had continued as eventfully as it began, no part of his biography would have remained unwritten. (Russell, 1903, §340)

The point is that if Tristram were to write about his nth day all through his nth year, then indeed every day would be chronicled. Like Galileo's line segment and the line, the days of eternal time are equinumerous with the years.

The cartographer's paradox

Whereas the diary of Tristram Shandy is stretched out temporally, Jorge Luis Borges suggests a spatial analogue, a supposed extract from a fictional seventeenth-century work:

> ...In that Empire, the Art of Cartography attained such Perfection that the map of a single Province occupied the entirety of a City, and the map of the Empire, the entirety of a

Province. In time, those Unconscionable Maps no longer satisfied, and the Cartographers Guilds struck a Map of the Empire whose size was that of the Empire, and which coincided point for point with it. The following Generations, who were not so fond of the Study of Cartography as their Forebears had been, saw that that vast map was Useless, and not without some Pitilessness was it, that they delivered it up to the Inclemencies of Sun and Winters. In the Deserts of the West, still today, there are Tattered Ruins of that Map, inhabited by Animals and Beggars; in all the Land there is no other Relic of the Disciplines of Geography.

–Suárez Miranda, *Travels of Prudent Men*, Book Four, Ch. XLV, Lérida, 1658.

Jorge Luis Borges (2004)

The great map of the province filling the city was itself a prominent geographical feature, depicted on the map as a reduced facsimile map-of-a-map, which was in turn displayed further reduced, and so on until details were lost in the fine grain of resolution. The Great Map of the Empire, however, depicting itself upon itself at the identity 1:1 scale, was self-depicted with no such loss of resolution.

But what of magnified scales? In keeping with Tristram Shandy, let me tell you of the competing sister empire that I encountered across the sea, having won the Great Contest of Cartography with its own grandiose map at a scale far exceeding 1:1, larger than life, so that the map of the city fills the province and that of the province fills the empire. Each square meter of land is depicted on 1 square kilometer of the Map, which displays the cellular structure of plants and the elaborately magnified details of the handwriting in the citizen's notebooks and diaries, written as though with enormous brushstrokes on a vast canvas. And indeed all the details of the Map are themselves geographic features to be writ large again, but farther off, and therefore writ even larger *again* still farther off, and so on continually. The cartographer's accidental drop of red ink on the vellum becomes recorded magnified first as a pool, and then a lake, and then an ocean, if one should venture far enough to find it. Nevertheless, if space is an unending Euclidean plane, then indeed there is space enough that every location on it is recorded infinitely often magnified in this self-similar way.

The Library of Babel

Borges also relates "The Library of Babel," a vast arrangement of identical hexagonal galleries, connected by narrow passages and spiral stairways, each gallery having four bookcases of five shelves; upon each shelf sits thirty-five books of uniform format, with each book having exactly 410 pages, each page having forty lines, and each line eighty letters, all from the same twenty-five-character set, if one includes spaces, periods, and commas. The librarians, one for each three galleries, are provided bare subsistence, but most have long abandoned the library or gone insane. One came to formulate the theory of the vast library: that it contained all possible books of that format, no two identical.

> From these two incontrovertible premises he deduced that the Library is total and that its shelves register all the possible combinations of the twenty-odd orthographical symbols (a number which, though extremely vast, is not infinite): Everything: the minutely detailed history of the future, the archangels' autobiographies, the faithful catalogues of the Library, thousands and thousands of false catalogues, the demonstration of the fallacy of those catalogues, the demonstration of the fallacy of the true catalogue, the Gnostic gospel of Basilides, the commentary on that gospel, the commentary on the commentary on that gospel, the true story of your death, the translation of every book in all languages, the interpolations of every book in all books. Borges (1962)

Most of the books, of course, are gibberish—one of them repeats "CVS" over and over for the entire book. But also, no one meaningful book is so precious so as not to be lost, for there are vast numbers of facsimiles, differing from it inconsequentially, perhaps by a single comma or space, or by shifting the text. Yet, there are only finitely many books in total. Each page displays $40 \times 80 = 3200$ characters, and so each book has $410 \times 3200 = 1312000$ characters, a little over one million. Since there are 25 possibilities for each character, the number of books is therefore $25^{1312000}$. A very large number indeed, but not infinite.

Quine writes on the "Universal Library," offering another conception via binary notation:

> There is an easier and cheaper way of cutting down. ...a font of two characters, dot and dash, can do all the work of our font of eighty. ...The ultimate absurdity is now staring us in the face: a universal library of two volumes, one containing a single dot and the other a dash. Persistent repetition and alternation of the two is sufficient, we well know, for spelling out any and every truth. The miracle of the finite but universal library is a mere inflation of the miracle of binary notation: everything worth saying, and everything else as well, can be said with two characters. It is a letdown befitting the Wizard of Oz, but it has been a boon to computers. (Quine, 1987, p. 223)

On the number of possible books

Since we are discussing the number of possible books, let me give the farcical argument that in fact, there are only finitely many possible books, even when one does not restrict the length. By *book*, let us understand something ordinary: a book consists of a finite sequence of pages, where each page has an ordinary form, consisting of, say, at most 100 lines of text, each line having at most 100 characters from a fixed finite character set, and with a page number appearing in the ordinary way at the bottom, incremented with each page. We do not assume anything about the number of pages in the book.

It might seem at first that there must be infinitely many possible books of this form, for we shall have the all-As book of any desired length, consisting of the letter A over and over again on each page, and similarly with B, and so on. Does this not lead to infinitely many different books? No. I claim that there are only finitely many possible books of the described form. The critical reason is the page number requirement, for once the number of pages becomes very large, there simply will not be enough room to write the page number at the bottom, if one should write it "in the ordinary way," as we said. Even if one allows the page number to spill onto more than one line, then eventually the page number will have to fill the entire page with its digits. So there is an absolute finite upper bound on how large the page number could be to legally appear on a page. And therefore there is an absolute upper bound on the number of pages that a "book" following my description could include. Since only finitely many characters appear on each page, it follows that there are only finitely many possible books of that type.

Meanwhile, the argument breaks down if one allows nonstandard page numbering, such as by writing 1 on the first page and +1 on all subsequent pages, or using the phrase "the next number after the previous page's number," or some such. If this is allowed, then clearly there would be infinitely many possible books by making them longer and longer.

3.8   Beyond equinumerosity to the comparative size principle

Although the Cantor-Hume principle provides us with means for expressing that two sets or classes have the same size—they are equinumerous—a robust concept of number and size would enable us also to make *comparative* size judgements, to say that one set or class is *at least as large* as another. For this, we would seem to need a principle going strictly beyond the Cantor-Hume principle, such as the one used by Cantor:

Comparative size principle. *One set is less than or equal in size to another if and only if the first set is equinumerous to a subset of the second.*

Let us write $A \lesssim B$ to mean that the set A is less than or equal in size to the set B. The principle asserts that $A \lesssim B$ exactly when there is an injective function from A to B:

$$A \lesssim B \quad \Longleftrightarrow \quad \exists f : A \xrightarrow{\text{1-1}} B.$$

The range of the function f is the subset of B with which A is equinumerous.

The comparative size principle amounts to a reflexive set-theoretic reformulation of Euclid's principle, since it is asserting that every set is at least as large as any of its subsets. The strict set-theoretic version of Euclid's principle, in contrast, asserts that every set is strictly larger than its proper subsets, and this is refuted using the Cantor-Hume principle by examples such as the natural numbers and the perfect squares, as observed by Galileo.

Our use of the $\lesssim$ symbol here carries some baggage in that it suggests that this comparative size relation will exhibit the properties of an order, or at least a preorder. We certainly want our comparative size relation to be orderlike—but is it? Well, the $\lesssim$ relation on sets is easily seen to be reflexive, meaning that $A \lesssim A$, as well as transitive, meaning that $A \lesssim B \lesssim C$ implies $A \lesssim C$. Thus, the comparative size relation $\lesssim$ begins to look like an order. A more subtle question, however, is whether the relation $\lesssim$ is antisymmetric. In other words, if two sets are at least as large as each other, must they have the same size? That way of saying it may lead some to expect that the answer is trivial or immediate. But it is not trivial. The inference, which is correct, is precisely the content of the Schröder-Cantor-Bernstein theorem.

Theorem 8 (Schröder-Cantor-Bernstein). *If $A \lesssim B$ and $B \lesssim A$, then the sets A and B are equinumerous: $A \simeq B$.*

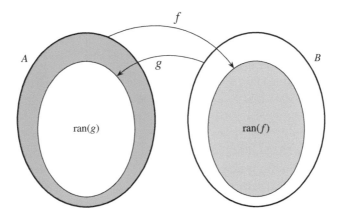

For sets A and B to obey $A \lesssim B$ and $B \lesssim A$ means that we have two injective mappings, one from left to right $f : A \xrightarrow{\text{1-1}} B$ and one from right to left $g : B \xrightarrow{\text{1-1}} A$. Each set injects into the other. The question is whether these two injective functions can somehow combine to cause a one-to-one correspondence between A and B, a bijective function from A onto B. The Schröder-Cantor-Bernstein theorem asserts that they can.

The theorem, whose proof is subtler than one might expect, has a convoluted history. It was stated by Cantor in 1887 without proof, proved but not published by Dedekind in 1888, proved by Cantor in 1895 but only by assuming the linearity of cardinals, proved by Ernst Schröder in 1896 as a corollary of a theorem of William Stanley Jevons and then again in 1897 independently (but this proof was later found to be faulty), and then proved independently by Felix Bernstein in 1897.

As Cantor had observed, the theorem is an immediate consequence of the assumption that every set can be well-ordered, and indeed, the theorem follows immediately from the assumption merely that all cardinals are linearly ordered by $\lesssim$. This linearity hypothesis, however, turns out to be equivalent to the axiom of choice, as proved by Friedrich Moritz Hartogs, and since one does not need this extra assumption to prove the theorem, it is better to do so without it.

Although my main point is that this theorem is subtler than a naive outlook would lead one to expect, nevertheless the theorem is not so terribly difficult to prove, and one can find a proof of it in any good set theory textbook. For example, I provide an elementary proof in (Hamkins, 2020c, theorem 117) based on the proof of Julius König (1906).

Let me illustrate an application of the Schröder-Cantor-Bernstein theorem by using it to prove that the real line $\mathbb{R}$ and the real plane $\mathbb{R} \times \mathbb{R}$ are equinumerous. What? Crazytown! The line and the plane have the same number of points? Yes, indeed. We clearly have $\mathbb{R} \lesssim \mathbb{R} \times \mathbb{R}$, since we can associate the real line, for example, with any given line in the plane. To prove the converse relation, observe first that the real line $\mathbb{R}$ is equinumerous with the positive real numbers $\mathbb{R}^+$ by the map $x \mapsto e^x$. Thus, the plane $\mathbb{R} \times \mathbb{R}$ maps bijectively to the positive quadrant $\mathbb{R}^+ \times \mathbb{R}^+$. Next, given a point (x, y) in that quadrant, let z be the real number that interleaves the digits of x and y in their decimal expansions:

$$(3.14159\cdots, 2.71828\cdots) \mapsto 32.1741185298\cdots$$

There is an irritating issue caused by the fact that some real numbers have two decimal representations, such as with $1.000\cdots = 0.999\cdots$, and so for the interleaving process, let us decide always to use the representations of x and y that do not end in all 9s. The resulting map $(x, y) \mapsto z$ is injective, since from z we can recover the decimal digits of x and y (note that z will never end in all 9s). This association is not onto $\mathbb{R}^+$, since the nonuniqueness issue means that we shall never realize the number $0.90909090\cdots$ by interleaving digits this way. What we have altogether, however, is an injection of $\mathbb{R} \times \mathbb{R}$ first into $\mathbb{R}^+ \times \mathbb{R}^+$ and then ultimately into $\mathbb{R}$, which injects back into $\mathbb{R} \times \mathbb{R}$. By the Schröder-Cantor-Bernstein theorem, therefore, these sets are all equinumerous. That is, although we have not exhibited an explicit one-to-one correspondence between the line $\mathbb{R}$ and the plane $\mathbb{R} \times \mathbb{R}$, we have exhibited injections in both directions, and so the theorem provides a bijective correspondence between them. An essentially similar argument shows that the real line $\mathbb{R}$ is equinumerous with the real number space $\mathbb{R}^n$ in any finite or indeed countable dimension. As mentioned earlier (page 9), this shocking result led Cantor to write to Dedekind in 1877, "I see it, but I don't believe it!" (Dauben, 2004, p. 941)

Place focus on reflexive preorders

I should like to discuss what I have observed as a common feature in several instances where a mathematical or philosophical analysis stumbles while extending ideas from the finite realm to the infinite. Suppose we are analyzing a preorder relation, such as the larger-than or at-least-as-large-as relation, or the better-than or at-least-as-good-as relation, or the relations arising with supervenience, or those arising with grounding or some other reduction concept. The problematic feature I have in mind occurs when one gives priority to the strict version of the order over the weaker reflexive order. For example, with the set-theoretic Euclid's principle, which asserts that every set is strictly larger than its proper parts, one has extended ideas that are true of finite sets for the strict order, but this principle stumbles in the infinite realm, as Galileo noticed with the natural numbers and the perfect squares. The corresponding principle for the reflexive order, meanwhile, is exactly the comparative size principle, which is totally fine and compatible with the Cantor-Hume principle. A similar situation occurs in the subject of infinitary utilitarianism, where the literature is full of versions of the strict Pareto principle, asserting that if every individual is made strictly better off (even in an infinite world), then the world as a whole is strictly better off. This principle, however, contradicts very natural isomorphism-invariance principles by admitting counterexample situations in a manner analogous to Galileo's examples (see Hamkins and Montero 2000a, 2000b). My view is that the strict Pareto principle is just as misguided in the infinite realm of utiliarianism as Euclid's principle is for infinite cardinality, for very similar reasons. Meanwhile, the reflexive version of the Pareto principle, which asserts merely that the resulting world is at least as good when all individuals are at least as good, avoids the counterexamples and remains compatible with those isomorphism principles. Similar issues arise in many other instances, including those concerned with various supervenience, grounding, and reduction relations. When working with a philosophical preorder relation, therefore, I tend to regard it as a fundamental mistake—often a kind of beginner's error—to take the strict version of the order relation as the primary notion for the purpose of forming intuitions or stating one's principles. One should instead base one's ideas and principles on the more robust reflexive order relation. One reason is that the same kind of problematic counterexample situations as Galileo observed will continually recur in these other infinitary contexts. A second reason is the mathematical observation that, in general, the reflexive version of a preorder relation carries more information than the strict version of the relation, in the sense that one can define the strict order from the reflexive order, but not conversely, since a given strict order can arise from more than one reflexive preorder (Hamkins, 2020c, p. 164); the basic problem is that the strict order cannot distinguish between indifference, $x \leq y \leq x$, and incomparability, $x \not\leq y \not\leq x$. Thus, the reflexive preorder relation is more fundamental than the corresponding strict order. To avoid the common pitfalls, therefore, kindly focus your philosophical analysis on the reflexive version of whatever preorder relation you are studying, rather than on the strict order relation.

3.9 What is Cantor's continuum hypothesis?

Cantor was intrigued by a curious observation. Namely, in mathematics we seem to have many natural instances of countably infinite sets:

> The set of natural numbers $\mathbb{N}$; the set of integers $\mathbb{Z}$; the set of rational numbers $\mathbb{Q}$; the set of finite binary sequences $2^{<\mathbb{N}}$; the set of integer polynomials $\mathbb{Z}[x]$.

We seem also to have many natural instances of sets of size continuum:

> The set of real numbers $\mathbb{R}$; the set of complex numbers $\mathbb{C}$; Cantor space (infinite binary sequences) $2^{\mathbb{N}}$; the power set of the natural numbers $P(\mathbb{N})$; the space of continuous functions $f : \mathbb{R} \to \mathbb{R}$.

But we seem to have no sets provably of intermediate cardinality between $\mathbb{N}$ and $\mathbb{R}$. Cantor struggled to settle the question of whether there was a set of intermediate cardinality between the natural numbers and the continuum.

The *continuum hypothesis* (CH), formulated by Cantor, is the assertion that indeed there are no infinities between the countable and the continuum, between the natural numbers $\mathbb{N}$ and the real numbers $\mathbb{R}$. Is it true? Is it refutable? The question consumed Cantor. Hilbert had listed the continuum hypothesis as the very first on his famous list of twenty-three open problems at the beginning of the twentieth century, problems that guided so much of the future of mathematical research.

Cantor was able to prove many instances of the continuum hypothesis, showing, for example, that it holds for closed sets of real numbers. He had wanted to work his way up to more complicated sets. This was ultimately shown to be a sound idea, at least in part. One can prove, for example, that it holds for the Borel sets, the sets arising in the hierarchy of sets starting with the open sets and closing under countable unions, intersections, and complements.

Going beyond this, in the late twentieth century, set theorists proved that if there are certain sufficient kinds of large cardinals that fulfill very strong axioms of infinity, then indeed the continuum hypothesis holds not only for the Borel sets, but for all sets in the projective hierarchy, which is the expansive hierarchy of sets that you can form by defining sets of real numbers in a language that allows basic arithmetic concepts and quantification over the integers and over the real numbers. In this way, Cantor's strategy was partially fulfilled.

Gödel proved in 1938 that if the axioms of ZF set theory are consistent, then they are also consistent with the continuum hypothesis and the axiom of choice. Thus, one cannot expect to refute these principles, and in this sense, it is safe to assume that they are true. This result explains why Cantor was not able to find a definitive set of intermediate cardinality between $\mathbb{N}$ and $\mathbb{R}$, since it is consistent with set theory that there is no such set.

Cohen proved in 1963 that if the axioms of ZF set theory are consistent, then they are also consistent together with the axiom of choice and the *negation* of the continuum hypothesis. In this sense, it is also safe to assume that the continuum hypothesis is false. This result may explain why Cantor was never able to find a proof that it was true, since it is consistent with the axioms of set theory that it is not; a counterpoint is that Cantor was not working in a formal system. We shall revisit these set-theoretic ideas in chapter 8.

3.10 Transfinite cardinals—the alephs and the beths

Let us now develop a little more of Cantor's theory of the transfinite. We have already seen that Cantor's argument shows that there are infinitely many different sizes of infinity, and indeed there are more different sizes of infinity than any given infinity.

Cantor defined the sequence of infinite cardinals by starting with the smallest infinity, $\aleph_0$, which is the size of the natural numbers or any countably infinite set, and more generally defining $\aleph_\alpha$ to be the αth infinite cardinality, leading to the sequence of infinite cardinalities:

$$\aleph_0 < \aleph_1 < \aleph_2 < \cdots < \aleph_\omega < \aleph_{\omega+1} < \cdots$$

The cardinal $\aleph_\omega$ is the supremum of the cardinals $\aleph_n$ for finite numbers n. This cardinal may seem quite large, as it is uncountable and larger than infinitely many other uncountable cardinals; indeed, it is a *limit cardinal*, which means that there is no immediately preceding infinity.

Meanwhile, the cardinal $\aleph_\omega$ has another curious property, which may make it seem comparatively accessible. Namely, because it is the supremum of the cardinals $\aleph_n$ for $n < \omega$, it is therefore a *singular* cardinal, the supremum of a comparatively small set of comparatively small cardinals. That is, $\aleph_\omega$ is the supremum of a set of size less than $\aleph_\omega$ consisting of cardinals each of size less than $\aleph_\omega$. With this ladder, we thus climb easily up to $\aleph_\omega$. Notice that the cardinal $\aleph_0$, in contrast, the countably infinite cardinality, is a *regular* cardinal, one that admits no such comparatively short ladder: every finite sequence of finite numbers has a finite supremum. In this sense, the cardinal $\aleph_0$ is inaccessible from below.

Cantor also considered the alternative related hierarchy of cardinals, the *beth hierarchy*, which arises by iterating the power set operation at each step:

$$\beth_0 < \beth_1 < \beth_2 < \cdots < \beth_\omega < \beth_{\omega+1} < \cdots$$

We start with the smallest infinite cardinal $\beth_0 = \aleph_0$, and at each stage, we apply the power set operation $\beth_{\alpha+1} = 2^{\beth_\alpha}$, taking suprema at limit stages.

The continuum hypothesis amounts to the assertion that $\aleph_1$ and $\beth_1$ are the same infinite cardinal, that our principal method of producing uncountable cardinalities—taking the power set of a countably infinite set—also happens to produce the very next uncountable cardinality. More generally, the *generalized continuum hypothesis* is the assertion that the two hierarchies are identical all the way up, that $\aleph_\alpha = \beth_\alpha$ for every ordinal α.

The cardinal $\beth_\omega$ is quite interesting, for it is a *strong limit* cardinal, meaning that it is closed under the power set operation, in that every set of size less than $\beth_\omega$ has its power set also of size less than $\beth_\omega$; this is because if a set has size less than $\beth_\omega$, then it has size at most $\beth_n$ for some natural number n, in which case its power set has size at most $\beth_{n+1}$, which is smaller than $\beth_\omega$. In this sense, the cardinal $\beth_\omega$ is inaccessible by means of power sets. Yet, like $\aleph_\omega$, it is singular; it admits a short ladder, a short cofinal sequence, for it is the supremum of the $\beth_n$ for $n < \omega$, making it the supremum of a small set of small cardinals.

Could there be a truly *inaccessible* cardinal? This would be an uncountable cardinal κ that was (1) inaccessible by power sets, in that every set of size less than κ also had its power set of size less than κ; but also (2) a regular cardinal, one admitting no short cofinal sequence, no sequence of fewer than κ many cardinals, each of size less than κ, whose supremum was κ. The smallest infinity $\aleph_0$, although countable, has both of those latter properties since the power set of a finite set is finite and the supremum of a finite sequence of finite cardinals is also finite. Thus, to ask for an inaccessible cardinal is to ask for a higher infinity, an uncountable cardinal that looks to the smaller cardinals, including $\aleph_0$, just as $\aleph_0$ looks to the finite cardinals. Can it happen?

The answer is complex and fascinating, a story that we shall tell in chapter 8. The hypothesis that there is an inaccessible cardinal is one of the first *large cardinal* axioms, a strong axiom of infinity, hypothesizing the existence of an infinite cardinality with certain precise largeness features. The large cardinal axioms are the strongest-known axioms in mathematics.

Lewis on the number of objects and properties

Expanding on Russell's observations concerning the number of propositions, David Lewis considers the dual problems: How many things are there? And then, how many properties of things are there? He offers the surprisingly precise answers of at least $\beth_2$ and $\beth_3$, respectively. Let me try to reconstruct his reasoning. Let us assume that the actual physical space and spacetime that we inhabit is well modeled, as in our physical theories, by the real numbers. (This is easily debatable; physicists need not take their physical models to be fully physically real; economists model human economic behavior with real-valued differential equations, even if they recognize that ultimately it is a discrete system, with finitely many people making finitely many choices; similarly, physicists might ultimately hold that space or time is finite and discrete.) The number of spatial or spacetime locations, therefore, would be the continuum, the cardinality of the real numbers, denoted by any of $\beth_1 = 2^{\aleph_0} = |\mathbb{R}| = \mathfrak{c}$. Any set of spatial points is a spatial *region*, which we may take to be a "thing." And so the number of things is at least the number of sets of real numbers, and this is $2^{\beth_1}$, the same as $\beth_2$. So there are at least $\beth_2$ many things. If any thing is determined by the collection of spacetime points that it inhabits, extended in both space and time, then we would also know conversely that the number of things was at most the number of sets of spacetime points, and so the number of things would be *exactly* $\beth_2$. Many philosophers would dispute that presumption, however, in light of classical examples such as the statue and the lump of clay, distinct things occupying the same location; and Kit Fine (2003) creates a statue by forming an alloy in a mold, so that "the alloy" and "the statue" coincide spatially and temporally, but they are different things, since perhaps one is well made and the other not.

As for properties, Lewis writes:

> Any class of things, be it ever so gerrymandered and miscellaneous and indescribable in thought and language, and be it ever so superfluous in characterising the world, is nevertheless a property. So there are properties in immense abundance. (If the number of things, actual and otherwise, is beth-2, an estimate I regard as more likely low than high, then the number of properties of things is beth-3. And that is a big infinity indeed, except to students of the outer reaches of set theory.) There are so many properties that those specifiable in English, or in the brain's language of synaptic interconnections and neural spikes, could be only an infinitesimal minority. (Lewis, 1983, p. 346)

Thus, he is arguing that if there are $\beth_2$ many things, and any set of things determines a property, then there are at least $2^{\beth_2}$ many properties, and this is precisely $\beth_3$. In other work, he considers the number of possible worlds, concluding that it also is $\beth_2$:

> We might ask how the inductively deceptive worlds compare in abundance to the undeceptive worlds. If this is meant as a comparison of cardinalities, it seems clear that the numbers will be equal. For deceptive and undeceptive worlds alike, it is easy to set a lower bound of beth-two, the number of distributions of a two-valued magnitude over a continuum of spacetime points; and hard to make a firm case for any higher cardinality. (Lewis, 1986, p. 118)

3.11 Zeno's paradox

Most of our discussion of infinity so far can be taken under a certain contemporary perspective on infinity, viewing infinite sets as an ordinary part of the mathematical landscape. But let us briefly mention some other older perspectives on infinity.

Zeno of Elea (c. 490–430 BC) argued that all motion is impossible because before you move from here to there, you must get halfway, but before you do this, you must get halfway to the halfway point, and before you do *this*, you must get halfway to that point, and so on, ad infinitum. Thus, he concluded, you can never begin. Similarly, Achilles can never overtake the tortoise in their footrace, because before doing so, he must get to where the tortoise was, but during this time, the tortoise has moved on, and when Achilles reaches *that* point, the tortoise has moved on yet again, ad infinitum. So Achilles can never catch the tortoise. What do you think of these paradoxical arguments?

Many mathematicians take them to be satisfactorily resolved by the ideas of calculus. The integral calculus, in particular, is founded on the idea that one may calculate a finite area by dividing it into successively tiny rectangles and taking a limit of those resulting areas. Similarly, the standard limit analysis of infinite series says, directly in opposition to Zeno, that one can add up an infinite series of numbers, such as

$$\frac{1}{2} + \frac{1}{4} + \frac{1}{8} + \frac{1}{16} + \frac{1}{32} + \cdots$$

and nevertheless achieve a finite sum. In this case, the series sums to 1, a fact that

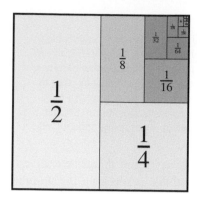

can be visualized with the help of the diagram here. The whole unit square has area one. This area is partitioned into smaller regions of area $\frac{1}{2}$, $\frac{1}{4}$, $\frac{1}{8}$, and so on. Since those regions exhaust the square, it follows that the sum of the series is 1.

Actual versus potential infinity

Another historical idea concerning infinity is Aristotle's distinction between actual and merely potential infinity. In ancient times, potentialists predominated, but now most mathematicians are actualists. The *potentialist* regards infinite collections as infinite only in a sense of unending potentiality. One can have more and more members of an infinite collection, as many (finitely many) as one likes, but it is not sensible to have them as a completed infinite totality. There are infinitely many natural numbers, in that one will never run out of them; we can never have all of them. The *actualist*, in contrast, claims to form infinite collections as completed totalities. The actualist considers the set of all natural numbers as a finished whole, and furthermore may go on to claim other infinite collections, such as the set of all real numbers, as completed totalities.

3.12   How to count

Let us conclude this chapter by learning how to count into the transfinite ordinals. This is a task that anyone can master. The ordinal numbers begin as you might expect, with the natural numbers:

$$0 \quad 1 \quad 2 \quad 3 \quad 4 \quad 5 \quad \cdots$$

But the ordinals continue beyond these finite numbers. The first infinite ordinal is known as ω, and it comes immediately after all the finite numbers, like this:

$$0 \quad 1 \quad 2 \quad 3 \quad 4 \quad 5 \quad \cdots \quad \omega$$

The ordinal ω is a *limit* ordinal, since it has no immediate predecessor, and the set of smaller ordinals is closed under the successor operation. After any ordinal, one can always add 1, and then another 1, and so on:

$$0 \quad 1 \quad 2 \quad 3 \quad 4 \quad 5 \quad \cdots \quad \omega \quad \omega+1 \quad \omega+2 \quad \omega+3 \quad \cdots$$

From this perspective, the ordinals look the same going forward, no matter where you start. After one has exhausted $\omega + n$ for every finite number n, one finally gets to $\omega + \omega$, which can also be written as $\omega \cdot 2$. This is the next limit ordinal after ω:

$$0 \quad 1 \quad 2 \quad 3 \quad 4 \quad 5 \quad \cdots \quad \omega \quad \omega+1 \quad \omega+2 \quad \omega+3 \quad \cdots \quad \omega \cdot 2$$

After this, one continues adding 1, eventually reaching $\omega \cdot 3$ and beyond:

$$
\begin{array}{lllll}
0 & 1 & 2 & 3 & \cdots \\
\omega & \omega + 1 & \omega + 2 & \omega + 3 & \cdots \\
\omega \cdot 2 & \omega \cdot 2 + 1 & \omega \cdot 2 + 2 & \omega \cdot 2 + 3 & \cdots \\
\omega \cdot 3 & \omega \cdot 3 + 1 & \omega \cdot 3 + 2 & \omega \cdot 3 + 3 & \cdots \\
\vdots \\
\omega \cdot n & \omega \cdot n + 1 & \omega \cdot n + 2 & \omega \cdot n + 3 & \cdots \quad \omega \cdot n + k \quad \cdots \\
\vdots \\
\omega^2
\end{array}
$$

Thus, we have counted up to ω^2, through all the ordinals of the form $\omega \cdot n + k$, where n and k are finite. The ordinal ω^2 is the first limit ordinal that is a limit of limit ordinals, since ω^2 is the limit of $\omega \cdot n$ as n increases in ω.

Counting up to ω^2 is rather like counting to 100. In counting to 100, one essentially counts to 10, but 10 times; a century is 10 decades, and counting within one decade is essentially the same as counting within another decade. Counting in the ordinals up to ω^2 is just like this, except that the "decades" have length ω and there are ω many of them. So, one counts to ω altogether ω times. The analogy is robust, for the ordinal numbers $\omega \cdot n + k$ are precisely the two-digit ordinal numbers, in base ω. So counting to ω^2 is counting through all the two-digit ordinals numbers, just like counting to 100.

But of course, the ordinals continue far beyond ω^2. They continue into the higher powers of ω like this:

$$
\begin{array}{llllll}
\omega^2 + 1 & \omega^2 + 2 & \cdots & \omega^2 + \omega & \cdots & \omega^2 + \omega \cdot n + k & \cdots & \omega^2 \cdot r + \omega \cdot n + k & \cdots \\
\omega^3 & \omega^3 + 1 & \cdots & \omega^4 & \cdots & \omega^m & \cdots & \omega^\omega & \cdots
\end{array}
$$

And then into the iterated powers and the fixed points of the powers:

$$
\omega^{\omega^\omega} \quad \cdots \quad \omega^{\omega^{\omega^\omega}} \quad \cdots \quad \epsilon_0 \quad \cdots \quad \epsilon_1 \quad \cdots
$$

The ordinal ϵ_0 is the smallest ordinal for which $\omega^{\epsilon_0} = \epsilon_0$, and it can be thought of as the infinitely iterated power, which explains why $\omega^{\epsilon_0} = \epsilon_0$:

$$
\epsilon_0 = \omega^{\omega^{\omega^{\omega^{\cdot^{\cdot^{\cdot}}}}}} \qquad \omega^{\epsilon_0} = \omega^{\omega^{\omega^{\omega^{\omega^{\cdot^{\cdot^{\cdot}}}}}}} = \epsilon_0.
$$

After this, one adds 1 and continues, and ϵ_1 is the next fixed point of this operation. Eventually, one gets to fixed points of the ϵ-process itself. The ordinal $\omega_1^{\mathfrak{Ch}}$ is the *omega one of chess*, which is the supremum of the game values that arise for positions in infinite chess, introduced in Evans and Hamkins (2014). The ordinal ω_1^{CK} is the Church-Kleene ordinal, which is the supremum of the ordinals that arise as the order type of a computable relation on the natural numbers.

Although all very large, these ordinals are all still merely countable ordinals, but the transfinite ordinals continue into uncountable realms:

$$\cdots \quad \omega_1^{\mathfrak{Cb}} \quad \cdots \quad \omega_1^{CK} \quad \cdots \quad \cdots \quad \omega_1$$
$$\cdots \quad \omega_2 \quad \cdots \quad \alpha \quad \cdots \quad \cdots \quad \cdots$$

The ordinal ω_1 is the first uncountable ordinal, ω_2 is the next one, and so on. Dots and more dots, of course, ultimately fail to convey the robust details of the subject, which I have merely hinted at here. Moving higher in the ordinals, one reaches much larger infinities, perhaps eventually reaching various large cardinals of set theory, such as inaccessible cardinals, Mahlo cardinals, measurable cardinals, supercompact cardinals, and more. The ordinals continue on and on and on without end, falling like grains of sand through the transfinite hourglass.

Since we discussed constructive and nonconstructive proof in this chapter, I should like to close with the playful claim of Twitter user Quarantine 'em (2020), in its entirety:

> There's a really great joke about non-constructive proofs.

Questions for further thought

3.1 Suppose that guests arrived at Hilbert's Grand Hotel in the manner described in this chapter: first one guest arrives, and then 1000 all at once, and then Hilbert's bus, followed by Hilbert's train, and finally Hilbert's half-marathon. If the manager followed the procedure mentioned in the chapter, describe who are the occupants of rooms 0 through 100. How did they arrive, and with which party? In which car or seat were they when they arrived if they arrived by train or bus? Which fraction did they wear in the marathon? If you were the very first guest to arrive, where do you end up in the end? And what is the first room above you that is occupied? How did that guest arrive?

3.2 Down the street from Hilbert's Grand Hotel is Hilbert's Cubical Co-op apartment complex, which is an infinite cubical building, like $\mathbb{N} \times \mathbb{N} \times \mathbb{N}$, where every occupant's residence can be described by a floor number n, a hallway number h, and a corridor r. Because the interior rooms have very little light, the entire cooperative wants to move to Hilbert's hotel. How can the manager accommodate them?

3.3 Can you give a direct argument that the set of all infinite binary sequences is uncountable?

3.4 Can you argue that space $\mathbb{R}^3$ has the same number of points as the real number line $\mathbb{R}$? And similarly in any finite dimension: $\mathbb{R}^n$ is equinumerous with $\mathbb{R}$. Can you argue this smoothly by induction, as a consequence from $\mathbb{R} \simeq \mathbb{R}^2$? [Hint: Show that if A and B are equinumerous, then so are $A \times C$ and $B \times C$.] What about infinite dimension?

3.5 How many continuous functions on the real numbers are there?

3.6 Assume that space is well modeled with the real numbers. How many regions of space are there? How many open regions of space are there? (If you are familiar with the Borel sets of real numbers, how many Borel sets are there?) Are your answers to these questions relevant to Lewis's calculation on the number of properties and the number of things?

3.7 In terms of cardinality, are there more integer-valued functions on the real numbers, or real-valued functions on the integers?

3.8 Galileo's Salviati seems to hold that size comparisons between infinite collections are not sensible, at least for the relations of equality, greater-than, and less-than. But what about less-or-equal and greater-than-or-equal size comparisons? Can Salviati hold that every set is at least as large as its subsets? And can Salviati hold that every set is the same size as itself?

3.9 How does Cantor reply to Galileo's suggestion (via Salviati) that size comparisons between infinite collections are not sensible? How does Cantor define such size comparisons?

3.10 Suppose that we have a supply of infinitely many billiard balls and an empty sack, and you undertake the following process: In the first minute, you place two balls into the sack, and remove one of them; in the next half-minute, you place two more balls in the sack, and remove a ball; in the succeeding quarter-minute, you place two more balls in the sack and remove one ball from the sack; and so on. After two minutes, you have completed infinitely many such exchanges. How many balls are in the sack at that time? Does it matter which ball is removed? Let us imagine that the balls are numbered 1, 2, 3, and so on, and that at each step, you always remove and discard the lowest-numbered ball currently in the sack. What then? Or suppose you always remove the largest-numbered ball? Or suppose you remove a random ball?

3.11 Argue that for any collection of fruit, there are more possible fruit salads to make than there are fruits. Identify the precise similarities amongst your argument, the argument in the text concerning the number of possible committees, and Cantor's proof that every set X is strictly smaller in size than its power set $P(X)$. If you are familiar with the Russell paradox (see chapter 8), also identify the precise similarity of these arguments with that.

3.12 Was it important in the number-of-committees argument that every person was associated with some committee? Was it important that committees were never named after multiple people? If not, rework the argument to show that there can be no surjective (onto) association of some of the people with all the committees. Similarly, prove that for any set X, there can be no surjective function from a subset of X to the power set of X.

3.13 Suppose that you have infinitely many \$1 bills (numbered 1, 3, 5, …) and upon entering a nefarious underground bar, you come upon the Devil sitting at a table piled high with money. You sit down, and the Devil explains to you that he has an attachment to your particular bills and is willing to pay you a premium to buy them from you. Specifically, he is willing to pay \$2 for each of your \$1 bills. To carry out the exchange, he proposes an infinite series of transactions, in each of which he will hand over to you \$2 and take from you \$1. The first transaction will take 1/2 hour, the second 1/4 hour, the third 1/8 hour, and so on, so that after 1 hour, the entire exchange will be complete. The Devil takes a sip of whiskey while you mull it over; should you accept his proposal? Perhaps you think that you will become richer, or perhaps you think that with infinitely many bills, it will make no difference. At the very least, you think that it will do no harm, and so the contract is signed and the procedure begins. How could the deal harm you? It appears initially that you have made a good bargain, because at every step of the transaction, he pays you \$2 and you give up only \$1. The Devil is very particular, however, about the order in which the bills are exchanged. The contract stipulates that in each subtransaction, he buys from you your lowest-numbered bill and pays you with higher-numbered \$1 bills. Thus, on the first transaction he accepts from you bill number 1, and he pays you with bills numbered 2 and 4. On the next transaction, he buys from you bill number 2 (which he had just paid you) and gives you bills numbered 6 and 8. Next, he buys bill number 3 from you with bills 10 and 12, and so on. When all the exchanges are completed, what do you discover?

3.14 Consider Laraudogoitia's beautiful supertask, Laraudogoitia (1996). Imagine infinitely many billiard balls of successively diminishing size, converging to a point. The balls are initially at rest, but then the first is set rolling. It collides with the next, transferring all its energy, and that ball begins to roll. Each ball in turn collides with the next, transferring all its energy. Because of the physical arrangements of the system, the motion disappears in a sense into the singularity; after a finite amount of time, all the collisions have taken place and the balls are at rest. So we have described a physical process in which energy is conserved at each step and in every interaction, but not overall through time. Discuss.

3.15 Another arrangement has the balls spaced farther and farther out to infinity, pushing the singularity out to the point at infinity. The first ball knocks the second, which sends the next flying, and so on out to infinity. If the balls speed up sufficiently fast, under Newtonian rules, it can be arranged that all the motion is completed in a finite amount of time. The interesting thing about this example and the previous one is that time symmetry allows them to run in reverse, with static configurations of balls suddenly coming into motion without violating conservation of energy in any interaction. Discuss.

3.16 Consider Thomson's lamp, which is on for one minute, off for a half minute, on for a quarter minute, off for an eighth, and so on. What do you think about the state of the lamp after two minutes, the total elapsed time?

3.17 Explain Cantor's proof of the existence of transcendental numbers, and discuss whether this proof is constructive. Does it provide merely a pure-existence proof of the existence of transcendental numbers, or does it provide a particular specific transcendental number?

3.18 Consider a dozen or so natural sets of real numbers, and determine in each case whether they are countable or size continuum, or whether they have intermediate cardinality between $\mathbb{N}$ and $\mathbb{R}$.

3.19 Describe a set that arises naturally in mathematics, in a subject other than logic and set theory, which has very large cardinality. How big is the largest such set you can find?

3.20 How many different cardinalities are there? How many countable cardinalities?

Further reading

Galileo Galilei (1914 [1638]). A classic; see especially the *First Day* discussion of infinity.

J. P. Laraudogoitia (1996). A brief but beautiful article, presenting the balls-to-infinity supertask example mentioned in the exercises.

Kurt Gödel (1947). Gödel's account of the continuum hypothesis, published prior to Cohen's independence result via forcing.

Jorge Luis Borges (1962), "The Library of Babel." A short work of literary genius, partaking of the finite and the infinite.

Joel David Hamkins (2018). This is my paper on arithmetic potentialism and the universal algorithm; the final section distinguishes between implicit and explicit potentialism.

Credits

Some parts of this chapter were adapted from my book, *Proof and the Art of Mathematics*, Hamkins (2020c). David Hilbert introduced his Grand Hotel in a 1924 lecture, "Über das Unendliche" ("On the Infinite")—see (Hilbert, 2013, p. 730). The presentation through Hilbert's bus, Hilbert's train, Hilbert's half-marathon and Cantor's cruise ship is my own invention from the early 1990s. The image of Cantor's ship is by Cunard (Postcards from the Early 1920s), available in the public domain via Wikimedia Commons. Regarding the issue of whether Cantor's proof of the existence of transcendental numbers is constructive, I was present at a mathematics colloquium public lecture where a highly distinguished mathematician made the claim that it was not constructive as part of his talk to a general audience; I spoke to him privately after the talk, and ultimately he agreed with me that Cantor's proof is actually constructive. Thanks to Joseph Moshenska for pointing me to Borges's Map of the Empire and related materials. The idea of a 1:1 map has evidently been considered by many authors, including Lewis Carroll (1894); see also Casey Cep (2014) for an informal contemporary summary. Wikipedia (2020) has a nice historical summary of the Schröder-Cantor-Bernstein theorem. Contemporary work on potentialism, such as Hamkins and Linnebo (2019) and Hamkins (2018), find the essence of potentialism in the idea of universe fragments and extensions, disconnecting the idea from infinity. The exercises on the deal with the Devil and the supertasks are adapted from presentations in Hamkins (2002). The fruit salad idea in question 3.11 is due to Martha Storey.

4 Geometry

Abstract. Classical Euclidean geometry is the archetype of a mathematical deductive process. Yet the impossibility of certain constructions by straightedge and compass, such as doubling the cube, trisecting the angle, or squaring the circle, hints at geometric realms beyond Euclid. The rise of non-Euclidean geometry, especially in light of scientific theories and observations suggesting that physical reality is not Euclidean, challenges previous accounts of what geometry is about. New formalizations, such as those of David Hilbert and Alfred Tarski, replace the old axiomatizations, augmenting and correcting Euclid with axioms on completeness and betweenness. Ultimately, Tarski's decision procedure points to a tantalizing possibility of automation in geometrical reasoning.

Classical Euclidean geometry is an ageless paragon of deductive mathematical reasoning, aiming to elucidate the fundamental truths of geometry, the mathematics and science of space, deducing them all from first principles. Starting with points and lines as primitive notions, Euclid gives us ten axioms—five essentially algebraic "common notions" and five essentially geometric "postulates"—and proceeds to build a monumental edifice, proving one statement and then another and another, each subsequently available for use in further proofs. In doing so, he taught us all the axiomatic method; he taught us how to do mathematics. He gave us a method for producing mathematics called the axiomatic method, as well as an unequaled example, *The Elements* (300 BC), by far the most successful and influential textbook ever written and the principal mathematics textbook for over two millennia.

4.1 Geometric constructions

In his deductions, Euclid makes fundamental use of two classical construction tools:

A *straightedge* or unmarked *ruler* enables you to construct a line containing any two given points and to extend that line in either direction arbitrarily.

A *compass* enables you to construct the circle having a given center and containing another given point.

These two tools suffice for all his applications.

Let us dive right into the Euclidean construction concept by constructing, according to the method of Apollonius, the perpendicular bisector and midpoint of a line segment. We are given a line segment *AB* (below at left), and we seek to construct the midpoint and perpendicular bisector. We begin by using the compass to construct the circle centered at *A* and passing through *B* (middle). Next, we construct the circle centered at *B* and passing through *A* (right).

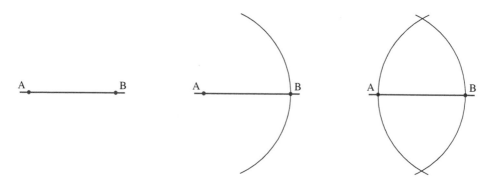

It is not necessary to draw these circles all the way around because we seek only the intersection points above and below; some geometers have the habit of marking only tiny arcs above and below, just enough to determine the intersections. Let *P* and *Q* be the points of intersection of the circles; with the straightedge, construct the line *PQ*, which is the perpendicular bisector intersecting the line *AB* at the midpoint *C*.

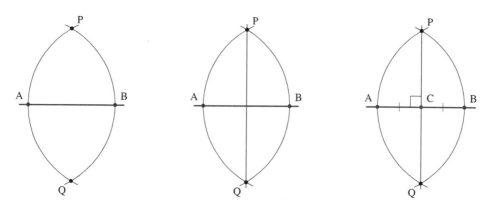

It is one thing to describe a construction, of course, and quite another to make mathematical assertions about it as we have. So let us now prove that *C* is the midpoint of *AB* and *PQ* is the perpendicular bisector.

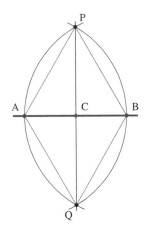

Consider the diagram at right. Because the points P and Q each lie on both circles, it follows that segments AP, AQ, BP, and BQ are all radii of those circles, and therefore congruent to AB, and hence also to each other. From this, it follows that triangle $\triangle APQ$ is congruent to $\triangle BPQ$, by the side-side-side congruence theorem. Since the corresponding parts of congruent triangles are congruent, we conclude that angle $\angle APQ$ is congruent to angle $\angle BPQ$. It follows that triangle $\triangle APC$ is congruent to $\triangle BPC$, by the side-angle-side congruence theorem. So segment AC is congruent to BC, since corresponding parts of congruent triangles are congruent, and therefore C is the midpoint of segment AB, as we claimed. Further, angle $\angle ACP$ is congruent to angle $\angle BCP$, and therefore these are right angles since together they make a straight angle. So line PQ is indeed the perpendicular bisector of AB, as we have claimed. QED.

Perhaps the reader has seen the two-column proof format, a format arising in twentieth-century American schools and used by millions of high schoolers, in which new assertions are made in the first column, with justification in every case provided in the second column. This format emphasizes the view that every assertion in mathematics requires justification. Meanwhile, Euclid wrote his *Elements* in a prose style, not far from what I wrote previously. But actually, to construct perpendicular bisectors in particular, Euclid proceeds a bit differently than Apollonius and I have: he first proves that one may construct an equilateral triangle on any segment (proposition I.1); he then uses this (and other developments) to prove that angles can be bisected (I.9); and finally, he combines these to bisect segments (I.10).

Contemporary approach via symmetries

Despite the enormous success of the Euclidean proof model, contemporary geometers might typically prefer an argument based on sweeping general symmetry principles, which are held to express more fundamental geometric ideas than the details of a particular construction. Perhaps one might argue like this: Given segment AB, reflect the plane through the actual perpendicular bisector, a process that swaps A and B and therefore swaps the two circles and therefore fixes the points P and Q of intersection. These points must therefore lie on the perpendicular bisector, and so PQ *is* the perpendicular bisector, and therefore C is the midpoint.

In this way, one understands the geometry by means of fundamental properties of congruence symmetries, rather than getting caught up in incidental details of a construction diagram. The symmetry principles allow one to see essentially at a glance that PQ is the perpendicular bisector.

Felix Klein (1872), in his professorial dissertation, established what has become known as the *Erlangen program* (named after the University of Erlangen-Nürnberg, where he worked), by which we are to understand and analyze a geometric space by means of its group of transformations and related invariants. Geometric properties are preserved by these transformations, and conversely (and importantly) any feature preserved by all transformations is deemed geometric. The Euclidean plane has congruences by translation, by rotation, and by reflection through a line, and these are all realized by a succession of at most three reflections. These transformations preserve metric properties, whereas geometric similarity means that we would also allow scaling transformations; oriented plane geometry rules out reflections, since those do not preserve handedness; and projective geometry has its own corresponding transformation groups. This approach unified the study of diverse geometries and led ultimately to an abstract structuralist approach to geometry, where one defines "a geometry" by exactly specifying its group of symmetries, which determine a corresponding congruence concept and accompanying geometric invariants.

Collapsible compasses

Let us turn to a certain controversial issue in the use of the compass (at least it was controversial at one time). Namely, in the usual manner of using a compass, one places the point of the compass at a point A and the drawing implement at B and then traces out the circle centered at A and having radius AB. It occasionally happens to be convenient, however, not to construct this particular circle, but to construct instead a circle of this same radius $|AB|$ centered at some other point C, perhaps some distance away:

In practice, this is an easy matter to achieve with a common type of rigid compass, one that holds its radius when lifted, since we simply measure the segment AB and then pick up the compass and place it at C, constructing the circle of radius $|AB|$ centered at C. Thus, we make a copy of the length AB at the point C. Is this legitimate? And what does *legitimacy* mean here? Such a compass use seems not directly authorized by Euclid's axioms, for while Euclid's postulate 3 states that every line segment is the radius of a circle, there is no axiom stating that every point C is the center of a circle of radius congruent to a given line segment AB. The worry, therefore, would be that allowing such a nonstandard construction method would invalidate the proofs in which it was used.

Cautious geometers in Euclid's day objected to what they viewed as a corruption of straightedge-and-compass constructibility and insisted upon the use of *collapsible* compasses, which hold their radius only while under the slight pressure of drawing a circle, collapsing when lifted. They did so specifically to prevent the objectionable manner of use.

But let us show that we can nevertheless always construct the desired circle at *C*, even if we use the compass only in the customary manner. This has become known as the *compass equivalence theorem*, and it was proved by Euclid in the *Elements*, proposition I.2, although Euclid's construction is different than the one I shall give here:

Given segment *AB* and point *C*, construct the line *AC*, the perpendiculars to this line at *A* and at *C*, and the circle centered at *A* containing *B*. Let *D* be the intersection of the circle with the perpendicular at *A*, and construct a perpendicular to *AD* at *D*. Let *E* be the intersection of the resulting line with the perpendicular line constructed at *C*. Since *ADEC* is a rectangle and *AB* is congruent to *AD*, it follows that segment *CE* is congruent to *AB*, as desired. (Meanwhile, there are sound reasons to prefer a different construction and argument for this theorem, which the reader will discover in question 4.1.)

The compass equivalence theorem shows that we are in fact enabled to copy segment *AB* to the point *C*, while employing only the customary use of the compass and without using it as a memory storage or measuring device. In light of this, the cautious objections concerning collapsible compasses might seem misplaced, as any construction undertaken with the forbidden compass use can be carried out with only the ordinary use. So why object?

Constructible points and the constructible plane

One naturally inquires about the power and limitations of one's construction methods. Suppose that some other nonstandard construction method, which might have seemed geometrically sound, turned out not to be simulable with the ordinary straightedge and compass. How would we regard it? And given a geometric figure, exactly which other points or figures are we able to construct with straightedge and compass?

Let us begin to carry out the universal construction process, systematically and exhaustively carrying out all possible constructions with straightedge and compass. We begin with two points *A* and *B*, establishing a unit distance:

With the straightedge, we may construct the line joining them; and with the compass, we may construct the two circles centered at each of them, having that segment as radius. These circles and the line intersect each other, creating 4 additional points of intersection and making 6 points in all:

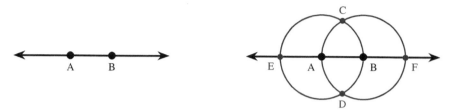

The next stage explodes with newly constructed points. Namely, using those 6 points, we may form 9 new lines and 14 new circles, and these figures intersect in many newly constructed points, providing 203 points altogether:

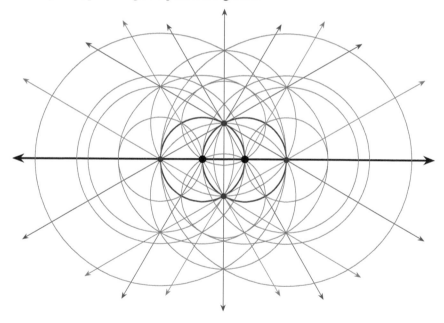

In a further subsequent stage, using the straightedge and compass with these 203 points, we again construct all possible lines and all possible circles and again find the resulting points of intersection. In an endless series of finite stages, we thereby generate all the *constructible points*, which constitute the *constructible plane*.

Because only finitely many new lines and circles are created at each stage, and these have at most finitely many new intersection points, it follows inductively that every stage in the universal construction process has only finitely many points. In Hamkins (2019), I introduced the *constructibility sequence*, beginning 2, 6, 203, and continuing at each stage with the number of points arising in the universal construction process that I have just described. At first, we didn't know the exact value of the next number on the sequence. We did not even know the exact number of lines and circles that arise at the next stage using the 203 points. Nevertheless, using exact computer algebra and many hours of computer processing time, Pace Nielsen (2019) found that the 203 points formed exactly 17562 lines and 32719 distinct circles; and with enormous further computational effort (6 continuous days of intense computer running time split amongst 6 processes on 3 machines), Teofil Camarasu found that these lines and circles produce in all 1723816861 distinct points of intersection. Thus, the constructibility sequence continues:

$$2, \quad 6, \quad 203, \quad 1723816861, \quad \cdots$$

The constructructiblity sequence now appears in the Online Encyclopedia of Integer Sequences, Hamkins (2020b).

In any case, the constructible points are formed by this process in a countable hierarchy of finite stages, and consequently, there are only countably many constructible points. In particular, since the real number plane $\mathbb{R}^2$ has uncountably many points, there must be some nonconstructible points. Not every point in the plane is constructible.

Meanwhile, the points in the constructible plane are naturally closed under straightedge-and-compass constructions, and from this, it follows that the constructible plane satisfies all the Euclidean axioms. One can thus view the constructible plane as an alternative geometric universe, one very like the real number plane $\mathbb{R}^2$, so far as Euclidean constructions are concerned, and yet missing some points. Nevertheless, with only the Euclidean tools, we shall remain unable to point to any particular location where a point is missing, since to identify such a location with those tools, of course, would be exactly to construct the point.

Constructible numbers and the number line

The classical concept of quantity, as mentioned in chapter 1, is fundamentally geometric; a quantity is a geometrical length, in relation to a standard of unit length provided by the line segment between two given points A and B. Let us rename these points as 0 and 1, for we shall think of the line segment joining them as representing a standard unit length, and the line containing these points is the *number line*.

A constructible number is a point on the number line that we are able to construct from the points 0 and 1 using the straightedge and compass only. For example, we can easily construct the number 2 by plac-

ing the point of our compass at 1 and constructing the circle with radius from 1 to 0.

And we can construct the irrational number $\sqrt{2}$, for this is the diagonal of the unit square constructed upon the segment from 0 to 1. But of course, we can construct far more than that. With the bisector construction, we can cut every line segment in half; with the segment-copying construction, we can add or subtract any two constructible numbers; by using similar triangles, we can multiply and divide them; by constructing certain similar triangles, we can take square roots (and furthermore, these methods alone are sufficient to construct all constructible numbers). In this way, the number line becomes densely filled with constructible numbers. The compass equivalence theorem implies that if we can construct any segment of length x, anywhere in the plane, then we can transfer that segment to the number line, with one end at 0, and thereby see that x is a constructible number. Thus, the constructible numbers are precisely the lengths of line segments that we are able to construct in the plane using the classical tools.

Descartes mounted the initial study of constructible numbers in *La Geometrie*, an appendix to *Discourse on the Method* (1637), introducing the foundational ideas of analytic geometry, synthesizing algebra and geometry by understanding geometric shapes in terms of their algebraic equations.

4.2   Nonconstructible numbers

On the contemporary perspective, mathematicians often identify the continuum of real numbers with the points on the classical number line, and in this sense, the collection of all constructible numbers forms a field, a subfield of the real numbers containing every rational number, and far more. Which real numbers are constructible? Is every real number constructible?

The answer is no. As we have mentioned, there are only countably many constructible numbers, but by Cantor's theorem, there are uncountably many real numbers. So there must be real numbers, points on the number line, that we cannot construct by straightedge and compass.

Although that is a soft proof via Cantor's theorem, nevertheless the existence of nonconstructible numbers was known even before Cantor as a consequence of deep results in algebra growing out of the field-extension ideas of Évariste Galois and others. Furthermore, this algebraic way of thinking about constructibility provided sophisticated information about the nature of constructible numbers. Let me try to explain a little about this theory. The main observation is that every new point constructed by straightedge and compass arises as a point of intersection of previously constructed lines or circles, and these have linear or quadratic equations. Thus, every newly constructed point is the solution of a certain linear or quadratic equation expressible in terms of previously constructed numbers. Each subsequently constructed number, therefore, has degree 1 or 2 over the previously constructed numbers. From this, using the theory of rational field extensions, it follows that every constructible number is algebraic over the field of rational numbers $\mathbb{Q}$, and furthermore, the degree of the constructible number over $\mathbb{Q}$ must be a power of 2, of the form 2^n for some n. In particular, we cannot construct numbers that solve an irreducible cubic (degree 3) or quintic (degree 5) equation, for such a real number will not have degree 2^n for some n.

Doubling the cube

For instance, we cannot construct the number $\sqrt[3]{2}$, since this number has degree 3, because it is a root of the irreducible polynomial $x^3 - 2$. This fact is pertinent for the confounding classical problem known as *doubling the cube*. The problem is, given a side length AB of a cube, to construct a side length CD for a cube of twice the volume. If one regards the original length AB as a unit length, then the first cube will have volume 1, and so the doubled cube will have volume 2, and consequently, side length $\sqrt[3]{2}$.

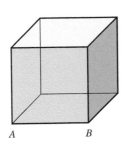

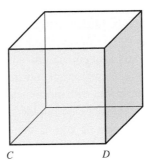

Ultimately, therefore, the problem of doubling the cube is exactly the problem of constructing the number $\sqrt[3]{2}$. Since this is impossible by the field-extension degree analysis mentioned previously, the conclusion is that it is impossible to double the cube by a straightedge-and-compass construction.

Trisecting the angle

Another long-standing open classical problem is that of *trisecting the angle*. The problem is, given an angle, to construct a trisection of it, an angle that is exactly one-third the size. In some cases, to be sure, one may easily construct the trisection. For

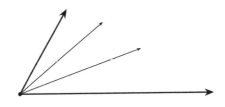

example, if the original angle is a straight angle, then the trisection would be a 60° angle, which arises in an equilateral triangle, and this we can easily construct; similarly, the trisection of a right angle has 30°, which we can construct by bisecting a 60° angle.

But consider an angle of 60°, whose trisection would be 20°. If you could construct such an angle, then you could construct a unit right triangle with this angle, whose leg was therefore of length cos(20°). The triple-angle formula of trigonometry shows that

$\cos\theta = 4\cos^3(\theta/3) - 3\cos(\theta/3)$, and consequently, $4\cos^3 20° - 3\cos 20° = \cos 60° = \frac{1}{2}$. So if $x = \cos 20°$, then $4x^3 - 3x = \frac{1}{2}$. This equation has no rational solutions, and from this, it follows that x has algebraic degree 3 over the rationals, which is not a power of 2, and so x is not constructible. Thus, it is impossible to construct an angle of 20° using the straightedge and compass only, and therefore it is impossible to trisect an angle of 60°.

Squaring the circle

Another classical construction problem is that of *squaring the circle*, the problem of constructing a square whose area is the same as the area of a given circle with a given radius AB. Since the unit circle has area π, this amounts to constructing the number $\sqrt{\pi}$, which would be the side length of a square of that same area. But

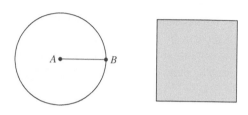

Ferdinand von Lindemann proved in 1882 that π is transcendental, and so also $\sqrt{\pi}$ is not algebraic, and therefore not constructible by straightedge and compass. So it is impossible to square the circle with those means only.

The transcendence of π implies that the converse problem of *circling the square*, to construct a circle whose area is the same as that of a given square, is also impossible by straightedge and compass, since a unit square has area 1 and the circle of this area would have radius $1/\sqrt{\pi}$, which is not algebraic because π is transcendental.

Circle-squarers and angle-trisectors

The resolution of the doubling-the-cube, trisecting-the-angle, and squaring-the-circle problems, having been open since antiquity and solved only in the nineteenth century, surely highlights the enormous power of the abstract developments in the theory of algebraic field extensions.

Yet, because the resolution in each case was a proof of impossibility rather than possibility, and furthermore because the proofs were not directly accessible to geometers but relied on the theory of algebraic field extensions, some stubborn souls refused to accept the impossibility proofs and carried on with the search for positive solutions. Mathematician and logician Charles Dodgson, better known as the author Lewis Carroll, wrote of his frustrations at trying to convince these "circle-squarers" of their errors; and logician Augustus De Morgan (1872) ridiculed them in his work *A Budget of Paradoxes*.

4.3   Alternative tool sets

The classical straightedge-and-compass tool set has been used for geometric constructions since antiquity. Meanwhile, there are other tool sets. Do they have the same power?

Compass-only constructibility

How surprising it must have been, after two thousand years of straightedge-and-compass constructions, to learn that any point constructible by straightedge and compass can also be constructed with a compass alone. You do not need the straightedge. This is the content of the Mohr–Moscheroni theorem, proved by Georg Mohr in 1672 and rediscovered by Lorenzo Moscheroni in 1797. The theorem can be proved by providing alternative compass-only procedures to simulate every possible use of a straightedge. These constructions are generally more complicated than straightedge-and-compass constructions, to be sure, but they do use only a compass.

For example, given two nonparallel line segments *AB* and *CD*, one seeks to find the point of intersection of their corresponding lines. One can do this with compass only in a construction involving a few dozen circles, including an initial choice of circle centered at an arbitrary point not on either line. Similarly, one must show that given a line segment and a circle, one can find the intersection points of the corresponding line with the circle (assuming they do intersect) using the compass only. This also can be undertaken with compass only in a construction that uses different methods, depending on whether the line passes through the center of the circle.

Straightedge-only constructibility

A dual version of the result calls for straightedge-only constructions, with the stipulation that one is provided at the outset a single circle with its center. The Poncelet-Steiner theorem, conjectured by Jean Victor Poncelet in 1822 and proved by Jakob Steiner in 1833, asserts that indeed, any point constructible by straightedge and compass can be constructed by straightedge alone, given such an initial circle.

The basic idea of the proof is to reflect any desired use of the compass back to the initial circle, undertake the construction with that circle, and then to transfer the result back up to the intended application. Note that the initial circle might be quite a different size or quite distant from the intended application. The theorem is similar to and strictly generalizes the sixteenth-century results of Lodovico Ferrari and Gerolamo Cardano concerning the power of "rusty" compasses, stuck at a fixed radius, showing that indeed these are constructively equivalent to an ordinary compass. Indeed, the rusty compass result is a consequence of the Poncelet-Steiner theorem, since one may use the rusty compass to make the initial circle and then carry on with the straightedge only. Meanwhile, the rusty compass theorem admits an easier proof, since for any desired use of another circle, one can construct a rusty-compass circle with the same center, and then use similar triangles to reflect the application to the rusty-compass circle and transfer the result back again, thereby omitting the need for the other circle.

Construction with a marked ruler

Meanwhile, let us now consider augmenting the classical tool set with additional construction tools. One might naturally hope to be able to construct figures that are not constructible using the straightedge and compass alone. Consider a *marked ruler*, for example, which is simply a straightedge containing two marks or notches. One is allowed to use a marked ruler by placing it so as to lie on a given point, while sliding and rotating so as to align the notches with two other figures.

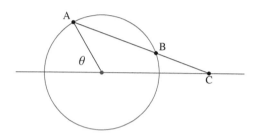

In the figure here, for example, given the blue line and the red circle and the point *A*, one is allowed to place the marked ruler at *A*, and then, by sliding and rotating, to find points *B* and *C* so as to align exactly with the notches. The point is that if one rotates the ruler downward, then the corresponding points would be too close, and if upward, then they would be too far apart.

So with the marked ruler, one can find points *B* and *C* that are collinear with *A* and exactly align with the notches, a unit distance apart. The importance of this is that if the circle also has unit radius, then it follows (as known in antiquity) that the angle at *C* is precisely the trisection of the indicated angle θ. With a marked ruler, therefore, one can trisect any given angle. And so the construction toolset of a marked ruler and compass is strictly more powerful than a straightedge and compass alone.

Origami constructibility

Another geometric construction method arises from the sublime Japanese paper-folding art of *origami*. In skilled hands, a single square of paper is folded successively to produce surprising works of exquisite beauty. In addition to this artistic aspect of origami, the process of folding is inherently geometrical, and one may take origami folding as an alternative method of geometric construction. How does it compare with the classical Euclidean straightedge-and-compass construction method?

For any of the allowed origami folds, we fold upon a certain line ℓ, determined by given information, and thereby construct not only that line, but also the reflection through that line of any figure we had previously constructed. Having made a fold, we may observe where a given point or figure is carried by the folding, transferring the point to its reflection.

It is allowed, for example, to fold along the line determined by any two given points. This origami construction is simulable with the straightedge and compass because with the straightedge, we may construct that line itself, and with the compass, we may construct the reflection through the line of any other point simply by constructing two circles. Another allowed origami fold is to fold so as to bring one given point to coincide with another. This is the same as folding along the perpendicular bisector of the segment between the two points, and this again is simulable by straightedge and compass.

Robert Geretschlager (1995) explains the equivalence of straightedge-and-compass construction with an axiomatic list of seven fundamental origami folds, making reductions in both directions. He also proves, however, that certain more complex origami folds go strictly beyond Euclidean constructibility. One kind of fold, for example, can be described as allowing one to fold along the common tangent to two parabolas, as specified by given focal points and vertex lines (one can in effect achieve this with actual paper). With the addition of this folding construction, it turns out, one can solve arbitrary cubic equations over the rational numbers, constructing $\sqrt[3]{2}$ and cube roots generally, which are not all constructible by straightedge and compass, as discussed previously. One can also trisect any given angle using these origami constructions. Thus, origami is realized ultimately to have a mathematical power surpassing the Euclidean tool set.

Spirograph constructibility

Another fanciful method of constructibility would be *spirograph* constructibility. Readers of a certain age may be familiar from their childhood with the Spirograph drawing toy, consisting of some geared rings and circles, with holes for a colored pen. By placing one gear inside another and turning the spiral with the pen, one produces a great variety of pleasing complex spiral figures.

Let us consider spirographs as a method of geometric construction. All the gears have the same size mesh, so they can be paired, and in the general case, let us suppose that we have circular gears and rings with any desired number of teeth. The circular gears have holes for the pen at an integer distance from the center. We are allowed to place the gears at any previously constructed points and then draw the resulting spirograph.

What is the power of spirograph constructibility? Are the resulting spirograph-constructible points also constructible with the straightedge and compass? One interesting observation is that one can draw straight lines with a spirograph if the inner gear has exactly half the diameter of the ring gear and the drawing point is on the circumference of the inner gear.

Meanwhile, it turns out that, like origami, spirograph constructibility extends strictly beyond that of the straightedge and compass. One way to see this is to observe that one may easily construct regular polygons via spirograph, with any desired number of sides; and indeed, essentially every spirograph pattern exhibits n-fold rotational symmetry, where n can be determined from the respective number of teeth in the two gears. For example, the red spirograph above has seven-fold rotational symmetry, and so its self-intersection points form a regular septagon. But it is known that with a straightedge and compass one cannot construct a regular septagon. According to the Gauss–Wantzel theorem, one can construct a regular n-gon by a straightedge and compass if and only if n is the product of a power of 2 and any number of distinct Fermat primes, primes of the form $2^{2^k} + 1$; and 7 is not realized in this manner. So spirograph constructibility exceeds the capabilities of the straightedge and compass.

4.4   The ontology of geometry

This brings us to the question of what we are achieving with geometric constructions. What are we doing when we carry out a geometric construction on paper, or perhaps as in the classical style, by etching in sand? Surely the figures that we actually construct are meant to be merely representational, imperfect suggestions of the perfect, ideal points, lines, and circles we intend to consider. In this case, it would seem silly to object to constructions that use the compass in a forbidden manner, for example, if we could have made the construction with only the allowed usage. Using a rigid as opposed to a collapsible compass would be like any other convenience that we make in our imperfect construction process: we need to take care to line up the straightedge perfectly and take care that it does not slip while we are drawing, and so on.

If geometry is meant to be about the nature and features of the ideal perfect geometric figures—points, lines, and circles—then there would seem to be no reason to object to augmenting our construction toolset with new tools, provided that we thought their use accorded with sound geometrical truths. For example, what could be the objection to marked rulers, used in the manner described here? One might object that in fact there might not be points B and C as in the figure on page 130, which are collinear with A and unit distance apart. Indeed, one cannot prove on the basis of the Euclidean axioms that there are such points B and C, precisely because (for some angles) these points are not constructible by straightedge and compass, whereas the collection of all constructible points is a geometrical universe that satisfies all the Euclidean axioms. Meanwhile, the stronger contemporary axiomatic systems of Hilbert and Tarski, using Dedekind's axiom of continuity, imply that those classical nonconstructibility problems do indeed have solutions. There is an angle trisector, as well as a cube duplicator and a squared circle; these points exist in space, even if we are not able to construct them with straightedge and compass.

In this way, the impossibility proofs identify flaws in our geometric theory; we are missing the axioms asserting that the points that these nonstandard constructions allow us to construct actually exist. The insistence on straightedge-and-compass constructions might be criticized as a form of dogmatic conservatism in mathematics. Are geometers with that outlook thinking, why should we presume to use other tools than the straightedge and compass, since these sufficed for Euclid to prove all his theorems? Is geometry about the theorems that we can prove from Euclid's axioms, or is it about an immutable geometrical reality about which we might realize further geometric truths? If the latter, we might seek stronger tools. The impossibility proofs hint instead at the idea of alternative nonclassical geometries, a possibility that later explodes forcefully into the subject with the discovery of non-Euclidean geometry.

4.5 The role of diagrams and figures

Let us consider a bit more closely the role of diagrams and figures in geometric proof arguments. From the time of Euclid to the present day, nearly every geometric proof has been accompanied by one or more proof diagrams. But are we to consider these diagrams to be part of the proofs in which they are used? Or are diagrams merely supplemental but inessential aids to proof?

One problematic issue is that diagrams are inherently more specific than the hypotheses of an argument entitle one to deduce. In a diagram, every line segment has a particular length and every angle is a particular angle, even when we have made no particular assumptions about those lengths or angles in the argument. Because of this overspecificity, the diagram might display accidental geometric relations, such as a point being inside a circle instead of outside, even when this feature is contingent on details of the initial figure and not a necessary consequence of the hypotheses. The worry is that this could lead one into deductive geometric error.

Kant

Immanuel Kant emphasized in his *Critique of Pure Reason* (1781) that geometrical reasoning can produce synthetic a priori truth judgements, to use the terminology of his classification of propositions. Mathematical conclusions for Kant can lie on the synthetic side of the analytic/synthetic distinction, for the concepts of a mathematical conclusion often grow strictly beyond the concepts of the hypotheses—mathematics synthesizes new concepts from prior concepts. An analytic conclusion, in contrast, would be one that is contained in the very meaning of the hypothesis. Yet, for Kant, mathematical reasoning is also a priori rather than a posteriori, since it is based on pure reasoning and not on experience.

Euclid's text can be seen largely as a work of synthesis rather than analysis, in that he is preoccupied with his construction tools and his methods for constructing new geometric figures. His axioms and even his proposition statements are directly concerned with exhibiting the power of the construction methods, with what we can construct or build with those tools, rather than with fundamental geometric truth. His propositions are stated—strangely, by contemporary standards—as construction goals rather than as statements of mathematical fact. For example, his tenth proposition reads like this:

Proposition 10. To bisect a given line segment.

This is synthesis, in that it is preoccupied with the process of building new geometric figures using our geometric construction tools. A geometrical analysis, in contrast, would be concerned with the geometric facts, for example that every line segment indeed has a midpoint, rather than with our ability to construct them. One might expect a geometric analysis to begin with geometric pure existence axioms, asserting that certain points and geometric figures exist, whether or not we are able to construct them, and proceed to investigate what further geometric truths there must be as a consequence.

   Kant further identifies the critical role of intuition in mathematics, providing the means by which we come to that geometric knowledge:

> Geometry is a science which determines the properties of space synthetically, and yet *a priori*. What, then, must be our representation of space, in order that such knowledge of it may be possible? It must in its origin be intuition. (1781, B40)

For Kant, an initial construction triangle can encompass validity for all triangles, based on this intuition, despite its specificity, because the relevant inferences and auxiliary constructions can be carried out for any given triangle:

> The single figure [a triangle] which we draw is empirical, and yet it serves to express the concept, without impairing its universality. For in this empirical intuition we consider only the act whereby we construct the concept, and abstract from the many determinations (for instance, the magnitude of the sides and of the angles), which are quite indifferent, as not altering the concept "triangle." (1781, B741–742)

Kant also finds synthetic a priori judgements elsewhere in mathematics, including arithmetic.

Hume on arithmetic reasoning over geometry

Hume (1739) writes on the fallibility of geometrical perception as a source of knowledge:

> I have already observ'd, that geometry, or the art, by which we fix the proportions of figures; tho' it much excels, both in universality and exactness, the loose judgments of the senses and imagination; yet never attains a perfect precision and exactness. Its first principles are still drawn from the general appearance of the objects; and that appearance can never afford us any security, when we examine the prodigious minuteness of which nature is susceptible. Our ideas seem to give a perfect assurance, that no two right lines can have a common segment; but if we consider these ideas, we shall find, that they always suppose a sensible inclination of the two lines, and that where the angle they form is extremely small, we have no standard of a right line so precise, as to assure us of the truth of this proposition. 'Tis the same case with most of the primary decisions of the mathematics. (Hume, 1739, I.III.I)

According to Hume, our geometric insight—that two distinct lines can have no segment in common, for example—is not based upon any figure that we might draw or even imagine, as any such figure will some angle of inclination, whereas other lines might have a much smaller inclination. Algebra and arithmetic, meanwhile, lead to a more certain form of knowledge than geometry:

> There remain, therefore, algebra and arithmetic as the only sciences, in which we can carry on a chain of reasoning to any degree of intricacy, and yet preserve a perfect exactness and certainty. We are possest of a precise standard, by which we can judge of the equality and proportion of numbers; and according as they correspond or not to that standard, we determine their relations, without any possibility of error. *When two numbers are so combin'd, as that the one has always an unite answering to every unite of the other, we pronounce them equal;* and 'tis for want of such a standard of equality in extension, that geometry can scarce be esteem'd a perfect and infallible science. (Hume, 1739, I.III.I, italics added)

The italicized phrase is Hume's statement of the Cantor-Hume principle, mentioned in chapter 1.

Manders on diagrammatic proof

Ken Manders (2008a, b) wrote influentially on the role of diagrams in Euclidean geometry, that is, in actual historical mathematical practice. Manders argues ultimately that proofs in Euclidean geometric practice have two parts, a diagrammatic part and a sentential part, which track different assumptions and information about the geometric situation. Diagrams are used to infer general spatial, or especially topological, information, such as incidence or intersection (or betweenness, as highlighted in the twentieth-century geometric theories), while exact metric concepts such as congruence are always made explicit in the proof text. One uses the diagram in effect to observe geometric information that is invariant under small deviations in the diagrammatic parameters, but exact claims about lengths and perpendicularity are made in the text.

Contemporary tools

Mathematicians commonly imagine their geometrical figures as indeterminate, flopping about, stretching, and collapsing, even to extreme cases of collinearity or degeneracy, in order to observe in their minds the effects of their hypotheses, and thereby to gauge whether their construction is fully general or dependent on some contingent feature. Modern technology enables this dynamism explicitly, such as with the interactive computer geometry system GeoGebra, which allows one to manipulate the input points of a construction while the constructed lines and circles are carried along in real time. Such visualizations, whether in the mind alone or on a computer screen, can reliably reveal accidental geometric relations for what they are.

How to lie with figures

In geometric argument, diagrams are seldom perfectly drawn. And even a very small error or imperfection in a geometric diagram can sometimes lead one to a false geometric conclusion. Let me illustrate how things can go wrong in the following false "proof" that every triangle is isosceles. (An isosceles triangle, if you recall, is one with two sides the same length.) We know, of course, that this claim is not true, since we may easily draw nonisosceles triangles; nevertheless, the argument appears solid. Can you spot any error?

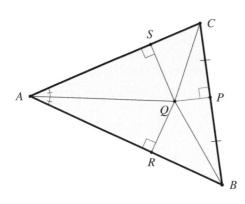

The claim is that every triangle is isosceles. To "prove" this, consider an arbitrary triangle $\triangle ABC$. Construct the angle bisector (blue) at $\angle A$ and the perpendicular bisector (green) of segment BC at midpoint P, and let Q be the point of intersection. Drop perpendiculars (red) from Q to AB at R and to AC at S. Because P is the midpoint of BC and PQ is perpendicular, we deduce $BQ \cong CQ$ by the Pythagorean theorem. Since AQ is the angle bisector of $\angle A$, the triangles AQR and AQS are similar, and since they share a hypotenuse, they are congruent. It follows that $AR \cong AS$, and also $QR \cong QS$. Therefore $\triangle BQR$ is congruent to $\triangle CQS$ by the hypotenuse-leg congruence theorem. So $RB \cong SC$. And therefore,

$$AB \cong AR + RB \cong AS + SC \cong AC,$$

and so the triangle is isosceles, as desired. It follows as an easy "corollary" that every triangle is equilateral, because in the proof we had proceeded from an arbitrary vertex A of triangle $\triangle ABC$, and so the argument shows that each pair of adjacent sides is congruent.

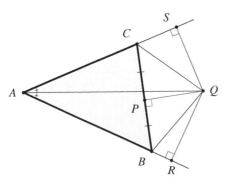

Perhaps someone might criticize the proof we gave above by saying that we do not necessarily know that the angle bisector at A intersects BC on the inside of the triangle. Perhaps the intersection point is outside the triangle, as in the diagram here. But the "proof" works just as easily for this case. Namely, we again let Q be the intersection of the angle bisector at $\angle A$ with the perpendicular bisector of BC at midpoint P, and again drop the perpendiculars from Q to R and S. Again, we get $BQ \cong CQ$ by the Pythagorean theorem, using triangles

$\triangle BPQ \cong \triangle CPQ$. And again, we get $\triangle ARQ \cong \triangle ASQ$ since these are similar triangles with the same hypotenuse. So again, we conclude $\triangle BQR \cong \triangle CQS$ by hypotenuse-leg. So we deduce $AB \cong AR - BR \cong AS - CS \cong AC$, and so the triangle is isosceles. The reader will uncover the flaws of the arguments in the end-of-chapter exercise questions.

Error and approximation in geometric construction

Meanwhile, let us analyze a bit more carefully the possibility of error in geometric constructions. Surely when carrying out a construction, even with extreme care, we inevitably make some small errors, perhaps by slightly misplacing the point of the compass or the position of the straightedge. Thus, we do not expect to implement a construction *exactly*, with infinite precision, but rather, the figure we actually construct will be merely an approximation of the more perfect ideal geometric object we imagine.

Furthermore, these errors might accumulate and magnify as they propagate through a construction. How sensitive are the familiar classical constructions to such small errors in the use of the straightedge or compass? Might we compare competing constructions in terms of accuracy?

Let us reconsider in this light the Apollonius construction of the perpendicular bisector of a line segment AB. We are given points A and B and seek to construct the midpoint and perpendicular bisector. We begin by placing the point of the compass at A and the writing implement at B and then constructing the circle centered at A with radius AB. When doing

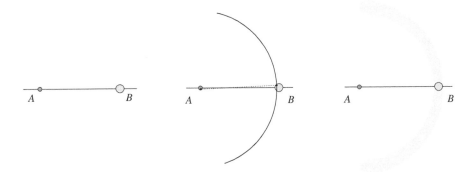

so, however, we are not perfectly accurate in placing the point of the compass at A—we may actually place it at some point within some small neighborhood of A (indicated in blue). Similarly, we do not place the writing or etching implement exactly at B, but instead at some point within a small neighborhood of B (indicated in orange). And then we trace out the arc of a circle centered near A, containing a point near B, as indicated in the middle panel. The space of all such arcs that could arise in conformance with those error bounds is indicated in the blurred orange arc, an image which was produced via computer by drawing dozens of such arcs.

Let us now similarly construct the circle centered at B and containing A; this is a new compass placement, with its own similar but independent errors, leading to the orange arc of possibilities shown here:

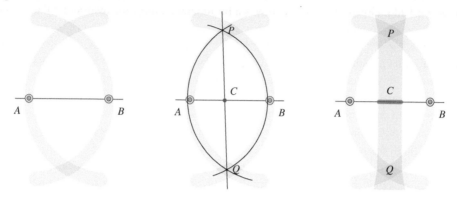

The two spaces of possibility intersect in the darker orange regions above and below, and the point is that any particular placement of the compass within the error bounds will lead to values of P and Q within that overlap. We might think of these overlap regions as representing the space of possible locations of P and Q, respectively.

Notice that in the particular sample case (black, in the middle figure), the resulting line PQ is noticeably nonperpendicular to AB, and the resulting point C is noticeably not the midpoint. The pink-shaded region (in the figure at right) indicates the space of all the bisectors that might arise in conformance with our error bounds on A and B, with the resulting possibilities for C shown in red. (A subtle complication is that the choices of P and Q from those overlap regions are not actually independent.)

In light of the small size of the original error bounds on the points A and B, it may be surprising that even such a standard simple construction as this—constructing the perpendicular bisector and midpoint of a segment—appears to have comparatively large error propagation: the shaded red region C is quite large and includes many points that one would not say are possibly the midpoint. In this sense, the Apollonius perpendicular bisector construction appears to be sensitive to the errors of compass placement.

Is there a better construction? For example, to improve the accuracy of the perpendicularity of PQ to AB, it would seem to help to use a much larger circle, which might lower the variation in the resulting "right" angle. But perhaps that expectation would seem reasonable only because we have so far assumed that compass error arises only with the placement of the points of the compass, not during the course of actually drawing the arc. One can imagine, for example, that errors also arise from a flexing of the compass during use, particularly with a very large compass, causing it to deviate from circularity, or from slippage, which might reasonably be expected to cause increasing error with the length or degree of the arc, and so on, and such a model might have greater errors with large circles. There would seem to be many different sources of error in actual geometric constructions.

Constructivist mathematician Andrej Bauer mentioned to me that in his school days, the classroom had a large compass which was a bit unwieldy, used for instructional constructions on the chalkboard. Andrej had noticed that one could reliably achieve more accurate midpoint results by using the compass to mark off segments of the same length from each end, nearly to the center and then simply guess the midpoint of the resulting small segment:

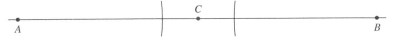

His teacher objected, despite Andrej's reliably superior results.

One could in principle analyze the accuracy of any of the usual geometric constructions, looking for sensitivity to various kinds of error, and indeed there is a literature of work doing precisely this. A somewhat more sophisticated model of error estimation, used in some of this work, views error as a probability distribution, analyzing how the distribution is propagated through a construction. For example, a distribution concentrating in neighborhoods of points A and B would lead through the Apollonius construction to a comparatively diffuse distribution near the midpoint.

Meanwhile, one might hope to improve the accuracy of a construction simply by repeating it several times, aiming to converge to a more accurate answer. For example, in the bisector algorithm, we might simply perform the Apollonius bisector construction twice, producing midpoint candidates C_0 and C_1, and we could proceed to find the midpoint of $C_0 C_1$ as a further presumably more accurate midpoint. Or we could iterate in some other manner and hope to converge to the actual midpoint. For example, we could produce many midpoint candidates and use the median point as our final answer. With reasonable assumptions on the probability distribution of the input, this method would indeed converge in a meaningful sense to the actual midpoint.

To my way of thinking, the issue of error propagation leads directly to a frank philosophical discussion of the role of geometric constructions in mathematics. In geometry, one may consider platonism with respect to the geometric objects—the ideal points, lines and circles, which our figures only imperfectly represent—but there is also platonism with respect to the constructions themselves, where we take the construction as a mathematical object, an algorithm or procedure, subject to mathematical analysis, much like one analyzes algorithms in complexity theory. Shall we imagine an ideal geometer in the platonic realm, indeed, why not an ideal Apollonius himself, with flowing white robes, carrying out a perfect instance of the bisector construction? How lovely that would be to see. We may analyze the nature of an algorithm or geometric construction, first under the assumption that it is carried out with absolute perfection; and then second, departing the platonic realm by analyzing as we did above how well it performs when carried out imperfectly.

Constructing a perspective chessboard

I should like to illustrate further the concept of toolset-limited constructions with a fun example; let me teach you how to construct a chessboard in two-point perspective, using only a straightedge, with no measuring or calculation of any kind. I view this activity in two lights. On the one hand, it shows how one can achieve surprising results with limited tools, in this case just a straightedge. But on the other hand, it is simply a fun activity that anyone can enjoy. This is something that you can

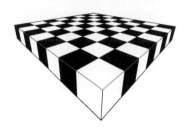

actually do, right now; all you need is a piece of paper, a pencil and a straightedge. Use a ruler or a chopstick or the edge of a notebook—anything that is straight. I shall wait right here while you gather your materials.

The construction follows the standard two-point perspective method. To begin, give yourself a horizon at the top of the page, with two points (orange), which we call the points at infinity, and an arbitrary point (blue) to serve as the front corner of the chessboard. Extend the front corner point to infinity at each side, and then also draw lines to form the front corner square of your board. This information will now determine the entire grid.

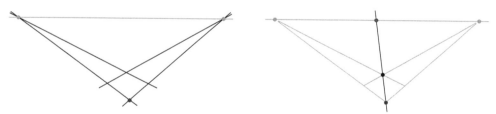

Construct the diagonal of the front-corner square and extend it to find the resulting diagonal point at infinity (brown) as shown above at right. After this, extend the right corner of the front square to the diagonal infinity, which determines a corner point for the second square, as shown below at left:

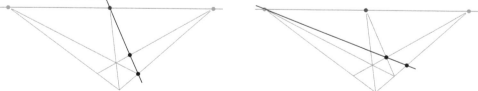

By extending that point to infinity at the left, one forms the second square, as shown above at right.

That line in turn determines a new point on the leading edge, which when extended to the diagonal infinity determines a point on the second rank, as below at left. By continuing in this way, one can produce as many first-rank squares as desired. Furthermore, the lines determine many additional points in the grid, a triangle of intersection points between the files and the diagonals:

The collinearity of these points is essentially the content of Desargue's theorem, often discussed in the context of projective geometry. These points determine further ranks of the board by extending to infinity at the right. Thus, we have constructed an entire chessboard:

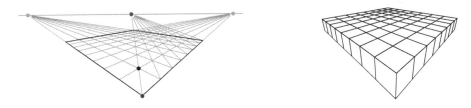

By adding a point at infinity down below (do so directly below the diagonal point at infinity, but a good distance down), one can extend the leading faces downward to give the board a sense of thickness, and then simply shade or color the chessboard in the usual chessboard pattern.

One can use this method to make chessboards of any size, with any number of tiles in any of a variety of perspective views. One finds innumerable fascinating instances of the method in Renaissance paintings, showing tiled floors and great plazas or steps receding into the distance with perfect perspec-

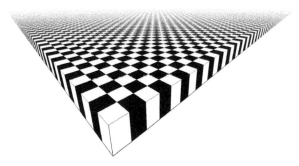

tive. Many of the mathematical ideas on perspective originated in art. With practice, I hope you may come to learn how to complete an infinite chessboard in finite time.

4.6   Non-Euclidean geometry

Let us consider next one of the most interesting and profound developments in geometry—
the discovery of non-Euclidean geometry. At issue is the famous parallel postulate, Euclid's fifth postulate in the *Elements*, which asserts that if a line is transverse to two lines
and forms angles with them summing to less than two right angles, then the lines can be
extended so as to intersect:

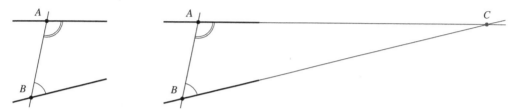

John Playfair's formulation of the axiom, which is equivalent given certain other geometric assumptions, asserts that for any line ℓ and any point P not on line ℓ, there is at most
one line through P parallel to ℓ. A strengthened form of Playfair's axiom asserts that there
is exactly one line through P parallel to ℓ:

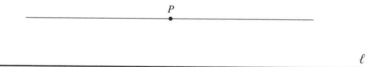

Euclid had put off the need for this axiom for some time, using it for the first time
only in his twenty-ninth proposition, after having developed quite a bit of his geometric
theory. Since antiquity, mathematicians have noticed the manifestly different character of
this postulate, which is technically detailed and particular in comparison with the other
axioms, which are often seen as asserting fundamental geometric truths. Does the parallel
postulate strike you as more like a theorem than an axiom? Perhaps one might naturally
expect simply to prove this assertion from the other axioms.

Indeed, during the more than two thousand years since the time of Euclid, geometers
have strived to do exactly that. There is a long history of attempts to prove the parallel postulate, with notable work in the theory of parallels by Ptolemy, Proclus, Giovanni Girolamo
Saccheri, Johann Heinrich Lambert, Adrien-Marie Legendre, and others. The early work
tended to proceed from a perspective of Euclidean geometry as the one true geometry, with
the goal of proving the parallel postulate from the other axioms.

Eventually, this perspective began to shift. János Bolyai had aimed to prove the parallel
postulate by contradiction. He accordingly assumed the negation of the parallel postulate and proceeded to make geometric deductions. But rather than finding the sought-after

contradiction, what he found developing instead was a beautiful new geometric theory—a theory that led him to strange new geometric conclusions, many of which stood in conflict with classical geometry, but a theory displaying such an internal coherence and geometric sense that it seemed consistent and without contradiction. He wrote in 1823, "out of nothing I have created a strange new universe" (see Greenberg 1992, p. 163).

Johann Carl Friedrich Gauss also had been working privately for some decades on his own theory of non-Euclidean geometry, but he planned to publish only posthumously, reportedly fearing "the howl from the boeotians" (Greenberg, 1993, p. 182), a remarkable fear in light of Gauss's enormous stature in mathematics. Gauss had discovered that certain non-Euclidean geometries have an inherent and absolute notion of scale, such that one may define a unit length in absolute terms rather than always only relative to a fixed unit length. Gauss had speculated that if the physical universe were non-Euclidean, we could hope to discover this unit scale and use it as an absolute standard of measurement.

The first published account of non-Euclidean geometry is that of Nikolai Lobachevsky in 1829, and he had given public lectures reportedly as early as 1826. Thus was born a fascinating new subject in mathematics. These early developments in non-Euclidean geometry proceeded on the axiomatic system, developing the theory by proving theorems from the axioms. The resulting theory seemed coherent and consistent, but how was one to know whether it really was? The consistency question was finally resolved in 1868, when Eugenio Beltrami provided semantic interpretations of the non-Euclidean geometries within the classical Euclidean theory, thereby showing that non-Euclidean geometry is as consistent as Euclidean geometry. He provided models of the non-Euclidean theories by reinterpreting the basic notions of point, line, circle, and angle in such a way so as to fulfill the non-Euclidean axioms.

Spherical geometry

In *spherical* geometry, for example, we walk about on the surface of a large sphere as though on the surface of a perfectly polished moon, a self-contained non-Euclidean geometrical universe of its own. We carry out our geometric constructions, perhaps with colored chalk, directly on the surface of the sphere. The points of this new geometric world are the points on the surface of the sphere, but the "lines" are *geodesics*, paths of shortest length, which on a sphere are precisely the *great circles*, arising from the inter-

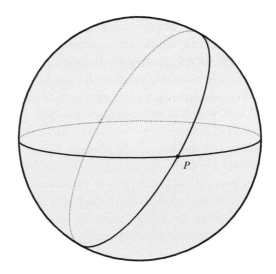

section with a plane through the center of the sphere; these are the paths that a jet might fly on a long-distance flight. These circles, of course, are not straight lines in the ambient Euclidean space in which the sphere is embedded, but they count as "straight" lines within the geometric universe of spherical geometry, which knows only of the surface of the sphere and nothing of the embedded space. Living on the surface of the sphere, we construct triangles and lines and circles, just as Euclid did, and when our figures are small in comparison with the size of the sphere, spherical geometry is nearly Euclidean; we might not even notice the deviation.

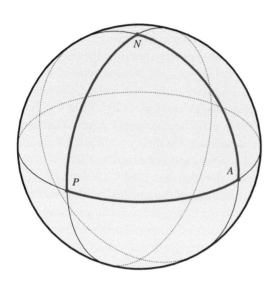

But on a large scale, the non-Euclidean nature of spherical geometry becomes obvious. In spherical geometry, for example, the angles of a triangle can add up to more than 180°. To see this, start a journey at the North Pole (*N*) and travel due south along the prime meridian to the equator at *A*; turn right and follow the equator one quarter of the way around to *P*; turn right again and head north back to the North Pole. Thus, you have formed a triangle △*NAP*, with three right angles. Indeed, in spherical geometry, all triangles have more than 180°, although precisely how much more depends on the size of the triangle. Because of this, the geometry admits an absolute unit of scale inherent in the geometry—a feature lacking in Euclidean geometry, which is invariant under scaling. In spherical geometry, we need not keep a standard meter iron rod locked in a museum case in Paris, for we could define the fundamental unit of distance as an inherent feature of space itself.

Consider how light would propagate in spherical geometry as a model of physical space. If light were emitted in all directions from a point source, it would travel outward locally, much as in Euclidean space. The curious thing to notice, however, is that the light would eventually regather on the opposite side of the sphere, at the dual point after traveling around the world. Waves in a watery world would similarly reconverge: if you jumped into the water at one point, circular waves would emanate outward and reconverge at the dual point on the opposite side of the world. But next, one must imagine a higher-dimensional analogue of this kind of space. A space traveler traveling long enough in one direction would eventually return to their starting point.

In spherical geometry, all lines intersect, since any two great circles have two antipodal intersection points. On the sphere, you might construct what you think are parallel lines, with a common perpendicular, and although locally it might appear that they never intersect, nevertheless, if you extended them halfway around the world, you would find that they meet. Thus, the strong form of Playfair's axiom fails. Although spherical geometry is suggestive of the non-Euclidean case, nevertheless not all of Euclid's other axioms are true in spherical geometry as I have described it. For example, the north and south poles are distinct points, but they lie on many different "lines," contrary to Euclid's first postulate.

Elliptical geometry

Felix Klein's version of the sphere, a model of *elliptical* geometry, addresses this issue by redefining *point* to refer to a pair of antipodal points. Thus, the north and south poles together count as a single point in this geometry, and similarly any pair of opposite points counts as one point. For any two pairs of antipodal points, there is a unique great circle containing them, fulfilling the first postulate for elliptical geometry. But meanwhile, any two lines intersect in elliptical geometry, and so the parallel postulate fails in that form.

Hyperbolic geometry

Consider next the case of *hyperbolic geometry*, with the particular model of the Poincaré disk. The points of this geometric universe are the points interior to a fixed Euclidean disk, and the "lines" are circle arcs, which meet the boundary of the fixed circle at right angles. This geometry has abundant parallel lines. Given any line *AB* and a point *C* not on that line, there are infinitely many lines through *C* and not intersecting *AB* (shown in colors). In this geometry, all triangles have angle sums strictly less than 180°, as illustrated by triangle △*ABC*. The concept of distance in the Poincaré model is a little subtle, and consequently also the

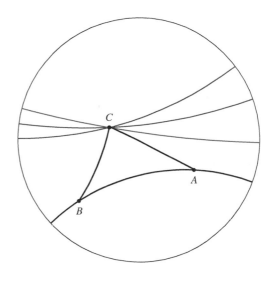

congruence relation, for one must understand that the Euclidean distances in the disk are to be greatly amplified (in a very precise manner) as one approaches the boundary. One can find numerous fascinating videos online showing what it is like to walk around in hyperbolic space, and based on such virtual experiences, one can gain a robust intuition for the nature of hyperbolic geometry.

The consistency proofs work by providing an interpretation of non-Euclidean geometry within Euclidean geometry, a way of seeing what non-Euclidean geometry could be like from the standpoint of an ambient Euclidean space. One concludes that if Euclidean geometry is consistent, then so is non-Euclidean geometry.

But one should not imagine non-Euclidean geometry only as a simulation within Euclidean geometry, for this would miss out on the view of non-Euclidean geometry as its own robust independent geometric concept; it has its own essential geometric nature, with its own geometric character that is fundamentally different from classical Euclidean geometry in important respects. One should imagine inhabiting hyperbolic space as it is, not merely as it is simulated inside Euclidean space.

Conversely, one may also find simulations of Euclidean geometry inside non-Euclidean geometry. Imagine an alien species taking hyperbolic space as fundamental; they construct within it a simulation of Euclidean space. There are such interpretations in both directions.

Curvature of space

Let us explore a little more the differing character of non-Euclidean geometry. Spherical geometry, for example, has positive curvature, and part of what this means is that if one selects a point on the sphere and draws circles of increasing radius on the surface of the sphere, then the circumferences of these circles do not grow quite as fast as they do in the Euclidean plane. The circles are bent in slightly because they lie in the surface of the sphere. The formula for the circumference, therefore, is no longer $2\pi r$, but rather something less. The circumference of a circle in spherical geometry is less than expected in Euclidean space; there are simply fewer locations within a given distance of you in spherical geometry than in Euclidean space.

Swing your arm around and contemplate the locations that exist at arm's length from you. In Euclidean space, this is a certain number of locations, having to do with the circumference of a circle, or in higher dimensions, with the area of the corresponding sphere. But in spherical geometry, the number of such locations at arm's length is somewhat less than in Euclidean space.

In hyperbolic space, in contrast, the opposite is true; there are more such locations at arm's length than in Euclidean space, and indeed, prodigiously more when the space is strongly negatively curved. Hyperbolic space is negatively curved, and so the circumference of a circle grows more rapidly with the radius than in Euclidean space.

If you are a resident of a city in hyperbolic space, then the number of shops and restaurants within a few blocks of your apartment would be enormous because in hyperbolic space, there are so many more locations that are this close to you. If the space is extremely negatively curved, then there could be whole undiscovered civilizations within walking distance, simply because a strong negative curvature means that there are such a vast number of locations quite nearby. Criminals escape easily in hyperbolic space, simply by walking away a short distance; it is too difficult to follow them far, since at every moment, one must choose from amongst so many further directions to continue. For the same reason, you or your loved ones may easily become lost in hyperbolic space, for it is so difficult to find your way exactly back home again; so please be careful and hold them close.

Let me emphasize again that both elliptical and hyperbolic geometry appear increasingly Euclidean on very small scales. If the geometric universe were vast, then a person living in such a space might believe the universe to be Euclidean based on their experience at comparatively small scale.

4.7   Errors in Euclid?

Euclid's *Elements* are often held up as a paragon of deductive mathematical reasoning. But how perfect is it mathematically? Did he make any mistakes? Does his mathematical development have any mathematical or logical flaws? The work is remarkably robust, although not perfect, and one should distinguish several types of error. We have no original copies of the *Elements*, and everything we know of the text is based on copies and translations that came some centuries later. One kind of error in Euclid could be described as the omission of axioms by implicit assumption. In his arguments, Euclid often implicitly assumed certain features of his constructions, features which are perfectly reasonable or even "obvious," but which mathematicians today do not regard strictly as consequences of his axioms. To begin, for the pedants, there are what can be seen as trivial omissions, for Euclid does not state explicitly that there are any points and lines, that every line has at least two points on it, or that not all points are collinear. But let us go ahead and give him these additional incidence axioms.

Implicit continuity assumptions

Euclid implicitly assumes that all his geometric figures are continuous without stating so. In his very first proposition, for example, he provides the construction an equilateral triangle on a segment (a construction that is very similar to the Apollonius bisector construction), and when doing so, he constructs two circles having segment AB as the radius, one centered at A and the other at B. Then, he lets P be the point of intersection of these circles and proves that triangle $\triangle APB$ is equilateral. But how do we know that the two circles actually intersect?

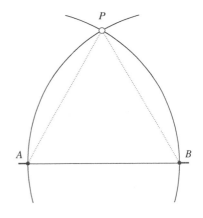

Why is there such a point P? Perhaps you find it obvious that they should intersect. But we are claiming to prove the theorems from the axioms here, not on the basis of unstated additional "obvious" facts, and we seem to have no axiom that implies that two circles on the same segment must intersect. Might we imagine that the circles simply pass through each other somehow, with no precise point of intersection? Perhaps each passes through a gap in the other.

To see how it could be that the circles fail to intersect, let us consider the *rational plane*—the set of points (x, y) in the real number Cartesian plane for which both coordinates x and y are rational numbers. Let us temporarily consider the rational plane as an entire geometric universe, a model of geometry, a world in which we can form lines and circles; but the only points that exist on them will be the rational points. In this rational world, if we have points A and B, then they are rational, and because the height of the triangle is $\sqrt{3}$ times half the base, it follows that the corresponding point P would not be rational. So in the rational plane, there simply is no such point P; the circles do not intersect in that world.

Indeed, in the geometric world of the rational plane, there simply are no equilateral triangles at all. There are also no regular pentagons or regular hexagons, or regular polygons other than the square—see (Hamkins, 2020c, chapter 9). The assumption that circles on the same radius must intersect is therefore an extra assumption, an implicit continuity assumption, clarified by the continuity and limit ideas that arose with Dedekind, Cauchy, and Weierstrass in the nineteenth century, as discussed in chapter 2. Dedekind proposed an explicit continuity axiom for geometry, related to the Dedekind completeness of the real numbers, and this was incorporated into Hilbert's formal system of geometry.

The missing concept of "between"

Another omission in Euclid is a proper theory of the concept of "between," which appears as an undefined and unaxiomatized concept in many of Euclid's constructions and arguments. In many instances, Euclid constructs collinear points A, B, and C and observes that B, say, is between A and C, and therefore AB is less than AC, and AC is the sum of AB and BC, and so forth. Similarly, he assumes that the angle bisector of angle $\angle Q$, as at right, will intersect the segment PR at a point between P and R.

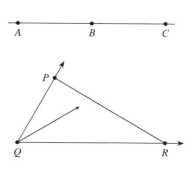

But the concept of between is subtler than may be expected, and a proper axiomatization would explicate the fundamental properties and assumptions about this concept. To appreciate the subtlety, notice that in spherical and elliptical geometry, for example, the concept of between goes awry, as there is a sense in which any of three collinear points A, B, and C could be regarded as between the others if one simply goes the other way around the sphere. Of the cities New York, Prague, and Sydney, which one is between the others? Part of what is involved here is an ambiguity about segments, since two points P and R on the sphere determine two line segments: you can fly from San Francisco to London on a great circle either by flying east over the Atlantic (the direct route), or by flying west over the Pacific and Asia (a longer route). In spherical geometry, it does not make sense to refer to *the* segment *PR* without elaboration.

Hilbert's geometry

Hilbert proposed a specific new axiomatization of geometry, augmenting and correcting Euclid and addressing all the flaws I have mentioned in this chapter and others as well. His system includes several axioms on betweenness, such as the assertion that if a point *B* is between *A* and *C*, then it is also between *C* and *A*; and given any three collinear points, exactly one of them is between the other two (so this excludes spherical and elliptical geometry).

A more substantial point-existence betweenness axiom asserts:

> For any two points *B* and *D*, there are points *A*, *C*, and *E*, such that point *B* is between *A* and *C*; point *C* is between *B* and *D*; and point *D* is between *C* and *E*.

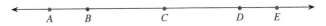

According to this account, a line segment *AB* consists of the points that are between *A* and *B*, and points *A* and *B* are defined to be on the "same side" of a line ℓ if no point of ℓ is between *A* and *B*. Hilbert then includes axioms to the effect that every line partitions the space into exactly two sides, which has the effect of making the geometry two-dimensional, a plane geometry, as desired. The Hilbert system also makes explicit several axioms detailing the notion of the incidence of a point on a line, the concept of congruence and axioms on parallelism. Marvin Jay Greenberg (1993, chapter 4) refers to *neutral geometry* as the geometry of the Hilbert theory without the parallel postulate, that is, with the axioms of incidence, betweenness, congruence, and continuity. This is the theory that is common to Euclidean and hyperbolic geometry, as Hilbert's other axioms already rule out elliptical geometry.

Tarski's geometry

Tarski developed an alternative axiomatization, called the *elementary theory of geometry*, based solely on points, with all the other geometric concepts expressed in terms of points by axiomatizing the undefined trinary betweenness relation and the relation of equidistancy, defined on pairs of pairs of points, and including a first-order continuity scheme.

4.8   Geometry and physical space

Is geometry the mathematical study of the properties of physical space? Geometers and natural philosophers throughout history have indeed thought so. According to this view, we understand points, lines, and circles as mathematical idealizations of features in physical space.

In her account of the philosophical tradition in non-Euclidean geometry, Joan Richards (1988) argues that it was Bernhard Riemann and Hermann von Helmholtz who triggered philosophical discussions of the nature of geometrical truth. Each analyzed the basic spatial concept, seeking to discover which geometric concepts were logically necessary and which parts were contingent on experience, asserting ultimately that the decision between Euclidean and non-Euclidean geometry was experientially determined. She writes that "thus, their analysis raised the question of whether Euclidean geometry was necessarily or only contingently true" (Richards, 1988, p. 64).

In light of the possibility of non-Euclidean geometry, one might take the Euclidean theory of space as a scientific hypothesis to be scientifically tested. How could we determine experimentally whether the physical universe was Euclidean or non-Euclidean? One way would be simply to form a large triangle and measure the angles to see if they add up to 180°. Gauss reportedly did so using three distant mountain peaks; the result conformed with Euclidean geometry to within the margin of error in the measurement. Lobachevsky called for such experiments to be undertaken with triangles on an astronomical scale.

Greenberg (1993) emphasizes the point that

> because of experimental error, a physical experiment can never prove conclusively that space is Euclidean—it can prove only that space is non-Euclidean.

The measurement will be such as it is, with its margin of error, and if 180° is excluded from that interval, then it shows that space is non-Euclidean; but if 180° is within the margin of error, then also other sums-of-degrees very near but not equal to 180° will also be within the margin of error. Thus, any experimental result compatible with Euclidean space is also compatible with non-Euclidean (but nearly Euclidean) space.

Suppose that the physical universe turns out not to be Euclidean—would this somehow refute any aspect of Euclid's theory of geometry? Would not all the theorems and constructions still be correct for the kind of geometry that Euclidean geometry was about? If so, doesn't this counterfactual already refute the idea that geometry is the mathematical study of physical space? Euclidean geometry is about the mathematical realm of objects—points, lines, and circles—that satisfy the axioms of Euclidean geometry, whether or not the physical universe does so.

Our best current physical theories indeed have grown to conflict with the view that the various classical geometric theories, whether Euclidean, hyperbolic, or elliptic, are the geometry of physical space. According to current theories, the nature of space (or spacetime as it is regarded by the physicists) is affected by the presence of mass, which warps and curves space in the vicinity of the mass. Light bends under the influence of gravity, as predicted by relativity theory and confirmed by experimental observation, such as observations of gravitational lensing of distant light sources around intermediate massive objects. Physical space is much more complex geometrically than any of the geometric theories we have been discussing.

There are further problematic issues with taking geometry to be the study of physical space. Perhaps it will turn out, as explained by some future improved version of quantum mechanics, that space (like energy) is quantized into discrete indivisible units; perhaps it is finite. And consider the homogeneity of space, the question of whether the local nature of physical space is everywhere the same. Perhaps a future improved physics will reveal that space can become torn or degraded somehow locally, making it inhomogeneous.

4.9   Poincaré on the nature of geometry

One is led ultimately to a view of geometry as concerned with purely mathematical questions, divorced from any necessary connection with physical space. Henri Poincaré responded to the issues above and to Kant by saying:

> A difficulty remains, however, and it is unsurmountable. If geometry were an experimental science, it would not be an exact science but rather one subject to continual revision. No, it would be sure of being incorrect from now on, since we know that there are no perfectly rigid solids.
>
> *Geometric axioms are then neither synthetic* a priori *judgments, nor experimental facts.*
>
> They are *conventions*. Among all possible conventions, our choice is *guided* by experimental facts, but it remains *free* and is only limited by the need to avoid all contradiction. In this way, the postulates can remain *strictly* true even when the experimental laws contributing to their adoption are only approximate. In other words, *the axioms of geometry* (I am not speaking of those of arithmetic) *are only definitions in disguise.*
>
> At this point, how should we understand this question: Is Euclidean geometry true? The question does not make any sense. One might just as well ask if the metric system is true and the old units of measure false; if the Cartesian coordinates are true and the polar coordinates false. One geometry cannot be truer than another, it can only be *more useful*. (Poincaré, 2018, p. 42, emphasis in the original)

For Poincaré, the existence of the alternative geometries means that the geometrical truths of physical space cannot be known a priori. The various geometrical frameworks can be studied mathematically, but they amount to conventions when put forth as a model of physical reality. The axioms themselves define the subject that they are about.

4.10   Tarski on the decidability of geometry

Let me conclude this chapter by returning to Tarski's elementary theory of geometry and something he proved about it, which to my way of thinking has profound implications for the subject of classical geometry as a mathematical activity. Namely, what Tarski proved is that his elementary theory of geometry is both complete and decidable. That is, his axiomatization of geometry logically settles every question that can be asked in the language of elementary geometry, and furthermore, he provides a computable procedure for determining the truth of any such statement. Amazing! After two millennia of mathematical struggle, we have reduced geometry to a rote procedure.

Tarski's geometric decision procedure is a consequence of his deep theorem on real-closed fields, in which he identified a complete set of axioms for the ordered real number field structure $\langle \mathbb{R}, +, \cdot, 0, 1, < \rangle$. Tarski proved that every assertion in the language of ordered fields is equivalent over his real-closed field axioms to an assertion not using any quantifiers, whose truth is therefore easily decided. For example, the existential assertion $\exists x \, ax^2 + bx + c = 0$ is equivalent in the real numbers to the assertion that the discriminant is nonnegative, $b^2 - 4ac \geq 0$, and this latter assertion is quantifier-free. Tarski proved, amazingly, that *every* assertion in the language of the ordered real number field is similarly equivalent to a quantifier-free assertion. This has consequences for geometry because the full language of ordered fields is extremely rich—rich enough to refer to all the classical geometric objects (points, lines, planes, circles, parabolas, conic sections, hyperboloids, angles, and so on) by referring to the algebraic equations that describe them. In this formal language, therefore, one can make sweeping assertions about all triangles, circles, or cones, requiring that there are further geometric objects with certain properties and so forth, for any algebraically expressible properties. And thus, by applying the Tarskian decision procedure, one can decide the truth of any such geometric assertion.

From an optimist perspective, the Tarskian decision procedure is a theoretical completion of the subject of geometry, a project begun over two thousand years ago and now completed with a decision procedure that automates the process of geometric reasoning, determining by rote the truth of any geometric statement, in any desired dimension, in the Cartesian model of analytic geometry provided by the real numbers. We have built a geometric truth-telling machine.

A counterargument to this grandiose vision, of course, is the observation that Tarski's decision procedure is of only theoretical use, since although we have a concrete account of the procedure, it simply takes far too long in practice to be useful. The time complexity of Tarski's original algorithm is so huge and fast-growing that it cannot be bounded by any elementary recursive function. Improved versions of the algorithm have brought the time complexity down to doubly exponential, a vast improvement but still infeasible for any practical purpose. The double exponential bound has been proved optimal, in certain senses, for any procedure based, as Tarski's method is, on quantifier elimination.

Another serious counterargument is that the algorithm does not accommodate reference to integers or natural numbers in the formal statements to which it applies, although classical geometry does make such references. In light of Gödel's theorems (see chapter 7), it follows that there can be no decision procedure for the elementary theory of arithmetic, and so it will be impossible to modify the Tarskian decision procedure to include arithmetic. In particular, we cannot use Tarski's procedure to decide questions of the form, "Will this specific constructive procedure achieve a certain configuration in finitely many steps?" or "is there a constructive procedure with these specific properties?" In this sense, this fuller conception of geometry is not completed by the Tarskian decision procedure.

Questions for further thought

4.1 In the proof we gave for the compass equivalence theorem, we used both a straightedge and a compass. In addition, because the proof relied on rectangles, it also used the parallel postulate. Nevertheless, there are proofs of this theorem using only a compass and not using the parallel postulate. Find such a proof (for example, in Euclid or by searching online) and undertake the construction yourself. Explain why this manner of arguing might be superior.

4.2 Show that the sum and product of any two constructible numbers are constructible.

4.3 Construct a regular hexagon with a straightedge and compass, given a segment to be used as a side. Which other regular polygons can you easily construct? (But look into the Gauss–Wantzel theorem.)

4.4 Discuss the ontology of nonconstructible numbers on the classical number line. Do these points exist if we cannot construct them?

4.5 In the text, we described how one can in principle exhaustively construct all the constructible points in the plane by starting with two points A and B, and then at each stage form all possible lines and circles using those points and finding their points of intersection before proceeding to the next stage. Does this process depend on which particular points A and B you start with? If so, in what sense have we defined "constructible point"? Is there a similar objection to the definition of "constructible number"?

4.6 In what sense are geometric constructions invariant under scaling? If I can construct a figure using a straightedge and compass, can I construct the same figure scaled by a factor of x, for arbitrary x? Are the constructible numbers invariant under arbitrary scaling?

4.7 Discuss the sense in which the compass equivalence theorem is a positive instance of a situation that arises negatively with some non-Euclidean construction tools, such as a marked ruler or the use of origami or spirographs.

4.8 Any physical straightedge has some finite length and any physical compass can open to a maximum radius. Do these physical limitations cause theoretical limitations in their capacity in principal to undertake arbitrary straightedge and compass constructions? Consider first the cases where only the compass is limited or only the straightedge is limited, and then consider fully the case where both are limited. See Hamkins (2020a), Goucher (2020).

4.9 Does the constructible plane satisfy all the same geometric truths as the real number plane $\mathbb{R}^2$? In other words, is the constructible plane an elementary substructure of the real number plane, with respect to assertions expressible in the language of geometry? [Hint: The real number plane has solution points realizing the trisection of the angle and the doubling of the cube. Are these verifiable there?]

4.10 We cannot double the cube with a straightedge and compass, but can you double the square? Given a square, construct a square with exactly twice the area.

4.11 Consider some other classical construction, such as the bisection of an angle, and analyze the propagation of error in the placement of the compass or straightedge.

4.12   Is every point on the number line constructible by straightedge and compass? How do you know?

4.13   Is the Cantor argument that there are nonconstructible numbers constructive? Does it provide the existence of a particular nonconstructible number?

4.14   Consider another classical construction, such as the angle bisector construction, and implement it in the GeoGebra system.

4.15   Argue that any construction that can be undertaken with the straightedge, compass, and the two basic origami folds mentioned in the text can also be undertaken with a straightedge and compass only.

4.16   Criticize the proof that every triangle is isosceles. Is it flawed?

4.17   Is geometry the mathematical study of physical space? If not, what is the subject matter of geometry?

4.18   Discuss the question of whether one could determine whether physical space is Euclidean experimentally. Could a measurement show that space is Euclidean? Or that it is not Euclidean?

4.19   Is Playfair's formulation of the parallel postulate true in spherical geometry? If not, why is this geometry considered non-Euclidean?

4.20   To what extent can the intuitions that we might have for the truths of Euclidean geometry also support non-Euclidean geometry, especially if the curvature is very near zero?

4.21   How is the completeness of Tarski's axiomatization related to the existence of a computable decision procedure? Can you argue that every complete theory having a computably enumerable axiomatization is computably decidable? (We shall discuss this issue further in chapter 6.)

4.22   Discuss the significance of Tarski's decision algorithm for geometry as a mathematical practice; what role does the infeasibility of the algorithm play?

Further reading

Marvin Jay Greenberg (1993). A fantastic account of the subject, rich in mathematics, history, and philosophy; highly recommended.

John Stillwell (2005). A completely solid mathematical development of the subject of geometry.

Dave Richeson (2018). A fun GeoGebra instance implementing a perspective chessboard construction, allowing one to see how changes in the setup of the initial data affects the perspective of the result.

Robert Geretschlager (1995). A fascinating and accessible account of the geometry of origami, including an analysis of origami as a method of geometric construction.

Viktor Blåsjö (2013). Follows through on the ideas of von Helmholtz concerning what it would be like to be a creature living in hyperbolic space, concerning visual experience and illusions in comparison with Euclidean space, and vice versa.

Credits

Thanks for the comments and suggestions of Deborah Franzblau and Ilya Kofman. The false proof that every triangle is isosceles is due to W. W. Rouse Ball, *Mathematical Recreations and Essays*, 1892 (Ball, 1905, p. 38), and the discussion here was adapted from my book *Proof and the Art of Mathematics* Hamkins (2020c). The chessboard construction using the straightedge alone is described, along with a discussion of its connection to art history, in (Stillwell, 2005, pp. 88–94). The impossibility of doubling the cube and trisecting the angle was first proved by Pierre Wantzel in 1837. The material on the history of non-Euclidean geometry, including the discussion of Bolyai and Gauss, was adapted from Greenberg (1993). Joan Richards's remarks are taken from Richards (1988). Tarski's account of the theory of elementary geometric constructions appears in Tarski (1959). His decision procedure for real-closed fields appears in Tarski (1951).

5 Proof

Abstract. What is proof? What is the relation between proof and truth? Is every mathematical truth true for a reason? After clarifying the distinction between syntax and semantics and discussing various views on the nature of proof, including proof-as-dialogue, we shall consider the nature of formal proof. We shall highlight the importance of soundness, completeness, and verifiability in any formal proof system, outlining the central ideas used in proving the completeness theorem. The compactness property distills the finiteness of proofs into an independent, purely semantic consequence. Computer-verified proof promises increasing significance; its role is well illustrated by the history of the four-color theorem. Nonclassical logics, such as intuitionistic logic, arise naturally from formal systems by weakening the logical rules.

What does it mean to prove a mathematical assertion? Mathematicians, it is said, are sometimes rejected by prosecutors during jury selection in the US criminal justice system because they do not have the ordinary understanding of what it means to prove something "beyond a reasonable doubt," the standard of evidence that juries follow for conviction; the mathematician's standard of proof is high. Let us distinguish sharply between truth and proof. A statement is true when the idea it expresses is the case. (There is, to be sure, a vibrant philosophical debate on the precise nature of truth, explicating what it might mean to have a truth predicate or to make a truth assertion.) To prove a statement, in contrast, means to provide a reason why it is true, a reason for believing it, a justification that the statement is indeed the case. And there is a similarly rich proof theory, which we shall discuss in this chapter. Perhaps you may imagine that a statement could be true independently of whatever reasons we may be able to provide for it? Is every mathematical truth true for a reason, such as a proof? Can there be accidental mathematical truths?

5.1 Syntax-semantics distinction

Proof and truth, therefore, lie on opposite sides of the syntax-semantics divide, for at bottom, a proof is a kind of argument, a collection of assertions structured syntactically in some way, while the truth of an assertion is grounded in deeply semantic issues concerning the way things are.

On the semantic side, we are concerned with meaning, truth, existence, and validity. If M is a mathematical structure and φ is an assertion in the corresponding language, possibly with parameters from M, then we define the satisfaction relation $M \models \varphi$, pronounced "M satisfies φ," to mean that φ is true in M. This relation is defined in any first-order language and many extensions by induction on the complexity of φ, in what is called a *compositional* manner, by reducing the truth of a compound assertion to the truths of instances of its constituent subassertions. A theory T, meaning a set of sentences in the language, is *satisfiable* if there is a model in which every sentence of the theory is true. Still on the semantic side, we have the notion of logical consequence, $T \models \varphi$, which means that φ holds in every model of the theory T. In the empty theory, we have the notion of validity $\models \varphi$, which means that φ holds in every model.

Proof, in contrast, lies solidly on the syntactic side, since ideally one can verify and analyze a proof as a purely syntactic object, without a concept of meaning and without ever interpreting the language in any model. For any theory T, we write $T \vdash \varphi$ to mean that there is a proof of φ from T. Similarly, $\vdash \varphi$ means that φ is provable from the empty theory, with no extra axioms beyond the logical axioms that might be built into the formal proof system itself.

Syntax	**Semantics**
Language	Meaning
Numeral	Number
Formula φ	Model M
Variable/constant symbol	Element $a \in M$
Relation symbol	Actual relation on M
Function symbol	Actual function on M
Theory T	Truth $M \models \varphi$
Proof $T \vdash \varphi$	Validity $T \models \varphi$
Consistency of T	Satisfiability of T
"Snow is white"	Snow is white
Mention	Use
Existence claim	Actual existence
Formalism	Platonism
Using "pants" as "trousers"	Using pants as trousers

We thus divide our logical conceptions along the syntax/semantic boundary, and many profound results of mathematical logic concern the interaction across this divide.

Use/mention

The distinction between semantics and syntax is also known as the *use/mention distinction*, the distinction between using a word and mentioning it. At high table recently in

Oxford (a regular gathering of college fellows and scholars at an enjoyable formal meal), my colleague, Alex Moran, citing "customary practice" in the north of England, said:

> I generally use pants as trousers.

Or perhaps he had said,

> I generally use "pants" as "trousers."

For the Americans, let me mention the common British usage of the word "pants" for what would be called "underwear," "underpants," or "panties" in the US. I made a discreet glance under the table and can report that he didn't seem to be using pants as trousers at the time. Since quotation marks are not usually pronounced, not even silently, with two curled fingers in the air, I couldn't be sure which sentence he had said. But as we had been discussing regional variations in language use, probably he had mentioned the words "pants" and "trousers" rather than used them.

My daughter Hypatia inquired,

> Does everything rhyme with itself?

I had replied that it did not, having wrongly (and perhaps teasingly) heard her to ask,

> Does "everything" rhyme with "itself"?

But then again, she might have meant

> Does "everything" rhyme with itself?

or possibly

> Does everything rhyme with "itself"?

5.2 What is proof?

It is a traditional mark of maturity when a fledgling mathematician is finally able reliably to recognize when they have got a proof and when they have not, and furthermore, to be honest with themselves about this. Less mature students are seen sometimes to allow nonsense into their arguments, perhaps without even realizing it—"it's not even wrong"— whereas the mathematician has generally either got a correct argument, or if not, would prefer to write nothing or to revise the claim to what had been proved. Mathematicians sharply distinguish between a proof and a proof idea.

But what is a proof? In high-school geometry, students often learn a standard two-column form of proof, in which certain kinds of statements are allowed in the first column, provided that they are justified by certain kinds of reasons in the second column. This form of proof highlights an all-important feature—namely, that a proof must provide a chain of reasoning that logically establishes the conclusion from the premises, while also emphasizing the idea that every mathematical assertion requires proof. In contemporary mathematical research, the rigid, two-column proof style gives way to an open, more flexible proof format. Most contemporary mathematical proofs are written in prose, essay-style, while retaining the defining feature that a proof must logically establish the truth of its conclusion.

Pressed for a precise definition, many mathematicians might have difficulty saying exactly what a proof is. In mathematical practice, a proof is any sufficiently detailed convincing mathematical argument that logically establishes the conclusion of a theorem from its premises. Successful proofs of new theorems often succeed by introducing new mathematical ideas or methods, and one gains mathematical insight from the proof beyond merely learning the truth of the theorem itself. We value such proofs because we might use the new methods to answer other questions that intrigue us. To prove a theorem by introducing a new method is often more valuable than to prove it with known techniques.

Proof as dialogue

Mathematical proofs often take place in a social context, and whether an argument counts as a proof can depend on the intended audience. For some audiences, perhaps more details will be needed, while in others, fewer will suffice. When a proof is offered and skeptics can probe the argument by asking for elaboration or fuller details in the difficult parts, then surely it counts as positive evidence of the robustness of the proof when clear explanations are readily and satisfactorily forthcoming. In contrast, when the proof-provider is unable to explain or justify a part of their argument more thoroughly, this is bad news. There may be a gap in the proof; perhaps the objections have broken the argument. In this way, we come to an understanding of proof as a kind of dialogue.

Catarina Dutilh Novaes (2020) emphasizes this dialogical nature of proof, explaining how this perspective meshes with the several interrelated roles for proof in mathematical practice, including the traditional (1) proof as verification, the logical truth-preserving function of proof, establishing the validity of the theorem, which she distinguishes from (2) proof as certification, meaning the community recognition of the validity of a proof, as it might arise via refereeing and community vetting:

> In Prover-Skeptic terms, we may say that a proof has been certified if it has been examined by a sufficient number of suitable Skeptics, and none of them has found errors in it. Of course, what counts as a "sufficient number" and a "suitable Skeptic" is to some extent a contextual matter. Inevitably, high-profile results by distinguished mathematicians will receive more attention from Skeptics, and thus will be certified to a higher degree if no mistakes are found than an unremarkable result published in an obscure journal. Presumably, the stakes are higher also because more people will rely on these "famous" results in their own work. Certification can thus also be conceived as a matter of degrees (perhaps similar to robustness and replication in empirical research): the more competent experts have scrutinized a proof and not found any significant mistakes, the more certified it is. (pp. 222–223)

She also distinguishes (3) proof for communication and persuasion and (4) proof as explanation, indicating not only *that* something is true, but *why* it is true; and also (5) proof as innovation, by which proof serves as a conduit of new ideas and perspectives. In several striking historic cases, flawed proofs were still seen as critically important because of the new mathematical ideas they offered. Finally, in the role of (6) proof as systematization, proofs are seen to unify related ideas that have not yet been joined together in mathematics.

Wittgenstein

Ludwig Wittgenstein (1956, III.§46) emphasizes our capacity to reproduce a proof exactly, as opposed to the difficulty of reproducing a shade of color or handwriting exactly:

> It must be easy to write down exactly this proof again. This is where a written proof has an advantage over a drawing. The essentials of the latter have often been misunderstood. The drawing of a Euclidian proof may be inexact, in the sense that the straight lines are not straight, the segments of circles not exactly circular, etc. etc., and at the same time the drawing is still an exact proof.

Wittgenstein also emphasizes that proof must be *understandable*; poor notation or obscure formalization can mean that proof is lost:

> I want to say: if you have a proof-pattern that cannot be taken in, and by a change in notation you turn it into one that can, then you are producing a proof, where there was none before.

For example, a smooth arithmetical notation may allow new insights that are not possible in a cumbersome formalism:

> if a man had invented calculating in the decimal system—that would have been a mathematical invention!—Even if he had already got Russell's Principia Mathematica.

Even when a foundational formalism has been established, nevertheless one can often express insightful mathematical ideas outside of it. Accordingly, he describes mathematics as "a MOTLEY of techniques of proof."

Thurston

The geometer Bill Thurston (1994) describes a vision of mathematics where mathematical "understanding" is a primary goal, and he argues that the definition-theorem-proof model is not necessarily the one best equipped to arrive at it. As an illustration, he rattles off seven ways of understanding the meaning of the derivative of a function—familiar conceptions using infinitesimals, slopes of tangent lines, ϵ formalism, linear approximation, and so on—and he imagines the list continuing; "there is no reason for it ever to stop." He imagines item 37 on the list: the derivative of a real-valued function on a domain is the Lagrangian section of a certain cotangent bundle:

> This is a list of different ways of thinking about or conceiving of the derivative, rather than a list of different logical definitions. Unless great efforts are made to maintain the tone and flavor of the original human insights, the differences start to evaporate as soon as the mental concepts are translated into precise, formal and explicit definitions.

In Thurston's account, tensions between formalization and mathematical understanding run to the very core of mathematics; formalization can erase mathematical meaning:

> The standard of correctness and completeness necessary to get a computer program to work at all is a couple of orders of magnitude higher than the mathematical community's standard of valid proofs. Nonetheless, large computer programs, even when they have been very carefully written and very carefully tested, always seem to have bugs.... When one considers how hard it is to write a computer program even approaching the intellectual scope of a good mathematical paper, and how much greater time and effort have to be put into it to make it "almost" formally correct, it is preposterous to claim that mathematics as we practice it is anywhere near formally correct.

But Thurston does not find mathematics unreliable. Rather, his point is that the reliability of mathematics arises not from formal proof, but from mathematical understanding:

> Mathematicians can and do fill in gaps, correct errors, and supply more detail and more careful scholarship when they are called on or motivated to do so. Our system is quite good at producing reliable theorems that can be solidly backed up. It is just that the reliability does not primarily come from mathematicians formally checking formal arguments; it comes from mathematicians thinking carefully and critically about mathematical ideas.

Formalization and mathematical error

Meanwhile, because many published proofs do have errors and many published theorems are incorrect, one wonders why mathematics does not simply collapse. Much like Thurston, Mike Shulman (2019) defends the idea that "the fundamental content of mathematics is ideas and understanding," rather than proof; even incorrect mathematical arguments can still be mathematically robust and valuable. He relates an instance:

> While working on a recent project I discovered no fewer than nine mistaken theorem statements (not just mistakes in proofs of correct theorems) in published or almost-published literature, including several by well known experts (and two by myself). However, in all nine cases it was simple to strengthen the hypothesis or weaken the conclusion in such a way as to make the theorem true, in a way that sufficed for all the applications I know of. I would argue that this is because the mistaken statements were based on correct ideas, and the mistakes were simply in making those ideas precise.

Shulman explains how a robust community of mathematicians responds effectively to error, even when the exact source of the error is not completely understood, such as when there is a complicated theorem proof and an equally complicated counterexample. Mathematicians rely on their well understood and tested ideas to lead them through the impasse.

> The point here is that a community of people developing ideas together is likely to have arrived at correct intuitions, and these intuitions can flag "suspicious" results and lead to increased scrutiny of them.

Part of Shulman's point is that formal mathematical statements may not actually capture the intended mathematical insight, and mathematicians accordingly may be less interested in the true/false or proved/refuted status of formal assertions, giving preference and priority to the mathematical insight and understanding that they might achieve without formalization.

Formalization as a sharpening of mathematical ideas

But does this let the mistakes slide off too easily? Let me illustrate by example how formalization works in ordinary mathematical practice. Suppose that we have a real-valued function on the real numbers that we have reason to think is very "wriggly," in a way that we think is relevant for a certain further consequence. We ponder: In what sense exactly is the function wriggly, and is this sufficient for the conclusion? Perhaps the function is wriggly in the sense that the derivative changes sign infinitely often. Thus, it very often goes up and then down, if only minutely. Or perhaps the function is wriggly merely in the sense that it has infinitely many inflection points? Or perhaps the function is differentiable, but on every interval, the derivative changes sign? Or perhaps the function is not differentiable, yet it is wriggly in the sense that it is not monotone on any interval? Or perhaps, it is crossed by secants with positive and negative slopes as steep as desired? Or perhaps it is nowhere differentiable, which is itself a form of wriggliness? Or perhaps there is still another conception of wriggliness that holds of our function and works for the application?

My point is that the process of formalization is not typically one of translating fully formed mathematical ideas into a sterile language, with punctuation and balanced parentheses being the main concern. Rather, the process of formalization is quite commonly also one of sharpening our ideas, giving substance and further meaning to initially vague conceptions or proof ideas; it is at the very heart of mathematical activity: figuring out and saying precisely what you mean. An initial idea about wriggliness admits dozens of precise formulations, each one a slightly different expression of that vague thought, and not all equivalent, although some of them may be. We might place this list next to Thurston's, for contrast. Whereas he had a list of different ways of thinking about the same underlying mathematical feature, equivalent when formalized, we have a list of inequivalent formal manifestations of the same underlying informal idea. To formalize the idea is to adopt one of these formulations, to have figured out what we meant and to express exactly that. In this way, formalization often gives more depth and meaning to our ideas and intuitions, rather than less.

Mathematics does not take place in a formal language

Some philosophers have the view that mathematical activity consists of arguing from axioms to theorems. These axioms and theorems are stated in a mathematical language of some kind, perhaps a formal or semiformal language, and so from this view, mathematical activity always involves a mathematical language. In particular, according to this view, mathematical activity does not take place in a mathematical structure, whether or not such structures enjoy a real existence, but rather in a mathematical theory and language, in which one makes claims and assertions about such a structure.

Meanwhile, mathematicians sometimes defend the idea that mathematical ideas are not always or even usually mediated in this way through a formal language, or indeed perhaps not through any language at all. A geometer might imagine a geometric shape, mentally manipulating it in order to "see" certain mathematical features as obvious. This process does not seem directly to involve any formal language, and yet the thoughts and analysis it involves are clearly mathematical. In these cases, the mathematical ideas are not naturally expressed in words or symbols, but in other conceptions, of a visual or spatial character or abstract, but not necessarily linguistic or symbolic. A dog thinks, but probably not in any linguistic or symbolic language as ordinarily understood. These alternative mathematical concepts are perhaps more difficult to communicate to others, but no less compelling to the mathematician himself or herself. And the phenomenon is not limited to geometry or the visual parts of mathematics; mathematicians in widely disparate mathematical specialities sometimes report that their internal conceptions of mathematical ideas are not identical to the expressions of those ideas in words or in a formal language. Many mathematicians report having idiosyncratic understanding of certain mathematical concepts, including very abstract notions of certain kinds of mathematical structure:

> How big a gap is there between how you think about mathematics and what you say to others? ... I've been fascinated by the phenomenon the question addresses for a long time. We have complex minds evolved over many millions of years, with many modules always at work. A lot we don't habitually verbalize, and some of it is very challenging to verbalize or to communicate in any medium. Whether for this or other reasons, I'm under the impression that mathematicians often have unspoken thought processes guiding their work which may be difficult to explain, or they feel too inhibited to try. One prototypical situation is this: there's a mathematical object that's obviously (to you) invariant under a certain transformation. For instance, a linear map might conserve volume for an "obvious" reason. But you don't have good language to explain your reason—so instead of explaining, or perhaps after trying to explain and failing, you fall back on computation. You turn the crank and without undue effort, demonstrate that the object is indeed invariant. Thurston (2016)

Thus, Thurston distinguishes between "thinking" and "explaining." One can see or think a mathematical fact for oneself, with one's own private conceptions, but since these are difficult or impossible to explain to others, one must find a different language for explanation. Thurston (2016) includes fascinating further instances of this phenomenon posted there by distinguished mathematicians such as Terence Tao and Timothy Gowers.

Eugenia Cheng (2004, p. 20) describes it like this:

> The answer is that a moral reason is harder to communicate than a proof.

> The key characteristic about proof is not its infallibility, not its ability to convince but its transferability. Proof is the best medium for communicating my argument to X in a way which will not be in danger of ambiguity, misunderstanding, or defeat. Proof is the pivot for getting from one person to another, but some translation is needed on both sides.

> So when I read an article, I always hope that the author will have included a reason and not just a proof, in case I can convince myself of the result without having to go to all the trouble of reading the fiddly proof. When this does happen, the benefits are very great.

Maddy (2017) identifies the *shared standard* role played by a foundation of mathematics, with agreed-upon standards of argument and proof. The existence of a formal proof system, as well as a common axiomatic framework, provides a final court of appeal for adjudicating mathematical claims. Thurston, however, seems to retreat from this, claiming that mathematical insight is often idiosyncratic, amounting to a Wittgensteinian private language, impossible or difficult to translate into a shared framework without the loss of some mathematical insight. John P. Burgess (2015) seeks to find room for informal proof, aiming to find necessary and sufficient criteria for what counts as proof, demarcating the informal proofs that count as legitimate. And yet, Burgess seems to dismiss claims that some proofs can be more "explanatory" than others.

Voevodsky

Coming ultimately to an opposite view as Thurston on the relevance of formalization, Vladimir Voevodsky (2014) relates how an instance of mathematical error was uncovered only after a long delay and then repaired with much more complicated proofs, which were also subsequently found to contain errors:

> This story got me scared. Starting from 1993, multiple groups of mathematicians studied my paper at seminars and used it in their work and none of them noticed the mistake. And it clearly was not an accident. A technical argument by a trusted author, which is hard to check and looks similar to arguments known to be correct, is hardly ever checked in detail.

The incident motivated him to find more secure formal foundations and led eventually to univalent foundations and homotopy type theory, discussed later in this chapter.

Proofs without words

Let me try to convey in an elementary way the kind of understanding that Thurston and others have mentioned. Although Thurston is concerned with deep understanding of advanced mathematical ideas, nevertheless the genre of mathematical proof called *proofs without words* offers, to my way of thinking, something of the spirit and feeling of what he intends while remaining on an elementary mathematical level. In these proofs without words, mathematicians attempt to communicate a mathematical argument using no words at all, conveying

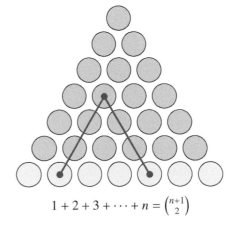

$$1 + 2 + 3 + \cdots + n = \binom{n+1}{2}$$

the proof purely by means of an insightful graphic or diagram. Often very clever, these arguments often aim to induce an *Aha!* moment of understanding when one grasps the idea. Consider the figure here, for example, as a proof without words of the indicated identity, where the notation $\binom{n+1}{2}$ means the number of ways of choosing two elements from a set of size $n + 1$. Can you see how the figure proves the identity?

Shall I explain the proof with a few words? The triangular arrangement of blue circles has one at the very top, and then two, and then three, and so on, making $1 + 2 + \cdots + n$ many blue circles in n rows. The key thing to notice is that every blue circle is determined by a choice of two green circles, and vice versa, in the diagonal manner indicated in red. Aha! It follows that the number of blue circles is equal to the number of ways of choosing two green circles, which therefore establishes the desired identity.

To my way of thinking—and despite the name of this proof genre—almost every so-called proof without words is improved by a few well chosen words. I find it a false virtue to offer a proof truly without explanation; I know of many cases where an intended proof-without-words was simply not understood. In such cases, therefore, I tend to regard these so-called proofs-without-words instead merely as poorly explained proofs. Use illuminating figures in your proofs, yes, but also words.

How to lie with figures

Of course, one must be on guard with figures, since they can easily lead one astray. Consider this diagram, offered as proof that

$$32.5 = 31.5.$$

The two colored-tile arrangements can be transformed into one another by rearranging the tiles, which each have exactly the same respective shapes and dimensions in both images. In the upper image, the tiles appear to

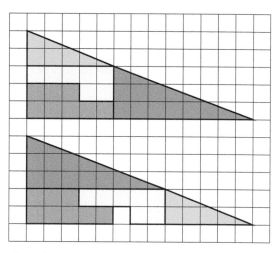

fill out a triangle with base 13 and height 5, having area 32.5; but in the lower one, there is a missing square, so the colored area is one less. Is this argument correct? The reader will criticize it in question 5.3.

Hard arguments versus soft

A hard argument is one that is technically difficult; perhaps it involves a laborious construction or a difficult calculation; perhaps it involves bringing disparate fine details together in just the right combination in order to succeed; or perhaps it involves proving various specific facts about a comparatively abstract construction, perhaps relating disparate levels of abstraction. With a hard argument, one must pay close attention. A soft argument, in contrast, is one that appeals only to very general abstract features of the situation, and one needs hardly to construct or compute anything at all. These are often my favorite proofs.

When a theorem has multiple proofs, one can often observe a certain conservation principle, the conservation of difficulty. One proof of the theorem may seem softer than another more difficult argument, but upon reflection, one realizes that the soft proof uses a background theory or theorem, whose development is itself difficult, often in just the same way. Thus, there is a conservation of difficulty; one needs to do the hard work one way or another. But of course, this conservation principle is merely a commonly observed phenomenon, not actually a universal principle. The history of mathematics is full of cases where a mathematician has found a genuinely easier proof, which then becomes the new standard of difficulty for that theorem.

Saharon Shelah, not known ever to shrink in the face of a mathematical difficulty, identifies in those difficulties the source of many mathematical pearls:

> Still others mourn the loss of the "good old days" when the proofs were with ideas and were not so technical. In general, I am not a great fan of the "good old days" when they treated your teeth with no local anesthesia, and the term technical is a red flag for me, as it is many times used not for the routine business of implementing ideas but for the parts, ideas and all, which are just hard to understand and many times contain the main novelties.
>
> My feeling, in an overstated form, is that beauty is for eternity, while philosophical value follows fashion.
>
> ... As for beauty, I mean the beauty in a structure in which definitions, theorems and proofs have their part in the harmony; but complicated proofs do not bother me....
>
> A disgusted reader may shout: "Beauty? You find in your mess some trace of beauty?" I can only say that I hear the music of those spheres or that every one likes his own dirt (the difference is small). (Shelah, 1993, Axis A)

Moral mathematical truth

Mathematicians sometimes speak of a mathematical statement being "morally" true, or a mathematical proof being "morally correct." A quick search just now revealed 431 instances of this language on MathOverflow, for example. What is meant? Well, it has very little to do with moral philosophy or notions of goodness. Rather, when a mathematician states that X is *morally* true, what they mean is something like, "X is an important and mathematically explanatory universal truth, nearly, except for some technical counterexamples that might arise from extenuating circumstances, which can be explained away or incorporated"; or, "X is true as a metaphor, not literally, but in a deep way that reveals mathematical insight." When X is morally true, then although it need not be literally true, strictly, nevertheless it (or something very like it) is true in the important relevant cases and furthermore, realizing these instances is important for mathematical understanding.

For an elementary example, consider the assertion that an nth degree polynomial over the rational numbers has n roots. This is not true over the real numbers, since $x^3 - x + 1$ has only one real number root, and so one should be working over the complex numbers or an algebraically closed field; and even then, it is not strictly true, since $(x - 1)^3$ has only one root even in the complex numbers. But many mathematicians would say that it is *morally* true that an *n*th degree polynomial has *n* roots, since in the generic case, roots are not repeated, and furthermore, if one introduces the concept of *multiplicity* for roots, where roots are counted more than once when they appear in the factorization with a higher degree, then the statement becomes true. And there are, of course, many more sophisticated examples.

But why the moral vocabulary? Eugenia Cheng (2004) explains it like this:

> Mathematical theories rarely compete at the level of truth. We don't sit around arguing about which theory is right and which is wrong. Theories compete at some other level, with questions about what the theory "ought" to look like, what the "right" way of doing it is. It's this other level of "ought" that we call morality. So what is morality? What counts as a moral reason for something to be true? Does it mean anything other than "an argument that isn't rigorous enough to count as a proof"? And does it really play a meaningful role in mathematics? ...Morality is about how one should behave, not just knowing that this is right, this is wrong. Mathematical morality is about how mathematics should behave, not just that this is right, this is wrong. Mathematicians do use the word "morally" for this idea. (pp. 7–8)

She distinguishes moral mathematical truth from elegance, saying that,

> An elegant proof is often a clever trick, a piece of magic..., the sort of proof that drives you mad when you're trying to understand something precisely because it's so clever that it doesn't explain anything at all. (p. 13)

And she distinguishes it from constructive proof and from explanatory proof, and from a host of other kinds of proof as well. Ultimately, however, she offers no positive full account of the term, even though she is quite right that this word is widely used by mathematicians. What does it mean?

> I am a mathematical moralist. I am more interested in moral truth than provable truth. But nevertheless I am a law-abiding mathematician. I grit my teeth and grudgingly chug through writing a proof even though I already know the result is true, and will not be any more convinced of its truth when I've written the proof. If only I could find a mathematical system in which moral truth and provable truth were always equivalent. Then I'd never have to prove anything again. (p. 17–18)

Let me try to give an account of the use of moral language in mathematics. As I see it, a mathematician says that statement X is morally true to mean that X encapsulates a great truth—a fundamental truth important for mathematical insight and progress—even if it may be one that is difficult to capture exactly; it may be fuzzy in the edge cases. This is a status of truth shared by many moral claims. They do not mean to assert that the truth of X is an absolutely proven literal fact, a derived universal truth. The statement X may be slightly informal or, if stated precisely, is not meant to be taken that way; precise versions of X will inevitably allow extenuating circumstances that technically admit violations. In the claim that one has the morally correct proof, there is a sense of "ought"—it is the right proof, the one that ought to work for that theorem, by using and expressing a profound mathematical insight, following it as a guiding idea through the argument. This kind of use of "morally correct" often thereby expresses value judgements about the quality of mathematical reasoning. One can imagine that combinatorialists might find counting arguments morally correct, while category theorists might find arguments morally correct when they identify and use a diagrammatic universal property.

In a separate research line connecting morality and mathematics, Justin Clarke-Doane (2020) explores the uncanny parallelism of morality and mathematics. On a surprising number of philosophical fronts, the two subjects face common problems and analogous solutions and rebuttals, whether the issue is realism, a priori justification, objectivity, naturalism, or pluralism. Clarke-Doane looks into what is truly common and what is not.

5.3 Formal proof and proof theory

So, what is proof? A truly robust answer arises with the concept of *formal proof* provided by mathematical logicians in the subject of proof theory. Let us concentrate on formal proof for the rest of this chapter. Formal proof is exacting and precise, written in a formal language and according to rigid rules concerning the allowed axioms and rules of inference.

Namely, a *proof* of an assertion φ from a theory T in a given formal proof system is a sequence or ordered system of assertions in the formal language, each of which is either a logical axiom, an axiom of T, or deducible from earlier assertions in the proof by an allowed rule of inference, and where the final assertion is φ, the theorem being proved. When there is such a proof of φ, we write $T \vdash \varphi$, pronounced "T proves φ."

To specify such a kind of formal proof system, one needs to specify the formal language, the logical axioms of the system and the rules of inference. The logical axioms, for example, might include all assertions having the form of a propositional tautology, as well as axioms dealing with quantifiers and equality. Rules of inference may include such rules as *modus ponens*, which asserts: from φ and $\varphi \to \psi$, deduce ψ. Such a rule is usually indicated by a line separating the premises and the conclusion, like this:

$$\frac{\varphi \qquad \varphi \to \psi}{\psi}$$

Another rule might be *disjunctive syllogism*, which asserts: from $\varphi \lor \psi$ and $\neg\varphi$, deduce ψ:

$$\frac{\varphi \lor \psi \qquad \neg\varphi}{\psi}$$

Some proof systems have many axioms and just a few deduction rules, and other systems have few or no axioms and a lot of deduction rules.

In some systems, proofs are arranged linearly in a sequence, but in other proof systems, the convention is to organize the proof in the form of a tree, highlighting the particular deduction rules that were used in the proof. There are a wide variety of formal proof systems. When studying proof theory, one would typically introduce and analyze a specific proof system. But in this chapter, let me rather stand back a bit and discuss matters generally. I would like to discuss the features that one might desire in a formal proof system, without introducing a specific system.

Soundness

We surely want our concept of proof to be truth-preserving in order to certify logical consequence. That is, if we can prove φ from a theory T, then we want it to be the case that φ is true whenever T is true. In other words, we want

$$T \vdash \varphi \quad \Longrightarrow \quad T \models \varphi.$$

This is called *soundness*. The notation $T \models \varphi$, pronounced "T logically implies φ," means that φ is true in every model satisfying T. A proof system is sound if whenever T proves an assertion φ, then φ is indeed a logical consequence of T; it is true in every model of T. To establish that a proof system is sound, it suffices to observe that the logical axioms are all valid—they are true in every model—and that the rules of inference are themselves truth-preserving. It follows from this by induction on the length of the proof that the entire proof system will be sound.

Notice that we are engaged here with the *theory-metatheory* distinction. Namely, in the object theory, we use the theory T; we assume the axioms of T and reason in the world described by T, the world in which the assertions of T are true. In the metatheory, in contrast, we mention the theory T; we reason about the assertions that can or cannot be proved in T or about the nature of the various models of T. Because of this theory-metatheory interaction, the subject of mathematical logic is also known as *metamathematics*.

Completeness

Conversely, we might want our proof system to be strong enough to prove all the valid inferences. That is, we might hope for the property that if φ is true in every model of T, then in fact there is an explicit proof of φ from T. A proof system with this property is said to be *complete*.

$$T \models \varphi \quad \Longrightarrow \quad T \vdash \varphi.$$

But is this a reasonable requirement? After all, validity is a sweeping universal property concerning all models of a theory, including uncountable models of vast unimaginable cardinalities, with all possible interpretations, a huge realm of possibility. Provability, in contrast, is a finitistic syntactic notion engaging with the finite combinatorial details of the formal system. Why should we expect to find a proof of a statement, a finite certificate of validity, just because the statement is true in all models? It may seem too much to ask. Is every mathematical consequence of a theory provable from that theory? Is every mathematical truth true for a reason?

I am astounded by the fact that indeed we are able to design comparatively simple sound and complete proof systems. These systems bridge the canyon between syntax and semantics, between proof and truth. In such a system, a statement is provable exactly when it is true in all models:

$$T \vdash \varphi \quad \Longleftrightarrow \quad T \models \varphi.$$

Thus, with a sound and complete proof system, there are indeed no accidental truths; every statement that is true in all models of a theory T is actually provable from the theory; every universal truth in this realm is true for a reason, a proof, a finite reason that we can express.

Compactness

Let me highlight a key consequence of completeness for a proof system, a consequence specifically of the finiteness of proofs. Namely, if a statement φ is a logical consequence of a theory T, then by completeness, there is a proof of φ from T. Since this proof is finite, it uses only finitely many axioms of T, and because it is a proof, this finite fragment of the theory is already sufficient to logically imply φ. This finiteness property is known as the *compactness theorem*.

Theorem 9 (Compactness theorem). *If $T \models \varphi$, then there are finitely many assertions $\psi_0, \ldots, \psi_n$ in T such that $\psi_0, \ldots, \psi_n \models \varphi$.*

It follows, equivalently, that a theory is satisfiable exactly when every finite subset of the theory is satisfiable.

I should like to emphasize that the compactness theorem is a purely semantic claim, making no mention at all of any proof system; it is expressed with $\models$, the semantic validity notion, rather than $\vdash$, the syntactic provability notion. The compactness theorem distills the semantic essence of the finiteness of proofs into a purely model-theoretic property, of inestimable fundamental importance in model theory, used in thousands of arguments. Although Gödel had originally proved the compactness theorem in the way we just described, as a consequence of his completeness theorem and the finiteness of proofs, it is far more common now for mathematical logicians and model theorists to dispense with the proof theory and prove the compactness theorem directly. The traditional Henkin proof of the completeness theorem, for example, works even more easily to prove the compactness theorem directly: given a finitely satisfiable theory, one extends it to a complete finitely satisfiable Henkin theory, and every such theory has a model built from the Henkin constants. One simply replaces consistency ideas in the usual Henkin completeness proof with finite-satisfiability, which is the semantic analogue of consistency.

Verifiability

A third key feature we want in our proof systems is *verifiability*. We want it to be clear and determinate whether what is in front of us is a proof or not. For a given proof candidate, the question of whether it is actually a proof or not should be something that can in principle be checked by a rote computational procedure. Because of this, proof systems are related to the concept of computability, discussed in the next chapter.

Strangely, verifiability often goes unmentioned in discussions of proof theory. One might find extensive talk of soundness and completeness, with nary a peep about verifiability. Yet the reader can be sure that even in such cases, there is an implicit insistence that proof systems should be verifiable. The ordinary way to define a proof system, after all, is explicitly to list the axioms and deduction rules, and if you are able to recognize the axioms and instances of the deduction rules in a system, then you can verify proofs simply by checking each step. For this reason, most proof systems that are proposed are indeed verifiable.

Nevertheless, one can easily imagine strange systems that might not be verifiable. For example, perhaps the axioms or rules of a system are merely computably enumerable but not computably decidable; in such a system, we could computably certify positive instances of proofs, but we would not necessarily be able to certify that something was not a proof, since we might not be able to know for sure that a certain axiom would not eventually be allowed. Another example would be cases where people speak of proving an assertion "in second-order logic." There is no verifiable sound and complete proof system for second-order logic, however, since we can prove that it is impossible to enumerate the second-order validities of arithmetic by any computable procedure. Yet, there are sound and complete unverifiable proof systems for second-order logic, defined semantically as in one of the systems considered next. Philosophers also sometimes provide various proof systems for second-order logic that are sound and verifiable, but not complete.

Sound and verifiable, yet incomplete

To emphasize the importance of having all three of these properties, let us see how one may easily design proof systems that have two but not all three of the properties. These systems are in each case either trivial or useless, which serves to underline the point that we do want all three properties.

To begin, let us provide a trivial proof system that is both sound and verifiable but not complete. For this, simply consider the empty proof system, the system with no logical axioms and no rules of inference. From a given statement φ, in this system you can deduce only φ itself. This is certainly sound, and also verifiable, but it is clearly incomplete. This proof system obviously does not capture the full extent of logical validity.

Complete and verifiable, yet unsound

Next, let us provide another trivial proof system, which is complete and verifiable but not sound. Consider the system in which every statement is allowed as a logical axiom, and every possible conclusion is allowed from any premise as a rule of inference. This system is clearly complete, since whenever Γ logically implies φ, then indeed φ is deducible from Γ in the system, since in fact every φ is deducible from nothing. The system is also verifiable. But it is clearly unsound—absurdly so—since absolutely every statement is provable in this system, including invalid statements.

Sound and complete, yet unverifiable

Let us now provide a first-order proof system that is sound and complete but not verifiable. This is perhaps the most interesting of these examples, for it shows that soundness and completeness are not the whole story in a proof system. This particular proof system, although sound and complete, is useless because it is not verifiable.

To define the system, let us take as our logical axioms all validities, that is, all statements true in all models; and let us have *modus ponens* as the only rule of inference: from φ and $\varphi \to \psi$, deduce ψ. These axioms and rules are clearly sound. The system is also complete. To see this, suppose that $T \models \psi$. By the compactness theorem, it follows that there are finitely many assertions $\varphi_0, \ldots, \varphi_n$ in T that logically imply ψ. It follows that

$$\varphi_0 \wedge \cdots \wedge \varphi_n \to \psi$$

is a logical validity. This assertion is logically equivalent to

$$(\varphi_0 \to (\varphi_1 \to (\cdots \to (\varphi_n \to \psi)))) \,.$$

By modus ponens applied successively to this axiom, using the fact that each φ_i is in the theory T, we conclude that $T \vdash \psi$ in this proof system, as desired.

In his popular logic lecture notes, Arnold Miller introduced a version of this system, which he called the *MM proof system* (the "Mickey Mouse" proof system). What is wrong with this system? Miller writes:

> The poor MM system went to the Wizard of Oz and said, "I want to be more like all the other proof systems," and the Wizard replied, "You've got just about everything any other proof system has and more. The completeness theorem is easy to prove in your system. You have very few logical rules and logical axioms. You lack only one thing. It is too hard for mere mortals to gaze at a proof in your system and tell whether it really is a proof. The difficulty comes from taking all logical validities as your logical axioms." The Wizard went on to give MM a subset Val of logical validities that is recursive and has the property that every logical validity can be proved using only Modus Ponens from Val." (Miller, 1995, p. 57)

Miller then proceeds to outline in a series of exercises how one may construct the set Val, a decidable set of validities from which all other validities can be proved. The resulting proof system becomes sound, complete, and verifiable.

The empty structure

Traditional proof systems often have the property that the assertion $\exists x \, x = x$ is explicitly formally provable. But do we regard this statement as valid? Can mathematical existence be proved purely from logic? The assertion fails, of course, in the empty structure, the structure having no elements in its domain. On its face, therefore, the provability of $\exists x \, x = x$ is a violation of the soundness of these proof systems; it is a statement that is provable but not true in every structure, for it is not true in the empty structure. This problem is addressed traditionally simply by excluding the empty structure from the subject, as a matter of definition: one defines validity $T \models \varphi$ to mean that every *nonempty* model of T is a model of φ. In this way, the proof systems can still be seen as sound. Right?

Is this satisfactory? No, not really, not to my way of thinking. In earlier decades, at the dawn of mathematical logic, it may have seemed innocent to exclude the empty structure. After all, we are not ever genuinely confused about the empty structure, and restricting the subject to the class of nonempty structures might seem to offer certain conveniences, such as with the treatment of the normal forms or surrounding the fact that every model admits expansions of the language, which is problematic for the empty structure when the larger language has constant symbols. But in contemporary times, mathematicians generally insist that well designed mathematical theories treat all their boundary cases robustly; and indeed, when a theory stumbles upon a trivial edge case, such as the empty structure, it is often taken to indicate flawed definitions or concepts. The failure to treat the empty structure properly in contemporary formal proof systems is a clear instance of "emptysetitis," the disease we discussed in chapter 1, occurring when a traditional mathematical development declines to take the empty structure seriously. This disease has been eradicated in most parts of mathematics, but it lives on in proof theory. For many mathematicians today, it is a nonstarter, a deal-breaker, to propose the use of a proof system that does not handle the empty structure properly. To my way of thinking, the older proof systems are simply mistaken on this issue, and it is past time to fix them.

Thankfully, some contemporary proof systems do handle the empty structure properly. One important difference concerns how the proof system behaves upon expansions of the language, especially with regard to new constant symbols, because although every nonempty structure is the reduct of a structure in any expanded language, this is not true of the empty structure. Specifically, since a language with constant symbols admits no interpretation in the empty structure, the relative validity relation $T \models \varphi$ can depend on the background language, even when T and φ use only a part of it. If the background language has constant symbols, then all models of T in that language are nonempty, but if not, then the empty structure may be relevant. In this sense, part of treating the empty structure robustly is recognizing that one's validity concept depends on the background language. In a language with constant symbols, $\exists x \, x = x$ is valid but not otherwise.

Formal deduction examples

Perhaps it will help to clarify the concept of formal proof by giving some example deductions. Without fully specifying the proof systems, let us imagine that we are working in formal proof systems that allow the logical axioms and rules of inference that I use in the two sample formal deductions provided next.

Theorem. $\exists x\, \varphi(x), \quad \forall x\,(\varphi(x) \to \psi(x)) \quad \vdash \quad \exists x\, \psi(x)$

Let us first give a traditional sequence proof. This kind of proof system typically has axioms for various quantifier manipulations and for various tautological forms, often with modus ponens as the only rule of inference:

1. $\exists x\, \varphi$, given.
2. $(\exists x\, \varphi) \to \neg \forall x\, \neg \varphi$, quantifier axiom.
3. $\neg \forall x\, \neg \varphi$, modus ponens.
4. $\forall x\,(\varphi \to \psi)$, given.
5. $\forall x\,((\varphi \to \psi) \to (\neg \psi \to \neg \varphi))$, logical axiom (contrapositive).
6. $[\forall x\,((\varphi \to \psi) \to (\neg \psi \to \neg \varphi))] \to [\forall x\,(\varphi \to \psi) \to \forall x\,(\neg \psi \to \neg \varphi)]$, quantifier axiom.
7. $\forall x\,(\varphi \to \psi) \to \forall x\,(\neg \psi \to \neg \varphi)$, modus ponens.
8. $\forall x\,(\neg \psi \to \neg \varphi)$, modus ponens.
9. $\forall x\,(\neg \psi \to \neg \varphi) \to (\forall x\, \neg \psi \to \forall x\, \neg \varphi)$, quantifier axiom.
10. $\forall x\, \neg \psi \to \forall x\, \neg \varphi$, modus ponens.
11. $[\forall x\, \neg \psi \to \forall x\, \neg \varphi] \to [\neg \forall x\, \neg \varphi \to \neg \forall x\, \neg \psi]$, logical axiom.
12. $\neg \forall x\, \neg \varphi \to \neg \forall x\, \neg \psi$, modus ponens.
13. $\neg \forall x\, \neg \psi$, modus ponens.
14. $\neg \forall x\, \neg \psi \to \exists x\, \psi$, quantifier axiom.
15. $\exists x\, \psi$, modus ponens. QED

And here is a natural deduction proof of the same theorem. In natural deduction proof systems, there are usually few or no logical axioms, but deduction rules are provided for the introduction and elimination of each of the logical constants:

$$
\cfrac{
\exists x\, \varphi(x) \qquad
\cfrac{
\cfrac{
[\varphi(u)] \qquad
\cfrac{\forall x\,(\varphi \to \psi)}{\varphi(u) \to \psi(u)}\ (\forall\ \text{Elim})
}{\psi(u)}\ (\to\ \text{Elim})
}{\exists x\, \psi(x)}\ (\exists\ \text{Intro})
}{\exists x\, \psi(x).}\ (\exists\ \text{Elim})
$$

This proof uses $\forall$ elimination and $\to$ elimination (modus ponens) at the top right, with the assumption of $\varphi(u)$, discharged in the final $\exists$ elimination; we should take care that the constant u does not appear in φ or ψ.

The value of formal deduction

How shall we view such formal deductions? On the one hand, formal proofs proceed via minute, primitive steps of reasoning that can be precisely verified, leaving no doubt about their correctness. This high standard of explicit logical rigor is certainly convincing. On the other hand, do we believe the logical validity of this example because of the deductions? No, we believed the validity before seeing the deductions; this validity is trivial: if there is an object with a certain property, and this property necessarily implies another property, then there is an object with this second property: the very same object. This validity requires no formal elaboration, and certainly the formal proofs are in no way more convincing than the informal elementary explanation I just gave.

This situation is typical of formal deductions: a formal deduction that can actually be carried out by hand is inevitably proving something trivial or essentially elementary. The conclusion can usually be seen as valid more easily, or more directly, without the cumbersome formal details. When working in a newly designed proof system, I am inclined to take the existence of a deduction of the kind given here principally as evidence that the proof system itself is working properly rather than as evidence of the validity of the theorem that it proves, which was not in doubt. Deductions focus our attention on the minutia of logical reasoning, such as the logical accounting of quantifiers perhaps with accompanying assumptions to be made and discharged or the details of distributing a logical connective. The proofs in our mathematics journals and lecture halls, in contrast, skip over those trivialities and focus our attention on the higher-level mathematical concepts and ideas that constitute the framework of our mathematical understanding. For this reason, we do not generally learn memorable mathematical ideas about the object theory from formal proofs directly. We do not usually learn any mathematics from a formal proof at all. I would find it absurd, for example, for someone to start writing down formal deductions in order to explore a new mathematical concept.[3] We learn by playing with and exploring mathematical ideas, not by manipulating formalizations of those ideas in a formal system.

Alexander Paseau (2016) makes the point that replacing a proof argument with an atomized formal version can reduce credence because whatever insightful ideas might have existed in the original proof are of necessity mixed in with innumerable banal atomic deductive steps in the formal proof; the key idea of the proof is lost in a sea of uninteresting details. He concludes for this and other reasons that "atomization is not in itself epistemically valuable" (p. 204).

---

[3] That is, in order to explore a concept in the object theory. Proof theorists, of course, might embark on formal proofs to explore an issue within proof theory, but this would be a fundamentally different kind of instance, in which the proofs themselves are the mathematical objects being studied.

In particular, in current ordinary mathematical practice, we rarely use formal deductions to convince ourselves of the truth of a mathematical conclusion; almost never.[4] Rather, proof theory is a theoretical tool that logicians use to analyze the underlying interaction of language and meaning in mathematics. We study the proof system abstractly, noting features that enable us to make conclusions about what may or may not be provable in it, or about the nature of those proofs.

I find the situation of formal proof and proof theory to be fundamentally similar to the use of Turing machines in computability theory (see chapter 6). We do not study Turing machines in order to use them to undertake actual computation; rather, we study Turing machines in order to understand the nature of computation and analyze its inherent capabilities and limits. Similarly, we study formal proofs not because we aim to undertake actual formal proofs of our mathematical arguments; rather, we study formal proofs to understand the nature of proof and analyze its inherent capabilities and limits.

5.4 Automated theorem proving and proof verification

A strong counterpoint to this perspective, however, which furthermore in the comparatively near future may completely overturn it, arises with the increasing power of computer-verified proof. This may simply be a game-changer in the realm of formal proof. Our generation is witnessing the coming-of-age of automated theorem proving and proof assistants, which seem destined to transform mathematical practice. Mathematicians of the future may be commonly using proof assistants and automated proof verifiers, with large and growing databases of formally verified proofs, to undertake their mathematical investigations. One imagines a vast repository of formal mathematical knowledge, all with verified proofs, aspiring to be the sum total of human mathematical insight, expressed in a formal system. The beginnings of that effort are now seriously underway.

Four-color theorem

Consider as an illustration the history of the four-color theorem, first conjectured in 1852 by Francis Guthrie, a student of Augustus de Morgan. The theorem asserts that every map can be four-colored, meaning that for any map you can draw in the plane, with various countries bordering one another (subject to certain natural requirements about connect-

---

[4] One might identify a few exceptions. The Russell refutation of general comprehension, for example, is so short that it might be regarded as a formal deduction, and therefore as an instance where we learned something important by a formal argument. Williamson (1998, p. 261) gives a semiformal derivation of the Barcan formulas in S5, afterward stating, "In light of these derivations, one might be inclined to regard it as simply a *discovery* that **BF** and **BFC** are truths of logic." Cranks sometimes write to me with "proofs" that the real numbers are countable, and I sometimes reply that if they claim to refute such a well established mathematical result, they must construct a formal proof in order to be taken seriously. Is this hypocritical if I do not insist on this standard in other cases?

edness), the countries can each be colored, using at most four colors, in such a way that adjacent countries get different colors.

After fitful false starts and a widely acclaimed proof by Alfred Kempe that had stood for over a decade (1879–1890) before it was realized to be wrong, the theorem was finally proved in 1976 by Kenneth Appel and Wolfgang Haken, who proved the theorem in part by using a computer to undertake a computational analysis of a large number of cases. Reducing the infinite problem to a finite one was itself a mathematical achievement.

The Appel-Haken proof engages with several philosophical issues concerning the nature of mathematical proof. Because of its computational nature, it was not possible for a person to check the details of the proof by hand, even in a lifetime of daily work. Yet, the proof was nevertheless widely accepted as a proof by the mathematical community. Does this violate an expectation that proofs should be surveyable? What good is a proof, after all, if you cannot verify or check it? How could it be convincing? Well, we could check that the program was correct (as indeed people did), in the sense that if the algorithm was undertaken exactly in accordance with it, with the right result, then it would indeed mean that the theorem was true. In this case, our confidence in the theorem would seem to be relying on knowing that the computer did carry out that algorithm correctly with that result. Because the operation of the computer is a physical process, subject to disturbance by gamma rays and quantum fluctuation, and so on, this would seem to give the proof of the theorem the character of a scientific experiment. This would seem very strange if one is accustomed to thinking of mathematical proof as a purely a priori deduction. Is the four-color proof an instance of mathematical knowledge arrived at empirically? We would seem here to have a posteriori mathematical reasoning.

Amazingly, the historical story of the four-color theorem continues beyond Appel and Haken to the 2005 development of Georges Gonthier, who was able to produce a *formal* proof of the four-color theorem, one which could be validated as correct by automated proof-checkers. This is a profoundly different kind of use of computers in the proof of the four color theorem because the formal proof object is logically robust and able to withstand detailed scrutiny as to its legitimacy. There can be no reasonable doubt now about the truth of the four-color theorem. Gonthier's verification gives the four-color theorem a greater level of certainty than is currently enjoyed by most of the other commonly accepted theorems of mathematics, since they have not yet generally been given formal proofs in a formal proof system.

Choice of formal system

Contemporary work on proof verification is aimed in part at finding the right formal system for this purpose. In the early twentieth century, mathematicians realized that set theory provided a robust and convenient mathematical foundation (see chapter 8); set theory seems to offer the capacity to express essentially arbitrary mathematical structure, and in this sense, one could view any mathematical argument as taking place in set theory. Set theory also had a simple, minimalist language, with just the set membership relation $\in$, and the Zermelo-Fraenkel (ZFC) axioms were first-order expressible. So it was implemented in several automated proof systems.

Meanwhile, some mathematicians chafed at the idea of translating their mathematical ideas into the formal language of set theory. Precisely because of its minimality, the resulting formal assertions are awkward or even absurd, drained of mathematical vitality, and the foundation does not automatically accommodate typing in a way that some mathematicians desire. Imagine composing a poem to your secret midnight lover, but having to convey it in Morse code by assembling piles of rocks, either short or tall, in a sequence. The set-theoretic issues seemed increasingly irrelevant to the mathematical ideas, and so some mathematicians have come to view set theory as the wrong formal foundation.

Michael Harris (2019) explains how formalization efforts can seem alien or limiting when undertaken with an unfamiliar foundation:

> A proof like the one Wiles published is not meant to be treated as a self-contained artifact. On the contrary, Wiles' proof is the point of departure for an open-ended dialogue that is too elusive and alive to be limited by foundational constraints that are alien to the subject matter.

What is desired is a formal system in which we can express our mathematical ideas naturally, a closer alignment of the formal foundation and the architecture of our mathematical conceptions and practice.

Attempting to address this, Voevodsky proposed univalent foundations and homotopy type theory as an alternative foundation explicitly aimed at facilitating the routine use of computer-verification of proofs. Synthesizing ideas of homotopy theory, type theory and category theory, this theory has been put forth as a promising formal system, one that has been described as closer in some domains to mathematical practice and amenable to proof formalization.

> Univalent foundations, like ZFC-based foundations and unlike category theory, is a complete foundational system, but it is very different from ZFC. To provide a format for comparison, let me suppose that any foundation for mathematics adequate both for human reasoning and for computer verification should have the following three components. The first component is a formal deduction system: a language and rules of manipulating sentences in this language that are purely formal, such that a record of such manipulations can be verified by a computer program. The second component is a structure that provides a meaning to the sentences of this language in terms of mental objects intuitively comprehensible to humans. The third component is a structure that enables humans to encode mathematical ideas in terms of the objects directly associated with the language. (Voevodsky, 2014, p. 9)

Thus, an important part of the goal of this new foundation is provide a formal system that is also close enough to the mathematician's way of thinking that it eases the process of formalizing actual mathematics. Advocates cite the ease of computational implementation as part of the attraction of univalent foundations.

Meanwhile, critics say that univalent foundations are oversold; many find the new foundation simply bewildering and alien, on a steep learning curve, far from meeting the naturalist usability goal. I suspect that the goal will be elusive for any formal system, since proof verification inherently involves mathematically uninteresting formalities of syntax that will inevitably stand apart from the ideas upon which mathematicians are focused, regardless of which formal system is used. Like Thurston, for this reason, I am skeptical that any formal system can be truly close to how mathematicians actually think and work. I find it likely that formal verification therefore will continue as its own speciality, with experts in various formal systems seeking to formalize the work of other mathematicians rather than as a routine process undertaken by every mathematician, as it is sometimes envisaged.

5.5 Completeness theorem

Let us discuss some of the ideas involved in proving the completeness theorem for a given formal proof system. The goal is to prove of a system that it is sound, complete, and verifiable:

$$T \vdash \varphi \quad \Longleftrightarrow \quad T \models \varphi.$$

To begin, it is generally easy to prove that one's system is sound, the forward implication, simply by observing that all the logical axioms and rules of inference are indeed valid. It follows by induction on the length of proofs that no proof can ever make an unsound deduction. It is also generally easy to see that one's system is verifiable if one has used a verifiable list of axioms and rules of inference that are uniformly verifiable in each instance. The difficult remaining part is to prove completeness. So let me discuss a few of the methods and ideas that play a role in proving completeness.

It is often useful to establish the *deduction theorem*, which asserts that $T \vdash \varphi \to \psi$ if and only if $T, \varphi \vdash \psi$. The forward implication of this equivalence, for example, is an immediate consequence if one has modus ponens as a rule of inference. The converse implication is also immediate in many natural deduction proof systems. In sequence systems, if $T, \varphi \vdash \psi$, then there is a proof $\theta_0, \ldots, \theta_n$. One then aims to show that $T \vdash \varphi \to \theta_i$ for each formula θ_i appearing in the proof, by considering exactly how θ_i itself had arisen in the proof. One can often proceed by induction, assuming that the claim has already been established for the earlier formulas. For example, if θ_i had been deduced by modus ponens, then $\theta_k \to \theta_i$ appeared earlier in the proof, and so by induction, we would have already known that $T \vdash \varphi \to \theta_k$, and also $T \vdash \varphi \to (\theta_k \to \theta_i)$. And so if one's proof system has the resources to move tautologically from those to $T \vdash \varphi \to \theta_i$, then we would have the desired conclusion. In this way, verifying the deduction theorem tells you a little about what your proof system must be like.

Another property that one often wants to establish is the *theorem on constants*, which asserts that if $T \vdash \varphi(c)$ and c is a constant not appearing in T, then $T \vdash \forall x \, \varphi(x)$, and furthermore, that it does so with a proof not mentioning the constant c. It is as though when you prove something about a totally new constant, you are really proving that this property holds universally, about an arbitrary object. Mathematicians use the theorem on constants routinely and without remarking on it, or perhaps without even realizing it. To prove that every group has a certain property, one says, "Let G be a group, ..." and then proves that G has the property, deducing on the basis of this fact that every group has the property. But let us be clear about what we are doing here. In proving the theorem on constants, we are not providing a reason to think that this process is legitimate, for indeed we already know that this is a valid form of reasoning. Rather, in proving the theorem on constants, what we are showing is that our formal proof system has succeeded in capturing this common form of mathematical reasoning. In some proof systems, such as natural deduction, it is easy to prove the theorem on constants via the $\forall$ introduction rule. In other systems, to prove the theorem on constants, one supposes $T \vdash \varphi(c)$, where c does not appear in T. The proof consists of assertions $\varphi_0, \ldots, \varphi_n$. One then proves by induction that $T \vdash \forall x \, \varphi_i(x)$, where x is a new variable that does not otherwise appear in any φ_i, and where we have replaced all instances of c by x. The point is that sufficient quantifier axioms in the proof system enable us to mimic the logical deductions that would have been made about the constant c, but now under the quantifier $\forall x$.

A theory T is *consistent* if it does not prove a contradiction. Otherwise, it is inconsistent. Thus, consistency is a purely syntactic notion, having to do with the existence of proofs of some kind, rather than a semantic notion concerned with the existence of certain kinds of models. The completeness theorem amounts to the claim that every consistent theory is satisfiable, that is, that the theory is true in some model. To establish completeness from this, assume that every consistent theory is satisfiable, and suppose that $T \models \varphi$. It follows that $T + \neg\varphi$ is not satisfiable. So, under our assumption, it must not be consistent. That is, $T, \neg\varphi \vdash \bot$. By the deduction theorem, $T \vdash \neg\varphi \rightarrow \bot$, and consequently $T \vdash \varphi$, as desired for completeness. Conversely, if a theory T is not satisfiable, then $T \models \bot$ vacuously, and so by completeness, T must prove this contradiction, and thus T is inconsistent.

The goal, therefore, becomes to prove that every consistent theory is satisfiable. To do this, one starts with a consistent theory T, and then gradually makes it a stronger theory, while retaining consistency. Following the Henkin proof, one adds to the theory sentences of the form $(\exists x \, \varphi(x)) \rightarrow \varphi(c)$, the *Henkin assertions*, where c is a new constant chosen specifically for this formula φ. This constant c, known as a *Henkin constant*, effectively names a witness x for the property φ, if there is any such witness. Using the theorem on constants, one can prove that these additions to the theory do not cause inconsistency. One then makes the theory complete by systematically adding either σ or $\neg\sigma$ for every sentence σ. At least one of these will be consistent. In this way, every theory T can be extended to a complete consistent Henkin theory $\bar{T}$.

Finally, at the heart of the Henkin completeness proof, one proves that every complete consistent Henkin theory is satisfiable. In a sense, such a theory itself tells you exactly how to build the model and what is true in it. Henkin's interesting idea calls for us to use the Henkin constants c themselves to build the model, essentially conflating the names with their referents, the objects that they name. Suppose that a mathematician has a large set A, perhaps uncountable, and he or she wants to invent a language containing names for every element of A. What is the easiest way to create a collection of names, such that every object $a \in A$ has a distinct name $\hat{a}$? Well, one easy thing to do is simply to regard each object $a \in A$ as its own name, to take $\hat{a} = a$. Thus, we have an intentional conflation of objects with their names. The Henkin proof basically amounts to turning that idea around backward. We start with a theory T and expand it by adding Henkin constants, which are effectively names for the witnessing objects for the existential assertions that the theory proves. We need to construct a semantic interpretation making the theory true, and we do so in the Henkin proof by simply taking the name c itself to be the object that c names. There is a slight complication, however, since we may have added different constant symbols c and d, which the theory ultimately proved to name the same object $\bar{T} \vdash c = d$. We do not want to have two objects with these names, but only one, so we define an equivalence relation $c \sim d$ on the names, which holds whenever the theory proves $c = d$, and we build the model out of the corresponding equivalence classes of names. The complete consistent Henkin theory $\bar{T}$ tells us everything we need to know about how to define the structure on the names, since it is a complete theory. One then proves that φ is true in the resulting Henkin structure if and only if $\bar{T} \vdash \varphi$. In particular, since $\bar{T}$ extends the original theory T, it follows that T is true in the Henkin model, and therefore T is satisfiable as desired. Thus, one proves Gödel's completeness theorem.

Theorem 10 (Completeness theorem, Gödel, 1929). *$T \vdash \varphi$ if and only if $T \models \varphi$.*

This was Gödel's celebrated dissertation result, forging the essential link between syntax and semantics, between proof and truth.

5.6   Nonclassical logics

When one lays out the precise logical rules of a formal system and begins to study them carefully, it is natural to wonder what would happen if one were to make changes in the system. What if we weakened the system by eliminating this rule or weakening that one; would the logic still work correctly? Perhaps we might sacrifice certain logical validities, such as the law of excluded middle or double-negation elimination. By making such changes to the proof system, we can easily create a diversity of alternative logics and nonclassical proof systems. In some cases, these alternative systems align in important respects with various nonclassical philosophical perspectives on mathematics, in a way that can help to clarify the meaning and fundamental principles of those perspectives. That is, in some cases, an alternative proof system can help to make clear exactly what the nonclassical logic is.

Take the case of *intuitionistic logic*, also known as *constructive logic* or *constructive mathematics*. Intuitionism is a philosophy of mathematics introduced by L. E. J. Brouwer that arose from a deep-seated desire to tie mathematical truth assertions more closely to their justifications and to the idea of explicit mathematical construction. When a constructivist asserts, "There exists x," it is meant that one can construct such an x, and to prove the statement constructively is to provide the construction explicitly. Similarly, a constructivist is entitled to assert a disjunction $p \vee q$ only when he or she is prepared also specifically to assert either p or q separately. For this reason, a constructivist does not necessarily assert all instances of the law of excluded middle $p \vee \neg p$, since perhaps neither p nor $\neg p$ is capable of assertion alone, in which case, according to constructivism, neither is $p \vee \neg p$. Similarly, a constructivist does not necessarily regard $\neg\neg p$ as logically equivalent to p, but rather as a weaker claim, the claim that $\neg p$ is not assertible. Thus, whereas classical logic has truth conditions for assertibility; intuitionistic logic has in effect proof conditions for assertibility. Arend Heyting, a student of Brouwer's and inspired by his intuitionistic vision, introduced a formal proof system aiming to implement it, and this formal logic has become known as *intuitionistic logic*. In this way, Heyting's formal system helped us to arrive at a precise understanding of what formal intuitionistic logic is.

Classical versus intuitionistic validity

This development inverts a certain interplay of syntax and semantics in comparison with classical logic. Specifically, in classical logic, one has a semantic validity concept that is independent of, and perhaps prior to, any proof system; one defines that an assertion is *valid* classically when it is true in all models. The goal of classical logic, then, is to capture this validity concept in a system of formal reasoning, and we judge a proof system by whether or not it does so. Namely, a proof system implements classical logic precisely if it is sound and complete with respect to semantic validity.

Intuitionistic logic, in contrast, does not seem to begin with a clear prior semantics or validity concept. Rather, intuitionism begins with general ideas about the nature of constructive reasoning; one then designs a proof system as Heyting did, so as to implement and formalize those guiding ideas. The result is intuitionist logic, provided as a formal proof system. In effect, the proof system itself helps us to clarify and express more fully what intuitionistic logic is in the first place. In particular, one can use the intuitionistic proof system to define the corresponding validity concept: to be valid intuitionistically means ultimately to be provable in the intuitionistic system, to be provable according to constructive principles of reasoning.

Perhaps this observation falls prey, however, to a confounding aspect of the dispute on the role of semantics in logic, namely, that much of the intuitionistic semantics appears like proof theory to the classical logician. After all, intuitionistic logic seeks meaning and semantics in the use of logical terms, that is, in the proof-theoretic rules governing their use, rather than in the truth conditions of those terms.

Critics of classical logic point out that classical validity was similarly unclear initially—classical logic also was born in its proof system and deductive rules, which were developed before the semantic notions were fully clear and before Tarski had defined the satisfaction relation for first-order logic. Mathematicians and logicians had informal concepts of truth and true-in-a-structure prior to Tarski's formal account, but were these sufficient to ground the classical validity notion, by which a statement is valid when it is true in all models? That the completeness theorem was recognized as important does seem to illustrate that one might judge a proof system in classical logic by whether it captures classical validity.

Various semantic conceptions did eventually emerge for intuitionistic logic, giving the subject greater depth and allowing a version of the syntax/semantics interplay. One prominent early idea was that intuitionistic logic is the logic of a Heyting algebra as opposed to the logic of a Boolean algebra, which underlies classical logic. A Heyting algebra is a distributive bounded lattice with relative pseudocomplements $a \to b$, implementing the intuitionistic form of implication and also negation $\neg a$, which is defined as $a \to \perp$. A Boolean algebra is a special kind of Heyting algebra, in which one has the law of excluded middle $a \vee \neg a = 1$, and also the law of noncontradiction $a \wedge \neg a = 0$, for every element a. In a Heyting algebra, one releases the law of excluded middle, but the law of noncontradiction still holds. One can form Heyting-algebra-valued models of set theory, for example, in analogy with the method of Boolean-valued models underlying the forcing method discussed in chapter 8. For any Heyting algebra A, one forms the A-valued set-theoretic universe V^A, where every object has a name, and truth assertions $\varphi(\dot{a})$ about those names have a truth value $[\![\, \varphi(\dot{a}) \,]\!]$ that is an element of A. One thus achieves a semantics of A-valued set theory, in a way that accords with intuitionistic logic, but not with classical logic if A is not a Boolean algebra. This construction is related to topos theory in category theory, where the internal logic of a topos can be intuitionistic.

Ultimately, it turns out that intuitionistic logic is sound and complete with respect to truth in all Heyting-algebra-valued models. This, therefore, provides a robust analogy between the two logics and their completeness theorems: intuitionistic logic is sound and complete with respect to Heyting-algebra-valued truth, and classical logic is sound and complete with respect to Boolean-algebra-valued truth, and indeed with respect to two-valued truth, which is to say, truth in all models.

While that is a satisfying mathematical symmetry between the logics, nevertheless my main point here is that there is also a certain important failure of this analogy. Namely, although we have a completeness theorem for intuitionistic logic via Heyting algebras, just as classical logic does via truth-in-all-models, nevertheless intuitionistic logic is not grounded in that validity notion. We did not begin in intuitionism with the validity concept of Heyting-algebra-valued truth, setting ourselves the goal of formalizing this validity and judging our proof systems by whether they did so. Rather, it was a mathematical discovery that intuitionistic validity aligned with Heyting-algebra truth. Indeed, to my way of thinking, we do not have a robust prereflective conception of true-in-all-Heyting-algebra-valued models as an independent motivating notion of validity for intuitionistic logic; we did not design the logic in order to capture that validity notion. Rather, the design of formal intuitionistic logic was guided by constructivist ideas concerning the nature of mathematical claims; the logic obeys and implements the idea of constructive proof. We then subsequently observe that this logic happens to align with Heyting-algebra validity, and this ultimately is why we care about Heyting-algebra validity; if it had turned out to be some other kind of algebra, then we would be talking about those algebras instead. In this sense, intuitionistic validity is grounded in intuitionistic logic, that is, in the proof system, whereas classical validity is grounded in truth and semantics—to be valid is to be true in all models—and our classical logic proof systems are designed to capture this notion.

Because intuitionism thus remains more deeply concerned with issues of mathematical reasoning, as opposed to mathematical model semantics and truth, it tends to remain more tightly connected with proof theory than is the rest of mathematical logic.

Informal versus formal use of "constructive"

In chapter 3 (see page 97), we discussed the idea of constructive arguments and proof, using this term with its common mathematical meaning, which is looser than in Brouwer's constructivism and constructive logic. Mathematicians will generally refer to an argument as constructive to mean simply that an existential claim is established by exhibiting a particular object realizing the property, in contrast to a pure-existence proof, where existence is established without identifying the particular witness. Ordinary uses of the word *constructive* in mathematics generally do not carry the implication that the construction conforms with the requirements of constructive logic.

But this sometimes causes confusion in mathematical discussions, when a classical mathematician and a constructivist mathematician interpret the terms differently. I have been a party several times to discussions on MathOverflow or Twitter, where a mathematician asks for a constructive argument in a certain mathematical case, is satisfied when an explicit construction is given, but then a constructivist mathematician objects that the explicit construction does not actually conform fully with constructivist logic; perhaps it had used double-negation elimination or excluded middle at some point. The discussion then becomes confused until it is realized that the participants disagreed on what it means to have a constructive argument.

Imagine a young writer proudly showing off the antique wooden desk she bought at an estate auction, but is told, deflatingly, that the piece actually is not antique; perhaps it is only 95 years old, whereas the trade definition of *antique* requires a century. Even so, it will be lovely to write at that desk, and I would happily call it antique along with her and most people. Similarly, the ordinary common meaning for *constructive* is a robust and useful mathematical concept, used throughout mathematics to distinguish comparatively constructive arguments from pure-existence proofs, whether or not those arguments conform fully to the requirements of constructive logic. To my way of thinking, it would be a loss for mathematical communication to erase this meaning and always allow only the stricter constructivist meaning.

But meanwhile, one can easily criticise this informal use of *constructive*, arguing that actually there is no robust meaning for this informal use, neither philosophically robust nor mathematically robust. Ultimately, if an argument is not constructive according to constructivist logic, then it will violate the constructivist principles given in defense of that logic. And so in the cases we are discussing, although it may seem that one has an explicit construction, actually one does not, for at the place in the construction that is not in ac-cordance with constructive logic, one must ultimately appeal to a nonconstructive method or process, such as division into cases or double-negation elimination. The constructivists really do understand what it means to have a constructive argument.

Epistemological intrusion into ontology

Some classical logicians resist calls for intuitionistic logic in ordinary mathematics as what they see as an intrusion of epistemological matters into discussions of mathematical truth. To say that we cannot assert $p \vee q$ because we are not prepared to assert either p or q separately, even in the case of simple arithmetic statements, is from the classical-logic point of view to confuse the truth of an assertion with our reasons for knowing that truth. In classical logic, of course, we can know a disjunction without yet knowing either disjunct. For the classical logician surveying a realm of classical structures and models, whose truths that he or she wants to discover, the intuitionistic objection seems to dissolve into these epistemological issues about our knowledge of those various structures and the uniformity of the procedures by which we might come to that knowledge. For the classical logician, the intuitionistic call to change the underlying logic thus amounts to a change in the subject from existence to construction, from definition to computability, from truth to proof.

The intuitionistic logician might reply that the classical logician is mistaken in the pre-sumption that there is indeed a robust realm of classical semantics, where every statement or its negation is true, but not both; perhaps this realm is merely a beautiful illusion.

No unbridgeable chasm

Meanwhile, the classical logician is, of course, able to discuss constructibility and com-putability and proof, without embedding those notions into the underlying logic. Classical logicians can investigate the nature of a topos or of Heyting-valued models, just as they investigate other kinds of mathematical structure, and thereby interact productively with the intuitionistic logician. For this reason, therefore, there is ultimately to my way of thinking no unbridgeable chasm between classical and intuitionistic logic. Logicians and mathematicians on each side can understand and communicate with each other across the canyon, provided that they recognize merely that assertions on the other side carry a some-what different meaning than on their own side.

Logical pluralism

According to the philosophical position known as *logical pluralism*, advanced for example by Jeffrey Beall and Greg Restall (2000, 2006), there is more than one correct notion of logical consequence. The view is that there is not just one true logic, but many. The spectrum of nonclassical logics is vast and includes intuitionist logic, relevance logic, linear logic, fuzzy logic, a wide assortment of other multivalued logics, diverse modal logics, and also paraconsistent logics, defended by Graham Priest and others (see Priest, Tanaka, Weber, 2018), in which some assertions can be seen as both true and false.

One prominent issue in the logical pluralism debate is the question of whether the meaning of the logical connectives is invariant across the various logics, or whether different notions of logical consequence require different meanings for the atomic logical constituents of the language. Quine (1986) famously argued in his discussion of deviant logics that a change in the logic amounts to a change in meaning—a change in the subject. For example, does the fact that one cannot prove $p \lor \neg p$ in intuitionistic logic, or the fact that $\neg\neg p$ is not necessarily logically equivalent to p, mean that intuitionistic logic gives a different meaning to disjunction or to negation? Does that fact that paraconsistent logic accepts some instances of $p \land \neg p$ mean that it is using a different meaning for conjunction or negation? Or is the meaning of the connectives the same across the different logics, although they have different notions of logical consequence?

Some of the nonclassical logics quite explicitly introduce different and incompatible notions of implication, disjunction, and negation, with different meanings and intension. Because these logics are often defined by specifying admissible deductive rules in a formal proof system, logical pluralists often find the meaning of their logical connectives in proof theory rather than in model theory. This is natural if one holds that meaning is determined by use, for it is precisely the rules of the formal proof system that would seem to govern their use, and the meaning of the logical connectives is therefore to be found in those deductive rules. I find it interesting to notice that this places the meaning of the logical connectives on what is traditionally considered the syntactic side of the syntax/semantics dichotomy, whereas the classical logic account of meaning resides in the truth conditions and the definition of satisfaction, on the semantics side of the dichotomy. Because logical pluralism finds meaning in the deductive rules of a proof system, therefore, it tends to blur, and even to undermine, the syntax/semantics distinction.

Classical and intuitionistic realms

Some philosophers have proposed that different parts of mathematics and of mathematical foundations have such different natures that they call for different logics. For instance, Nik Weaver (2005) argues that classical logic is appropriate in the realm of arithmetic, where he takes mathematical assertions to have a clear and distinct definite nature; but in the murkier set-theoretic realm of sets of numbers or sets of sets of numbers, which he takes to have a less definite nature, we should be using intuitionistic logic.

Perhaps illustrating this proposal, most intuitionistic logicians (but not all) undertake the analysis of their intuitionistic theories using classical logic in the metatheory. For example, one might say that either a given assertion is intuitionistically derivable or it is not, an instance of the law of excluded middle in the metatheory. This can be seen as an instance of Weaver's proposal to use classical logic in more definite realms if the metatheoretic issues are taken as more definite, for example, if they are regarded as essentially finitary or arithmetic.

5.7 Conclusion

The subjects of proof theory and mathematical logic bring out a certain virtuous cycle of interaction with the philosophy of mathematics, whereby essentially philosophical questions concerning the nature of proof and truth and mathematical reasoning inspire an essentially mathematical analysis of corresponding formal proof concepts, with results and insight that feed once again into the philosophical inquiry. I find it incredible that we have formal proof systems for which semantic validity, a sweeping concept ranging across all possible semantic contexts, aligns exactly with provability in the system. Universal truth is thereby reduced to a finite reasoning process—every universal truth is true for a finite reason—and this surely informs further philosophical questions and analysis.

Questions for further thought

5.1   Can you answer all four of Hypatia's questions about whether everything rhymes with itself? Is the rhymes-with relation an equivalence relation? Some people find it questionable whether anything rhymes with itself, although if one takes this position, then the rhymes-with relation will be neither reflexive nor transitive. Certainly it is a poor poet who rhymes a word with itself, say, *headache* with *headache*, rather than *heartbreak*. But does *headache* indeed rhyme with itself?

5.2   David Madore (2020) has suggested a necessary enlargement of the use/mention distinction to the case of use/mention/title. He says, "The Lord of the Rings" is 5 words long; *The Lord of the Rings* is hundreds of pages long; but the Lord of the Rings is Sauron. Altogether, the Lord of the Rings is a character in *The Lord of the Rings* whose title is "The Lord of the Rings." Discuss.

5.3   Criticize the proof-by-diagram given in the chapter, that 32.5 = 31.5. Namely, the argument is: The upper triangle in the diagram has base 13 and height 5, for an area of 32.5. The colored pieces can be rearranged as indicated to form the lower triangle, which has the same size, but it is missing exactly one square. Thus, 32.5 = 31.5.

5.4   Does the formalization of a mathematical idea impoverish the idea or enrich it?

5.5   Is every mathematical argument formalizable in principle?

5.6 Discuss whether different mathematical ideas can have the same formalization. Or whether the same mathematical idea can have different formalizations.

5.7 Are the mathematicians who describe a mathematical argument as "constructive," when it does not fully obey the rules of constructive logic, making a mistake?

5.8 Suppose that we have two proof systems, $\vdash_1$ and $\vdash_2$, and that all the logical axioms and inference rules of the first are included amongst the axioms and rules of the second. Show that the completeness of one of the proof systems (which one?) implies the completeness of the other; and similarly, the soundness of one of them (which one?) implies the soundness of the other.

5.9 Fix a particular sound and complete proof system. Show that a theory T is inconsistent if and only if $T \vdash \psi$ for all ψ. This is called the *blowup* principle of classical logic—that is, from a contradiction, you can prove anything.

5.10 Show that every formal proof system is equivalent to a system having no logical axioms.

5.11 Using the fact that validity is not computably decidable (see the Entscheidungsproblem in chapter 7), argue that every sound, complete, and verifiable proof system admits a "speed-up" phenomenon. Namely, for any sound, complete, and verifiable proof system $\vdash_1$, there is another sound, complete, and verifiable system $\vdash_2$, such that every $\vdash_1$ proof is also a $\vdash_2$ proof, and for at least one validity σ, there is a strictly shorter proof in the second system $\vdash_2 \sigma$ than in the first $\vdash_1 \sigma$.

5.12 Show that any formal proof system with no logical axioms and with nonempty hypotheses in all its rules of inference is incomplete. (Hint: Consider the proofs arising from the empty theory.)

5.13 What is the difference between having modus ponens as a rule of inference and having the tautology $\varphi \to ((\varphi \to \psi) \to \psi)$ as a logical axiom?

5.14 Show that no sound formal proof system having only logical axioms and no rules of inference is complete.

5.15 Although there are many formal proof systems, with different logical axioms and rules of inference, discuss the extent to which they must agree if they are sound and complete.

5.16 Suppose that we have two proof systems, $\vdash_1$ and $\vdash_2$, which are both sound and complete. If one of them is verifiable, does this imply that the other also is verifiable?

5.17 Argue that the sound and complete proof system considered on page 173, which was not fully verifiable, is nevertheless verifiable for positive instances, meaning that there is a computable procedure that will correctly validate all and only the proofs of this system (even if it makes no judgements on non-proofs). Is it actually important in a proof system that we are able to recognize and reject non-proofs, if we meanwhile have the capacity correctly to validate and accept all and only the actual proofs? (This question is fundamentally concerned with the distinction between computable decidability and semi-decidability or computable enumerability, considered in chapter 6.)

5.18   Consider the statement: Every mathematical truth is true for a reason. Discuss. Does your answer vary by mathematical subject, such as arithmetic versus geometry? Does your answer depend on whether we are talking about truth-in-a-particular-model or truth-in-a-theory?

5.19   With some minimal assumptions on features of the proof system, show that a proof system is sound and complete exactly when consistency in that system aligns with satisfiability, for every theory.

5.20   Prove the compactness theorem as a consequence of the completeness theorem. Discuss whether, conversely, there is a proof of the completeness theorem from the compactness theorem.

5.21   What would it mean for one proof system to be simulable by another?

5.22   Is intuitionistic logic sound for semantic validity? Is it complete for semantic validity?

5.23   Discuss the senses in which intuitionistic logic is or is not sound and complete.

5.24   Discuss the claim: "In classical logic, semantics are primary, whereas in intuitionistic logic, semantics are secondary."

5.25   Do the logical connectives have different meanings in intuitionistic as opposed to classical logic?

5.26   To what extent, if any, does logical pluralism require one to reject or blur the syntax/semantics distinction?

Further reading

Catarina Dutilh Novaes (2020). An important new work emphasizing the dialogical nature of proof and deduction, highlighting the adversarial nature of the prover/skeptic dialogue. She pours cold water on the simplistic view of proof as merely truth-preserving deduction, and finds riches in the view of proof as dialogue.

William P. Thurston (1994). A classic essay on Thurston's view of mathematical understanding. Some aspects are revisited twenty-five years later in his MathOverflow discussion on thinking and explaining, Thurston (2016).

Alexander Paseau (2015, 2016). Enjoyable articles discussing how mathematicians might gain knowledge even in the absence of proof and what epistemic value is added, if any, by complete rigor.

Timothy Williamson (2018). He raises doubts about the claims of advocates of nonclassical logics that the deviant logics are consistent with the use of classical logic within pure mathematics, specifically considering issues arising in the applicability of pure mathematics to natural and social science.

Vladimir Voevodsky (2014). The author explains how, after discovering errors in his own published mathematical work, he sought to establish the routine use of formally verified proof systems in mathematics and developed the univalent foundations to facilitate this.

Michael Harris (2019). A column for *Quanta Magazine* discussing the nature of proof, particularly in connection with Andrew Wiles's proof of Fermat's last theorem, arguing that a good proof like that is a point of departure rather than a terminus for mathematical inquiry.

Gisele Dalva Secco and Luiz Carlos Pereira (2017). An account of the four-color theorem with special regard to Wittgensteinian distinction between proof and experiment.

Univalent Foundations Program (2013). This is the main introduction to homotopy type theory and univalent foundations as a foundation of mathematics.

Credits

The claim about mathematicians and jury duty may be an urban legend. Some of the material in this chapter is adapted from my book, *Proof and the Art of Mathematics* (MIT Press), as well as from my book in progress, *Topics in Logic*. The triangular choices proof-without-words is due to Larson (1985); see also Suárez-Álvarez (2009). The "proof" that $31.5 = 32.5$ is due to New York City amateur magician Paul Curry in 1953; see O'Connor (2010). Weaver's view on classical logic for arithmetic and intuitionistic logic for set theory is also defended by Solomon Feferman.

6 Computability

Abstract. What is computability? Kurt Gödel defined a robust class of computable functions, the primitive recursive functions, and yet he gave reasons to despair of a fully satisfactory answer. Nevertheless, Alan Turing's machine concept of computability, growing out of a careful philosophical analysis of the nature of human computability, proved robust and laid a foundation for the contemporary computer era; the widely accepted Church-Turing thesis asserts that Turing had the right notion. The distinction between computable decidability and computable enumerability, highlighted by the undecidability of the halting problem, shows that not all mathematical problems can be solved by machine, and a vast hierarchy looms in the Turing degrees, an infinitary information theory. Complexity theory refocuses the subject on the realm of feasible computation, with the still-unsolved P versus NP problem standing in the background of nearly every serious issue in theoretical computer science.

What does it mean to say that a function is computable? Is there a hierarchy of computational power? Can every mathematical question be solved in principle by computation? We seem to have an intuitive grasp on computability—a computation is some kind of finite, mechanistic, rote procedure, each step of which is performed in accordance with certain clear instructions set out in advance—and a computable function is simply one that can be computed by such a procedure. But what exactly counts as such a procedure? Exactly what kind of steps are allowed? What kind of instructions? Can we reify the concept of computability by providing a precise model of it in a way that accords with our prereflective intuition about this concept? The aim may be an idealized notion of computability—what might be called "computability-in-principle"—a notion of what we could compute if we were not limited by a lack of resources, lack of time, space, paper, pencils, or memory. We want to know what it means to say that a function on the natural numbers is computable in principle, an account of when a mathematical question can be settled in principle by computation, given sufficient resources and without regard to efficiency or the limitations of time or space. Is there a mathematically rigorous concept of computability in this idealized sense? I would like to outline several possible approaches to answer this.

6.1 Primitive recursion

Gödel had sought initially to answer the question by proceeding from below, by identifying some operations on functions that, intuitively, preserve computability. If we should start with a collection of primitive functions, which we know are computable according to our prereflective concept of "computable," then we can generate new functions by systematically applying those operations and perhaps come in this way to produce a large class of computable functions, if not all of them. Our context here is functions on the natural numbers. It turns out to suffice for Gödel's class of functions that we begin with a collection of surprisingly simple primitive functions:

The constant zero function, $z(n) = 0$.
The successor function, $s(n) = n + 1$.
The projection functions, $p(x_1, \ldots, x_n) = x_i$.

From these primitive functions, we can generate many new functions through successive definitions by composition or recursion. The *primitive recursive* functions are precisely the functions that can be generated in this way.

Composition. If g and h are primitive recursive unary functions, then so is the function f defined by composition:

$$f(x) = h(g(x)).$$

More generally, for functions with more than one argument, if h and $g_1, \ldots, g_n$ are primitive recursive functions, then so is the function f defined by composition:

$$f(x_1, \ldots, x_k) = h(g_1(x_1, \ldots, x_k), \ldots, g_n(x_1, \ldots, x_k)).$$

We expect the class of computable functions to be closed under composition because if you are able to compute functions g and h, then we can see how to compute the composition function $h(g(x))$ as follows: on input x, first compute the value $g(x)$ and then feed that value into h, producing $h(g(x))$.

Recursion. If g and h are primitive recursive, then so is the function f defined by recursion:

$$
\begin{aligned}
f(x_1, \ldots, x_k, 0) &= h(x_1, \ldots, x_k) \\
f(x_1, \ldots, x_k, n + 1) &= g(x_1, \ldots, x_k, n, f(x_1, \ldots, x_k, n)).
\end{aligned}
$$

Do you see how recursion works? Suppressing the auxiliary arguments $x_1, \ldots, x_k$, each next value of the function $f(n + 1)$ is determined from the previous value $f(n)$, which in turn is determined from *its* previous value, and so on, terminating with the initial value $f(0)$. Recursion, therefore, seems at heart to be a computable process. In order to compute f at any particular n, we simply compute the value of the function at 0, and then at 1, at 2, and so on until we arrive at n. If the anchor function h and the recursive step function g are both computable, then by this recursive process, the solution function f also will be computable. Gödel thus identified recursion as a fundamental computable operation, a prescient observation in light of the pervasive and powerful use of recursion in contemporary computer programming.

Let us illustrate how composition and recursion enable us to add new useful functions to the class of primitive recursive functions. We do not initially have the addition function $m + n$ explicitly in the class, but we do have the successor function $s(n) = n + 1$, and we can view addition as repeatedly adding 1:

$$
\begin{aligned}
m + 0 &= m \\
m + (n + 1) &= (m + n) + 1.
\end{aligned}
$$

The point is that we can read this as a *definition* of addition, a recursive definition. The first line, the anchor case, specifies $m + 0$ to be m; and the second line, the successor case, defines $m+(n+1)$ in terms of the previous value, $m+n$. Thus, we have generated the (binary, two-argument) addition function from the (unary) successor function by recursion, and so addition itself is a primitive recursive function. Similarly, multiplication can be defined recursively as repeated addition, like this:

$$
\begin{aligned}
m \cdot 0 &= 0 \\
m \cdot (n + 1) &= (m \cdot n) + m.
\end{aligned}
$$

And one can continue: exponentiation can be defined recursively as repeated multiplication, tetration $2^{2^{\cdot^{\cdot^{\cdot^2}}}} \Big\} n$ as recursive exponentiation, and so on.

Recursion is an analogue or partner of the principle of mathematical induction. Whereas induction is a method of proof, used to prove arithmetic facts, recursion is a method of definition, used to define arithmetic functions. With each method, our knowledge concerning the truth of a statement or the value of a function at a number depends on corresponding knowledge about the previous number or numbers. Nearly every nontrivial theorem proved in Peano's system of arithmetic relies fundamentally on induction, and limiting the induction scheme in Peano arithmetic results in a strictly weaker system. Recursion is similarly powerful. One should not think of recursion in connection only with certain obviously recursive functions, such as the factorial function, but rather as a pervasive and fundamental tool that gives power to the class of primitive recursive functions.

Implementing logic in primitive recursion

To begin to see this, it will help to implement some ideas from logic. We are concerned with functions on the natural numbers, and so let us regard the number 0 as representing the truth value *false*, while positive numbers all represent *true*. For any primitive recursive function r, one may then consider the collection of inputs x, for which $r(x) > 0$, as defining a corresponding primitive recursive relation $R(x)$. The function r tells you of a given input x whether the relation $R(x)$ holds or not: if $r(x) > 0$, then $R(x)$ holds; if $r(x) = 0$, then $R(x)$

fails. The elementary logic of these truth values turns out to be computable. For example, multiplication $n \cdot m$ is an *and* operator, since

$$n \cdot m > 0 \quad \Longleftrightarrow \quad n > 0 \quad \text{and} \quad m > 0.$$

Thus, $n \cdot m$ is true if and only if n is true and m is true. Similarly, addition $n + m$ is an *or* operator, since for natural numbers

$$n + m > 0 \quad \Longleftrightarrow \quad n > 0 \quad \text{or} \quad m > 0.$$

One can define the negation operation recursively by $\neg 0 = 1$ and $\neg(n + 1) = 0$, which turns true into false and vice versa. The function defined by $\text{Bool}(0) = 0$ and $\text{Bool}(n + 1) = 1$ (named for George Boole, who emphasized such computational aspects of logic) normalizes all the truth values to be 0 or 1. We could equivalently have defined $\text{Bool}(n) = \neg\neg n$. Using this computational logic, one can easily show that the primitive recursive relations are closed under intersection, union, and complement. Thus, we have a kind of primitive recursive logical expressivity.

Extending this, one can see that the class of primitive recursive functions allows *definition by cases*, an instance of `if-then-else` programming logic, used pervasively in contemporary programming. That is, if we have primitive recursive functions g, h, and r, then we can define the function

$$f(n) = \begin{cases} g(n), & \text{if } r(n) > 0 \\ h(n), & \text{otherwise.} \end{cases}$$

So $f(n)$ has value $g(n)$ if we are in the first case $r(n) > 0$; and otherwise, the value is $h(n)$, the default value. This is primitive recursive because it can be defined by

$$f(n) = g(n) \cdot \text{Bool}(r(n)) + h(n) \cdot \text{Bool}(\neg r(n)).$$

The point is that if the condition $r(n) > 0$ holds, then $\neg r(n) = 0$, which kills the second term, and we shall have $f(n) = g(n) \cdot 1 + 0$, which is $g(n)$; and if, in contrast, $r(n) = 0$, then we get $f(n) = 0 + h(n) \cdot 1$, which is $h(n)$, as desired in each case.

Similarly, the class of primitive recursive functions is closed under the operation of *bounded search*, analogous to `for i=1 to n` programming logic. That is, for any primitive recursive function r, we consider the function f defined by

$$f(x, z) = \mu y < z\, [r(x, y) = 0].$$

The bounded μ-operator here returns the least $y < z$ for which $r(x, y) = 0$, if it exists, and otherwise returns the value $y = z$ if there is no such $y < z$. This can be defined by recursion like this: $f(x, 0) = 0$ and $f(x, n + 1) = f(x, n)$, if this is less than n; and $f(x, n + 1) = n$, if we are not in the first case and $r(x, n) = 0$; and otherwise $f(x, n + 1) = n + 1$.

Using these methods, it follows that with only primitive recursive functions, we can compute any property expressible in the language of arithmetic using only bounded quantifiers, and this includes the vast majority of finite combinatorial properties that arise in mathematics (although parts of mathematical logic are specifically focused on relations not captured in this way). For example, if you have an easily checked property of finite directed graphs, or some finite probability calculation, or an easily checked question about finite groups, then these properties will be expressible in the language of arithmetic using only bounded quantifiers, and so the corresponding functions and relations will be primitive recursive. In this way, we see that the class of primitive recursive functions is truly vast and includes most of what we would want to say is computable.

Diagonalizing out of primitive recursion

Has Gödel's strategy succeeded? Does the class of primitive recursive functions include every computable function? No. Gödel had realized, for extremely general reasons, that there must be computable functions outside the class of primitive recursive functions. What he realized is that we have an effective enumeration of the class of all primitive recursive functions, and because of this, we can effectively produce a function outside the class.

Let me explain. Every primitive recursive function admits a description, a kind of construction template—a finite set of instructions detailing precisely how it can be generated from the primitive functions using composition and recursion. These instructions also tell us precisely how to compute the function. Furthermore, we can systematically generate all such construction templates and effectively enumerate them on a list e_0, e_1, e_2, and so on. In this way, we can produce an effective enumeration f_0, f_1, f_2, ..., of the class of all primitive recursive functions, where f_n is the function described by the nth construction template e_n. This enumeration will have repetition, since as previously mentioned, different construction templates can result in the same function.

The key point is that whenever one has an effective enumeration of functions like this, then, like Cantor, we can diagonalize against it, defining the diagonal function

$$d(n) = f_n(n) + 1.$$

This function, by design, is different from every particular f_n, since by adding 1 we specifically ensured that $d(n) \neq f_n(n)$, so the two functions differ on the input n. Nevertheless, the function d is intuitively computable, since on input n, we may generate the nth construction template e_n in the effective enumeration of such templates, and then carry out the instructions in e_n in order to compute $f_n(n)$, and then add 1. So d is computable, but not primitive recursive.

Furthermore, this same argument would apply to any notion of computability for which we are able to provide an effective listing of all the computable total functions on the natural numbers, because in such a case, the diagonal function would be computable, but not be on the list. For this reason, it may seem hopeless to provide a rigorous theory of computability. Any precise notion of "computable" would seem to involve a concept of algorithm or instructions for computing, and for any such precise concept of algorithm, whatever it is, we would be able effectively to enumerate all the algorithms on a list e_0, e_1, e_2, ..., thus giving rise to an effective listing of all the computable functions f_0, f_1, ..., for this notion; and then we could define the diagonal function $d(n) = f_{e_n}(n) + 1$ as before, which would be intuitively computable, but not be on the list. Right?

Well, no. We shall see later in this chapter how Turing's ideas overcome this obstacle. Meanwhile, the diagonal function produced here is computable and not primitive recursive, but we seem to have very little additional information about it; furthermore, its exact nature depends highly on incidental details concerning the manner in which we enumerate the construction templates.

The Ackermann function

Wilhelm Ackermann provided a fascinating function, an extremely fast-growing one, which intuitively is computable in principle but which, it turns out, is not in the class of primitive recursive functions. The Ackermann function $A(n, m)$ is a binary function on the natural numbers, defined on two arguments according to the following nested recursion:[5]

$$
\begin{aligned}
A(0, m) &= m + 1 \\
A(n, 0) &= 1, \text{ except for } A(1, 0) = 2 \\
&\quad \text{ and } A(2, 0) = 0 \\
A(n + 1, m + 1) &= A(n, A(n + 1, m)).
\end{aligned}
$$

Let us play with it a little to see how this definition works. The first two lines define the anchoring values of $A(n, 0)$ and $A(0, m)$, when one of the arguments is 0; and the last line defines $A(n + 1, m + 1)$ when both arguments are positive.

Wait a minute—isn't this definition circular? You may notice that the third line defines $A(n + 1, m + 1)$ in terms of A itself. What is the nature of this kind of recursion? Is this definition legitimate? Yes, indeed, the definition is legitimate; it uses a well founded nested recursion, which is totally fine. To see this, let us think of the individual layers of the function, each becoming defined in turn. The anchor defines the 0-layer function as

$$A(0, m) = m + 1.$$

The next two lines define the $n + 1$ layer $A(n + 1, \cdot)$ by recursion, using the previous layer. So the layers are defined recursively.

---

[5] His actual function was slightly different, but with the same key recursion and fast growth phenomenon; I prefer this version because it has more attractive functions in each layer.

Let us see how it plays out. Can we find a closed-form solution for $A(1, m)$? Well, we start with the value $A(1, 0) = 2$, and then we must obey $A(1, m + 1) = A(0, A(1, m)) = A(1, m) + 1$. Thus, we start out at 2, and then for each increase in m by 1, we also add 1 to the answer. So it must be the function

$$A(1, m) = m + 2.$$

What about the next layer, $A(2, m)$? It starts out with $A(2, 0) = 0$, and then at each stage, we have $A(2, m + 1) = A(1, A(2, m)) = A(2, m) + 2$. So we are adding 2 at each stage, for each increment of the input. Consequently,

$$A(2, m) = 2m.$$

And how about $A(3, m)$? It starts at $A(3, 0) = 1$, and at each stage, $A(3, m + 1) = A(2, A(3, m)) = 2 \cdot A(3, m)$. So we start at 1 and multiply by 2 at each step, resulting in

$$A(3, m) = 2^m.$$

The essence of the Ackermann function is that each layer of the function is defined as an iteration of the function on the previous layer. And what will the iteration of exponentiation look like? The fourth layer starts at $A(4, 0) = 1$, and at each step, it takes the exponential $A(4, m + 1) = 2^{A(4, m)}$, which leads to iterated exponential towers:

$$A(4, m) = \left. 2^{2^{\cdot^{\cdot^{\cdot^2}}}} \right\} m.$$

Donald Knuth introduced the notation $2 \uparrow\uparrow m$ for this function, which is called *tetration*. Thus, the layers of the Ackermann function rapidly become extremely fast-growing, to a mind-boggling extent. Can you describe the number $A(5, 5)$? Please do so in question 6.5. These are powerful functions indeed.

Each layer of the Ackermann function is a primitive recursive function, a fact that one can show by the kind of analysis we have just performed on the first few layers. But meanwhile, it turns out that A itself is not a primitive recursive function. One can show this by proving that every primitive recursive function f is eventually bounded by some layer n of the Ackermann function; that is, $f(m) \leq A(n, m)$ for all sufficiently large m. It follows that the diagonal Ackermann function $A(n) = A(n, n)$ is not primitive recursive, since it is not bounded by any particular layer. But the Ackermann function is intuitively computable, since on any input, we can simply apply the recursive rule to compute it. The Ackermann example shows that although primitive recursion is capable of producing powerful functions, it is nevertheless weaker than the nested double recursion of Ackermann, which produces even faster-growing functions, which are still "computable" according to our intuitive idea of computability.

6.2   Turing on computability

Let us turn now to the ideas of Alan Turing, who in 1936, while a student at Cambridge, wrote a groundbreaking paper that overcame Gödel's diagonalization obstacle and provided a durable foundational concept of computability, while answering fundamental questions of Hilbert and Ackermann. His ideas ultimately led to the contemporary computer era. Wouldn't you like to write an essay like that? To arrive at his concept of computability, Turing reflected philosophically on what a person was doing when undertaking computation; he distilled computation to its essence.

In that era, the word "computer" referred not to a machine, but to a person—or more specifically, to an occupation. Some firms had whole rooms full of computers, the "computer room," filled with people hired as computers and tasked with various computational duties, often in finance or engineering. In old photos, you can see the computers—mostly women—sitting at big wooden desks with pencils and an abundant supply of paper. They would perform their computations by writing on the paper, of course, according to various definite computational procedures. Perhaps some of these procedures would be familiar to you, such as performing elementary arithmetic on a long list of numbers, but others might be less familiar. Perhaps they were undertaking Newton's method in an iterative calculation aimed at finding the zero of a certain function to within a certain accuracy; or perhaps they were finding numerically the solution to a certain differential equation. Undertaking such a procedure, they would make various marks (perhaps letters or numerals) on the paper according to its dictates; at times, the procedure might involve looking at the marks they had made previously on the paper in front of them, or perhaps looking at previous or later pages, using that information to change or augment their marks. At some point, perhaps the procedure comes to completion, producing a computational result.

Turing aimed to model an idealized version of such computational processes. He observed that we can make various simplifying assumptions for the processes without changing their fundamental character or power. We could assume, for example, that the marks are always made within a regular uniform grid of squares on the paper. And because the two-dimensional arrangement of those squares is unimportant, we can imagine that the squares are aligned in one dimension only—a long paper tape divided into square cells. We can assume that the marks come from a fixed finite alphabet; in this case, we might as well assume that only two symbols are used, 0 and 1, since any other letter or symbol can be thought of as a finite combination or word formed from these elemental bits. At any moment the human computer is in a certain mode of computation or state of mind, and the particular next step in their computational task can depend on this mode and on the particular marks appearing on the papers in front of them. Since those marks can change the state, the computer in effect can achieve a short-term memory for small numbers of bits, and therefore we may assume that the computational instructions depend only on their current state and the one symbol directly in front of them. A human computer, even an idealized one, however, has only finitely many such states of mind. In this way, by contemplating the operation of an idealized human computer carrying out a computational task, Turing was gradually led to his now famous machine concept.

Turing machines

Let me explain how the machines work. A Turing machine has an infinite linear paper tape divided into cells. Each cell can accommodate the symbol 0 or 1, which can be changed, and the tape is initially filled with 0s. Alternatively, one may imagine 1 as a tally mark and 0 as a blank cell, the absence of a tally mark.

Turing's original machines actually had *three* symbols, 0, 1, and blank; this allowed him to use binary notation for input and output instead of unary, as here. Meanwhile, all such variations turn out to give rise to the same concept of computability, and so I am presenting here the slightly more primitive notion.

The head of the machine can move on the tape from one cell to the next, reading and writing these symbols. Or perhaps one thinks of the head as stationary, with the tape passing through it. At any moment, the head is in one of finitely many *states q*, and the machine operates according to a finite list of instructions of the form:

$$(q, a) \mapsto (r, b, d),$$

which has the meaning:

> When in state q and reading symbol a, then change to state r, write symbol b, and move one cell in direction d.

A Turing-machine program is simply a list of finitely many such instructions.

In order to use such a machine for computation, we designate one of the states as the *start* state and another as the *halt* state. At the beginning of the computation, we set up the machine in the input configuration, which has the head on the leftmost cell in the *start* state and with the first n cells showing 1 for input n, and all other cells 0:

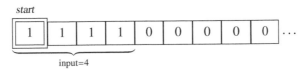

Computation then proceeds according to the instructions of the program.

If the *halt* state is ever achieved, then the output of the computation is the number of cells showing 1 before the first 0 cell appears:

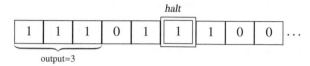

A partial function $f : \mathbb{N} \to \mathbb{N}$ is *Turing computable* if there is a Turing-machine program that can compute the function according to the procedure we have described. We handle binary functions and higher arities $f : \mathbb{N}^k \to N$ by separating the inputs with 0 cells.

Partiality is inherent in computability

Turing provides a notion of computability for *partial* as opposed to total functions on the natural numbers, since a computable function may not be defined on all input. After all, on some input, a computational procedure may run forever, never finding completion. Because of this, partiality is inherent in computability. Let us write $f(n)\!\downarrow$ and say the computation *converges* to indicate that function f is defined on input n; otherwise, we write $f(n)\!\uparrow$ and say that the computation *diverges*.

It is precisely this partiality feature that enables Turing to escape Gödel's diagonal conundrum. Although we can effectively enumerate the Turing-machine programs and thereby provide a list of all Turing-computable functions $\varphi_0, \varphi_1, \varphi_2, \ldots$, nevertheless when we define the diagonal function $d(n) = \varphi_n(n) + 1$, this is not a contradiction, since it could be that $d = \varphi_n$ for some n for which $\varphi_n(n)$ is not defined. In this case, the adding-1 part of the definition of d never actually occurs because $\varphi_n(n)$ did not succeed in producing an output. Thus, we escape the contradiction. The diagonal argument shows essentially that partiality is a necessary feature of any successful concept of computability.

Examples of Turing-machine programs

Let us illustrate the Turing machine concept with a few simple programs. The following Turing-machine program, for example, computes the successor function $s(n) = n + 1$:

$$(\mathit{start}, 1) \mapsto (\mathit{start}, 1, R)$$

$$(\mathit{start}, 0) \mapsto (\mathit{halt}, 1, R).$$

We use R to indicate a move to the right. We can see that this computes the successor function by observing what will happen when we execute the instructions on a machine set up with an input configuration. Namely, the program will successively move through the input, remaining in the *start* state while scanning the tape to the right until it finds the first 0, changes it to 1, and then halts. Thus, it has the effect of making the input one longer, and so it computes the function $s(n) = n + 1$.

Let us now compute the addition function $n + m$. The input configuration for a binary function places the two inputs on the tape, separated by a 0 cell:

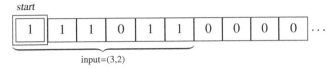

input=(3,2)

Our algorithm will be to go look for that middle 0, change it to 1, and then go look for the end of the input, and then back up and erase the very last 1, giving a string of $n + m$ many 1s, as desired:

$$
\begin{aligned}
(start, 1) &\mapsto (start, 1, R) \\
(start, 0) &\mapsto (q, 1, R) \\
(q, 1) &\mapsto (q, 1, R) \\
(q, 0) &\mapsto (e, 0, L) \\
(e, 1) &\mapsto (halt, 0, R).
\end{aligned}
$$

One can easily produce more examples, which you will do in the exercise questions. Although these particular functions are quite trivial, with greater experience one comes to see that essentially any finite computational process can be carried out on a Turing machine.

Decidability versus enumerability

A *decision problem* is a particular kind of computational problem, in which one wants to decide computationally whether a given input x is a positive or negative instance of a certain phenomenon. For example, given a finite directed graph, is it 3-colorable? Is it connected? Given a weighted network, does it have a Hamiltonian circuit with less than a given total cost? Given a finite group presentation, is the group torsion free? In the general case, a decision problem is a set A, encoded as a set of natural numbers, with the task to decide on a given input x whether $x \in A$ or not. Such a decision problem is computably *decidable* if indeed there is a Turing machine that can perform the task; it is convenient for decision problems to have two halting states, called *accept* and *reject*, such that the outcome of the decision is reflected in the halting state rather than on the tape.

A decision problem A is *computably enumerable*, in contrast, if there is a computable algorithm that enumerates all and only the elements of A. We can create a Turing-machine program that will, for example, write exactly the prime numbers on the tape; thus, the set of primes is computably enumerable. Are the computably decidable decision problems the same as the computably enumerable problems? If a problem is computably decidable, then it is also enumerable, since we can systematically test each possible input in turn, and enumerate exactly those for which we get a "yes" answer.

Conversely, however, there is a problem. If we can enumerate a decision problem, then we can correctly compute all the "yes" answers: on a given input, we simply wait for it to appear in the enumeration, saying "yes" when it does so. But we seem to have no way ever to say "no," and indeed, we shall see later that some problems are computably enumerable, but not decidable. The computably enumerable problems are exactly those for which we have a computable algorithm that correctly answers "yes," but is not required ever to say "no" or even to come to a completion on the negative instances.

Consider the following example, which highlights a subtlety concerning claims of computability. Define the function $f(n) = 1$ if there are n consecutive 7s somewhere in the decimal expansion of π, and otherwise $f(n) = 0$. Is f computable? A naive attempt to compute f would begin, on input n, to search through the digits of π. The number π is a computable number, and so we have a computable procedure to generate its digits and search through them. If we find n consecutive 7s, then we may quickly give output 1. But suppose that we have searched for a long time and not yet found a sufficient sequence of consecutive 7s. When will we ever output 0? Perhaps the function is not computable?

Not quite. Let us consider the extensional nature of f rather than the intentionality of the definition. Consider the possibilities: either there are arbitrarily long sequences of 7s in π or there are not. If there are, then the function f is always 1, and this is a computable function. If there are not, then there is some longest string of 7s, of some length N, in which case $f(n) = 1$ if $n < N$, and otherwise 0. But regardless of the value of N, all such step functions are computable. Thus, we have provided a list of computable functions and proved that f is one of them. So f is a computable function, although we do not know which program computes it. This is a pure-existence proof of the computability of f.

Universal computer

The first computational devices to be built were each designed to perform a single, specific computational task; the program was in effect engineered or hard-wired into the machine. To change the program, one needed to rip out the old wires and mechanics and organize them anew. Turing realized, however, that his Turing machine concept accommodated the construction of a *universal* computer—a computer that could in principle run any other program, not by rewiring it or changing its hardware, but rather just by giving it the new program as a suitable input. Indeed, he outlined the design of a universal Turing machine in his original paper. This was an enormous conceptual advance—to view a program as akin to other inputs—and was of huge importance for the development of actual computers. The idea has been implemented so pervasively that perhaps it may even seen banal now, as we load apps into our laptops and phones; we carry universal computers in our pockets.

The universal Turing machine is a Turing machine that accepts as input another Turing-machine program e and input n, and computes the result $\varphi_e(n)$ of that program on that input. For example, perhaps the universal machine inspects the program e and simulates the operation of e on input n.

The existence of universal Turing machines shows that having arbitrarily large mental capacity, as measured in the number of states, is not required to undertake arbitrary computational tasks. After all, the universal computer has only a certain fixed finite number of states (and not necessarily a very large one), and these suffice for it to carry out in principle any computational procedure if one should only write out the instructions for that procedure on the tape. Turing conceived of the machine state as analogous to a human mental state, and in this sense, the existence of universal machines show that a bounded mind—one with only finitely many mental states (and not necessarily very many)—can nevertheless undertake the same computational tasks as larger minds, if provided with suitable instructions and enough note-taking space. We humans can in principle undertake any computational task simply by equipping ourselves with sufficient paper and pencils.

"Stronger" Turing machines

One way of establishing the power of Turing machines is to prove that they can simulate other seemingly more powerful machines, thereby verifying the point that Turing made in his conceptual analysis of the ideal human computer. Turing machines with only 0 and 1, for example, can simulate Turing machines with much larger finite alphabets by grouping the 0s and 1s into fixed-length words, representing symbols in such an alphabet, like ASCII encoding. Similarly, a Turing machine with only one tape can simulate a Turing machine with multiple tapes, or with tapes that stretch infinitely in both directions, or with a larger head capable of reading several cells at once, or with several independent heads that move according to common coordinating instructions. In this way, one gradually shows the power of the primitive machines by ramping up to these stronger devices.

A Turing machine in a larger alphabet, for example, can undertake arithmetic using binary (or decimal) notation rather than the unary notation presented earlier in this chapter, and this is more satisfactory in certain respects. One cannot use binary notation for input and output if the only symbols are 0 and 1, because one cannot recognize the end of the input, as eventually the tape is all 0s; but perhaps the input simply had a long string of 0s before some additional 1s, and the machine needs to be able to tell the difference.

For each of the particular machine improvements, the new machines become easier to use and to program, but they do not actually become fundamentally more powerful in the sense of computing new functions that were not computable with the more primitive machines. The reason is that in each case, the primitive machines are able to simulate the more advanced machines, and so every computational process that you can undertake with the new machines can also be undertaken (in principle) with the primitive machines.

One might naturally seek to strengthen the Turing machine model to make the machines "better" in some way—perhaps easier to use or faster for some purpose; students often suggest various improvements. But we should be careful not to misunderstand the purpose of Turing machines. We do not actually use Turing machines for computation. No Turing machine has ever been built for the purpose of undertaking computation, although they have been built for the purpose of illustrating the Turing machine concept. Rather, the purpose of the Turing machine concept is to be used in thought experiments to clarify what it means to say that a function is computable or to provide a concept of computability that is precise enough for mathematical analysis. The Turing machine concept provides the concept of *computable by Turing machines*, and for this purpose, it is often better to have the simplest, conceptually cleanest model. We want a model that distills our intuitive idea of what a human computer can achieve in principle. Such a distillation is more successful and surprising when it reduces the concept of computability to more primitive computational capabilities rather than to a stronger one.

Meanwhile, the equivalence of the simple models with the much stronger-seeming models is nevertheless important, for the truly profound results of computability theory are the *undecidability* results, where we show that certain functions are *not* computable or certain decision problems are *not* computably decidable. In this case, not only are the problems not solvable by the primitive Turing machines, but they are also not computably solvable by the enhanced machines, precisely because ultimately these machines have the same power.

Other models of computatibility

At around the same time as Turing wrote his paper and after, several other logicians put forth some other models of computability. Let me briefly mention a few of them. Alonzo Church had introduced the λ-calculus and Emil Leon Post introduced his combinatory process, both in 1936, the same year as Turing's paper. Stephen Cole Kleene, also in 1936, extended the class of primitive recursive functions by recognizing that what was missing was the idea of *unbounded* search. The class of *recursive* functions is obtained by also closing the class under the following: if g is recursive, then so is the function

$$f(x_1, \ldots, x_k) = \mu n[g(n, x_1, \ldots, x_k) = 0 \wedge \forall m < n \; g(m, x_1, \ldots, x_k){\downarrow}],$$

where this means the smallest n for which $g(n, x_1, \ldots, x_k) = 0$ and $g(m, x_1, \ldots, x_k){\downarrow}$ for all $m < n$; it is left undefined if there is no such n. Thus, partiality enters into the class of recursive functions.

In the early days of the development of actual computing machines, which came a bit later, there was a rivalry of ideas in computer architecture between groups in the UK and the US; in a sense, the dispute was between Turing and von Neumann, with the von Neumann group basing their ideas on the *register machine* concept. Register machines, an alternative machine model of computability, are a bit closer to the machines actually being built (and much closer to contemporary computer architecture than are Turing machines). The machines have finitely many registers, each holding a single number, with particular registers designated for input and output. The programs consist of a sequence of instructions of a simple form: you can copy the value of one register to another; you can increment a register by 1; you can reset a register to 0; you can compare one register with another, branching to different program instructions depending on which is larger; and you can halt. The output of the machine at the halting stage is the number in the output register.

There are now many other models of computability, including dozens of varieties of Turing machines, allowing multiple tapes, two-way infinite tapes, larger alphabets, larger heads, and extra instructions; register machines; flowchart machines; idealized computer languages such as Pascal, C++, Fortran, Python, Java, and so on; and recursive functions. Logicians have found Turing-equivalent processes in many surprising contexts, such as in Conway's Game of Life.

6.3 Computational power: Hierarchy or threshold?

Given these diverse models of computability, the fundamental question is whether any of them captures our intuitive concept of computability. Let me present two fundamentally different visions of computability.

The hierarchy vision

According to this view, there is a hierarchy of computational power—a hierarchy of concepts of computability that are all intuitively computable, but that become increasing powerful and complex as one moves up the hierarchy to stronger machines. With the stronger machines, we have greater computational power; we are able to compute more functions and computably decide more decision problems.

The threshold vision

According to this view, there is a huge variety of different models of computability, but once they are sufficiently powerful, they are all computably equivalent, in the sense that they give rise to exactly the same class of computable functions. Any notion of computability, once it meets a certain minimal threshold of capabilities, is fully equivalent to any of the others.

Which vision is correct?

The hierarchy vision may seem attractive because it seems to accord with our experience of actual computers, which seem to have been getting stronger and stronger over the years. But that would be a distracting confusion of the idea of theoretic computational capability with speed-of-computation capability, which is a very different thing. When studying computability in principle, we do not concern ourselves with speed of computation or efficiency in terms of memory or space, but rather with the boundary between what is computable in principle and what is not, regardless of how long the computation takes. With this conception of computational power, it turns out, the evidence is overwhelming for the threshold vision, not the hierarchy vision. The fundamental fact, for all the diverse models of computability that have been proposed as a serious model of computability (including Turing machines, register machines, flowchart machines, idealized computer languages, recursive functions, and many others) is that they are all computably equivalent. In each case, the class of computable functions to which the models give rise is extensionally identical to the class of Turing-computable functions.

One can prove this by showing for each pair of models of computability that they can simulate each other, and therefore any function that can be computed with one of the models can be computed with any of the other models. Turing did this in the addendum to his original paper, for the notions of computability available at that time. It suffices to establish a great circle of simulation, by which every Turing machine can be simulated by a flowchart machine, which can be simulated by a register machine, which can be simulated by recursive functions, which can be simulated by an idealized C++ program, which can be simulated in idealized Java, which can simulate a Turing machine. So all the models are equivalent. We say that a model of computability is *Turing complete* if it can simulate the operation of Turing machines and vice versa, so that the class of computable functions to which it gives rise is the same as the class of Turing-computable functions.

Our personal experience with actual computers improving in power over the years with better and faster designs results not from any fundamental advance with respect to ideal computability, but rather from advances in the resources in memory and speed that are available to the machine. So this evidence, while extremely important for our actual computer use, is not actually evidence for the hierarchy vision of ideal computability.

6.4   Church-Turing thesis

The *Church-Turing* thesis is the view that all our various equivalent models of computability correctly capture the intuitive idea of what it means for a function to be computable. Thus, the thesis asserts that we have got the right model of computability. This is not a mathematical thesis, but a philosophical one. An important piece of evidence in favor of this view is that all the attempts at reifying the concept of computability have turned out to be equivalent to each other.

Jack Copeland has emphasized the distinction between what could be called the *strong* Church-Turing thesis, which asserts that the class of functions that it is possible to compute in principle by any mechanistic rote procedure is precisely the class of Turing-computable functions. The *weak* Church-Turing thesis, in contrast, asserts merely that the class of functions which it is possible in principle to compute, using the idealized picture of a human sitting at a desk, with unlimited supplies of paper and pencils and following a rote mechanistic procedure, is precisely the class of Turing-computable functions.

In addition to those two philosophical positions, there is a kind of *practical* Church-Turing thesis, which is simply the view that in order to know that an algorithmic process can be carried out by a Turing machine, it suffices simply to explain the process sufficiently so that it becomes clear that it is a computable procedure. After the experience of writing a few dozen Turing-machine programs, it becomes clear what kind of procedures Turing machines can implement, and for those who have this experience, it is no longer always necessary to actually write out the Turing-machine programs in detail if all that is desired is the knowledge that there is indeed such a program.

The situation is somewhat analogous to that in proof theory, where mathematicians rarely write out formal proofs of their theorems, but rather give their arguments at a higher level of explanation. Successful proofs, one generally thinks, could at possibly great effort and cost be transformed into formal proofs. In each case, we have a formal notion of proof or computation, which is very important and enlightening as a general concept, but tedious and unenlightening in particular instances. We may get as little mathematical insight from an actual formal proof as from an actual Turing-machine computation; rather, mathematical insight comes from understanding a mathematical argument or a computational algorithm at a much higher level of abstraction.

And there is clearly an analogue of Thurston's view on formalization, but for computability. Namely, just as Thurston argued that mathematical ideas and insights are not necessarily preserved to formalization (see chapter 5, page 161), similarly the ideas and insight of a computable algorithm, described in human-understandable high-level language, might not survive the implementation in a low-level but formal programming language, such as a Turing-machine program. A program written in machine code may seem impenetrable, even if it is actually just implementing a well known algorithm such as quick-sort.

Computation in the physical universe

The evidence we have in favor of the Church-Turing thesis is mainly evidence only in favor of the weak Church-Turing thesis. Turing's philosophical analysis of the nature of computation, which led to his machine concept, was explicitly concerned only with the weak Church-Turing thesis. Meanwhile, we live in a quantum-mechanical, relativistic physical universe, with bizarre physical phenomena that we are only beginning to understand. Perhaps we might hope to take computational advantage of some strange physical effect. Perhaps the physical world is arranged in such a way that allows for the computation of a non-Turing-computable function by means of some physical procedure.

To give you a sense of the range of possibility, let me mention that some researchers have proposed algorithms for computation that involve falling into a black hole and taking advantage of relativistic time foreshortening so that parts of the machine undergo an infinite time history while remaining in the light cone of another part of the machine that experiences only a finite time. In this way, the machine as a whole seems able to simulate an infinite unbounded search and thereby come to know in a finite amount time the answer to a question that a Turing machine could not.

Are we to take such algorithms as refuting the Church-Turing thesis? It is a crazy idea, I agree, but do we actually have good reasons to rule out the possibility of desktop machines employing such methods, perhaps with low-mass microscopic black holes contained inside them? Imagine a black hole on a chip—IPO coming soon! To establish the strong Church-

Turing thesis, after all, one needs not just to rule out this particular crazy idea, but to rule out all possible crazy ideas, to argue that there is in principle no physical means by which we might undertake a computational task not caught by Turing machines. Our best physical theories seem neither complete nor certain enough to support such a sweeping claim.

Sometimes people object to the various super-Turing physical procedures that have been proposed with practical physical concerns, such as energy resources, and so on. But of course, such objections can also be made against the Turing machine model itself. At a large scale, Turing machines will be using extremely long, vast paper strips. Eventually, they will become unmanageable on the surface of the Earth. Shall we put the paper tapes in orbit, floating about in space? Surely the paper will tear; if the tape is coiled somehow, perhaps like the DNA coiled in our genes, then once large enough, it will collapse on itself by the force of gravity, perhaps forming a small sun; and are there enough atoms in the universe to form a tape large enough to write out the number $A(5, 5)$? Do we take such objections as undermining the concept of idealized computability-in-principle? If not, then do we have the same attitude toward black-hole computation and the other fantastic or bizarre physics-based computational methods?

6.5   Undecidability

Let us turn now to the question of establishing noncomputability. Amazingly, we can identify concrete decisions problems that cannot be solved in principle by any Turing-computable procedure.

The halting problem

Consider the problem of deciding whether a given Turing-machine program p will halt on a given input n. This is known as the *halting problem*. Is the halting problem computably decidable? In other words, is there a computable procedure, which on input (p, n) will output yes or no depending on whether program p halts on input n? The answer is no, there is no such computable procedure; the halting problem for Turing machines is not computably decidable. This was proved by Turing with an extremely general argument, a remarkably uniform idea that applies not only to his machine concept, but which applies generally with nearly every sufficiently robust concept of computability. Let us give the argument.

The claim is that the halting problem for Turing machines is not computably decidable by any Turing machine. Suppose, toward contradiction, that it is. Consider the following strange algorithm. On input p, a program, our algorithm asks if program p will halt if given p itself as input; under our assumption that the halting problem is decidable, this is a question whose answer we can compute yes-or-no in finite time, and so let us do so. If the answer comes back that, yes, p would halt on input p, then in the strange algorithm that we are describing, let us jump immediately into an infinite loop, which will never halt. And if the answer comes back that, no, p would not halt on input p, then let us halt immediately in our new algorithm. Let q be the program that carries out the algorithm that we have just described. Consider now the question: does q halt on input q? Well, by inspection of the algorithm, you may observe that q halts on a program p exactly when p does not halt on p itself as input. Consequently, q halts on input q exactly when q does not halt on input q. This is a contradiction, so our assumption that the halting problem was decidable must have been false. In other words, the halting problem is undecidable. The argument is extremely flexible and shows essentially that no model of computability is able to solve its own halting problem.

Other undecidable problems

The undecidability of the halting problem can be used to show the undecidability of many other problems in mathematics. The question of whether a given diophantine equation (polynomials over the integers) has a solution in the integers is undecidable; the word problem in group theory is undecidable; the identity of two elementary algebraic expressions in a formal language of "high-school algebra," defined by Tarski, is undecidable; the mortality problem for 3×3 matrices (given a finite set of matrices, determine whether a finite product of them, repetitions allowed, is zero) is undecidable. There are thousands of interesting undecidable problems, and a typical pattern of argument is to show that a problem is undecidable by reducing the halting problem to it.[6]

The tiling problem

A favorite example of mine is the tiling problem: given finitely many polygonal tile types, can they tile the plane? For instance, can the two tiles shown here tile the plane? In the general case, this problem is undecidable, with a remarkable proof. For any Turing machine, one constructs a collection of tiles, such that the program halts exactly when those

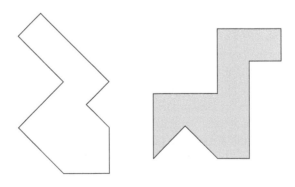

tiles admit no tiling of the plane. One designs the tiles carefully so as to simulate the computation—the tiling continues so long as the computation keeps running—and the halting configuration causes an obstacle. Thus, the halting problem reduces to the tiling problem.

---

[6] In a subtle error mistaking the logic of reducibility, one sometimes hears, "This problem is undecidable because it reduces to the halting problem." But this logic is backward, because a problem is proved difficult not by reducing it to a difficult problem, but rather by reducing a difficult problem to it.

Computable decidability versus enumerability

Although the halting problem is undecidable, it is nevertheless computably enumerable, and we thereby distinguish these two concepts. The halting problem is computably enumerable because we can systematically simulate all possible programs, enumerating the ones that halt. We simulate the first program for one step, the first two programs each for two steps, the first three programs each for three steps, and so on. Whenever we find that a program has halted, we enumerate it onto the tape. Any program that does halt will eventually be enumerated by this algorithm, and so the halting problem is computably enumerable.

6.6   Computable numbers

Alan Turing essentially founded the subject of computability theory in his classic 1936 paper, "On Computable Numbers, with an Application to the Entscheidungsproblem," in which he achieves so much: He defines and explains his machines; he describes a universal Turing machine; he shows that one cannot computably determine the validities of any sufficiently powerful formal proof system; he shows that the halting problem is not computably decidable; he argues that his machine concept captures our intuitive notion of computability; and he develops the theory of computable real numbers.

That last concern was the title focus of the article, and Turing defines, in the very first sentence, that a computable real number is one whose decimal expansion can be enumerated by a finite procedure, by what we now call a Turing machine, and he elaborates on and confirms this definition later in detail. He proceeds to develop the theory of computable functions of computable real numbers. In this theory, one does not apply the functions directly to the computable real numbers themselves, but rather to the programs that compute those numbers. In this sense, a computable real number is not actually a kind of real number. Rather, to have a computable real number in Turing's theory is to have a program—a program for enumerating the digits of a real number.

I should like to criticize Turing's approach, which is not how researchers today define the computable numbers. Indeed, Turing's approach is now usually described as one of the natural, but ultimately mistaken, ways to proceed with this concept. One main problem with Turing's account, for example, is that with this formulation of the concept of computable numbers, the operations of addition and multiplication on computable real numbers turn out not to be computable. Let me explain. The basic mathematical fact in play is that the digits of a sum of real numbers $a + b$ are not continuous in the digits of a and b separately; one cannot necessarily say with certainty the initial digits of $a + b$, knowing only finitely many digits, as many as desired, of a and b.

To see this, consider the following sum $a + b$:

$$0.343434343434\ldots$$
$$+\quad 0.656565656565\ldots$$
$$\overline{0.999999999999\ldots}$$

The sum has 9 in every place, which is fine, and we could accept either $0.999\ldots$ or $1.000\ldots$ as correct, since either is a decimal representation of the number 1. The problem, I claim, is that we cannot know whether to begin with 0.999 or 1.000 on the basis of only finitely many digits of a and b. If we assert that $a + b$ begins with 0.999, knowing only finitely many digits of a and b, then perhaps the later digits are all 7s, which would force a carry term, causing all those 9s to roll over to 0, with a leading 1. Thus, our answer would have been wrong. If, alternatively, we assert that $a + b$ begins with 1.000, knowing only finitely many digits of a and b, then perhaps the later digits of a and b are all 2s, which would mean that $a + b$ is definitely less than 1. Thus, in any case, no finitely many digits of a and b can justify an answer for the initial digits of $a + b$.

Therefore, there is no algorithm to compute the digits of $a+b$ continuously from the digits of a and b separately. It follows that there can be no computable algorithm for computing the digits of $a + b$, given the programs that compute a and b separately.[7] One can make similar examples showing that multiplication and many other very simple functions are not computable if one insists that a computable number is an algorithm enumerating the digits of the number.

So what is the right concept of computable number? The definition widely used today is that we want an algorithm not to compute exactly the digits of the number, but rather to compute approximations to the number as close as desired, with a known degree of accuracy. A computable real number is a computable sequence of rational numbers, such that the nth number is within $1/2^n$ of the target number. This is equivalent to being able to compute rational intervals around the target real number, of a size less than any specified accuracy. And there are many other equivalent ways to do it. With this concept of computable real number, then the operations of addition, multiplication, and so on, all the familiar operations on the real numbers, $\exp(x)$, $\log(z)$, $\sin\theta$, $\tan\alpha$, will all be computable.

Meanwhile, although I have claimed that Turing's original approach to the concept of computable real number was flawed, and I have explained how we usually define this concept today, the mathematical fact is that a real number x has a computable presentation in

---

[7] This latter claim is actually quite subtle and sophisticated, but one can prove it with the Kleene recursion theorem. Namely, let $a = .343434\ldots$, and then consider a program to enumerate a number b, which begins with 0.656565 and keeps repeating 65 until the addition program has given the initial digits for $a + b$, at which point our program for b switches either to all 7s or all 2s, in such a way so as to refute the result. The Kleene recursion theorem is used in order to construct such a self-referential program for b.

Turing's sense if and only if it has a computable presentation in the contemporary sense. Thus, in terms of which real numbers we are talking about, the two approaches are extensionally the same. Let me quickly argue this point. If a real number x is computable in Turing's sense, so that we can compute the digits of x, then we can obviously compute rational approximations to any desired accuracy simply by taking sufficiently many digits. And conversely, if a real number x is computable in the contemporary sense, so we can compute rational approximations to any specified accuracy, then either it is itself a rational number, in which case we can certainly compute the digits of x, or else it is irrational, in which case, for any specified digit place, we can wait until we have a rational approximation forcing it to one side or the other, and thereby come to know this digit. This argument works in any desired base. Notice that issues of intuitionistic logic arise here, precisely because we cannot tell from the approximation algorithm itself which case we are in.

So Turing has not actually misidentified the computable real numbers as a class of real numbers. But as I have emphasized, we do not want to regard a computable real number as a kind of real number; rather, we want to regard a computable real number as a program, a finite program for somehow generating that real number because we want to be able to feed that finite object in other computable algorithms and compute things with it. For this purpose, with Turing's approach, addition is not a computable function on the computable real numbers; but with the rational-approximation approach, it is.

6.7   Oracle computation and the Turing degrees

So the halting problem is not computably decidable. But suppose that we were to equip our computing machine with a black-box device, of unknown internal configuration, which would correctly answer all queries about the halting problem. What then? What could we compute with this newly augmented computational power? Are there still problems that we cannot solve? In this way, Turing arrives at the idea of *oracle* (or *relative*) computation. Turing's halting problem argument shows that even with an oracle, there are some problems that remain out of reach. This is because the great generality of Turing's undecidability argument means that it applies also to oracle computation: the halting problem relative to a particular oracle will not be solved by any computable procedure using that oracle.

Let us be more specific about our model of oracle computation. Imagine that we augment a Turing machine with an extra tape, the oracle tape, which the machine can read but not write on, and which contains the membership information of a fixed oracle, $A \subseteq \mathbb{N}$. The oracle tape has a pattern of 0s and 1s written out in exactly the pattern of membership in A. Thus, A itself becomes decidable by the machine with oracle A, because on input n, we can simply check the nth cell of the oracle tape to discover whether $n \in A$ or not. So this concept of relative computability is strictly stronger than Turing computability, when the oracle A is not decidable.

We say that a function is *A*-computable if it can be computed by a Turing machine equipped with oracle *A*. We define that *B* is Turing computable *relative to A*, written $B \leq_T A$, if the characteristic function of *B*, having value 1 for members of *B* and 0 otherwise, is computable from a machine with oracle *A*. Two oracles, *A* and *B*, are *Turing equivalent* if each is computable relative to the other. This is an equivalence relation, whose equivalence classes are called *Turing degrees*.

The undecidability of the halting problem shows that for every oracle *A*, there is a strictly harder problem *A′*, called the *Turing jump* of *A*, which is simply the halting problem for the machines with oracle *A*. So $A <_T A'$. It is traditional to use the symbol 0 to represent the Turing degree of all decidable sets, and $0'$ is the Turing degree of the halting problem. So we have $0 <_T 0' <_T 0'' <_T 0''' <_T \cdots$, and so on. Are there incomparable degrees? Yes. Are there degrees strictly between 0 and $0'$? Yes. Are there computably enumerable sets strictly between 0 and $0'$? Yes. As a partial order, are the Turing degrees dense? No, there are minimal noncomputable degrees, although the computably enumerable degrees are densely partially ordered. Are the Turing degrees a lattice? No, there are pairs of Turing degrees with no greatest lower bound. There is a hugely developed literature of research investigating the fundamental nature of the hierarchy of Turing degrees, including all these questions and many others.

My own perspective on computability theory is that it is a kind of infinitary information theory. The Turing degrees represent the possible countable amounts of information that one might have. Two oracles that can compute each other have the same information, and when $B \leq_T A$, then *A* has at least as much information as *B*. Studying the Turing degrees is to investigate the possible amounts of information that one can have in principle.

6.8   Complexity theory

Let us come finally to the topic of complexity theory, a huge area of contemporary research in theoretical computer science, which focuses not on the distinction between computable and uncomputable and on degrees of noncomputability, but rather on the distinction between feasible and infeasible computation and on degrees of feasibility. I find the topic to be rich with philosophical issues, which I would like to see explored much further than they have been by philosophers.

Complexity theory aims to refocus computability theory on a more realistic and useful realm. We want to attend to the limitations on computational power imposed by the scarcity of resources, such as time or space. Such limitations are often totally and unrealistically ignored in classical computability theory, but they are paramount concerns in actual computation. What good to us is an algorithm that on a small input takes the lifetime of the universe to provide its answer? In order to develop a robust theory of computational resources, what is needed is a robust concept of *feasible* computation. What does it mean to say that computational process is feasible?

One can easily introduce notions of feasibility with extremely hard edges by imposing specific bounds on resources, such as requiring all computations to stop in 1 billion steps, say, or to use at most 1 billion memory cells. But such a theory of feasibility would be ultimately unsatisfying, inevitably caught up in edge-effect details about those specific bounds. What is needed instead is a softer notion of feasibility, a relative feasibility concept, where the feasibility of a computation is measured by a consideration of the resources used by a computation in comparison with the size of the input. Larger inputs may naturally require more resources. This leads one to consider asymptotic bounds on resource use as the size of the input becomes large.

We may compare the efficiency of algorithms, for example, by the number of steps of computation that they involve. Consider the sorting problem: given n numbers, put them in order from smallest to largest. How would you do it? How many basic steps would your algorithm require? By a "step" here, we would mean something like the process of comparing two values or the operation of moving a value from one location to another. There are dozens of sorting algorithms now known, of varying computational complexities. Some of them work better in the average case, but worse in the worst case, while others are fast even in the worst case. Examples include:

Bubblesort. Repeatedly scan the list; swap adjacent entries that are out of order. Worst case n^2 steps.

Comb sort. An improvement to bubblesort, dealing more efficiently with "turtles" (entries that are far out of place), but still n^2 in the worst case.

Merge sort. A divide-and-conquer algorithm that sorts sublists first and then merges them; $n \log n$.

In-place merge sort. A variation that achieves $n \log^2 n$ in the worst case.

Heap sort. Iteratively take the largest element from the unsorted region; $n \log n$.

Quick sort. A partition-exchange sort, which in the worst case takes n^2 comparisons, but on average and in practice is significantly faster than its main competitors of merge sort and heap sort; average time $n \log n$.

I highly recommend the incredible AlgoRythmics channel on YouTube, which you can find at https://www.youtube.com/user/AlgoRythmics. This group exhibits all the various sorting algorithms via Hungarian dance and other folk dance styles. It's great!

To appreciate the vast differences in the efficiency of the algorithms, consider that when $n = 2^{16} = 65536$, then we have (using the base 2 log):

$$
\begin{aligned}
n^2 &= 4,294,967,296 \\
n \log^2 n &= 16,777,216 \\
n \log n &= 1,048,576.
\end{aligned}
$$

So there is a genuine savings with the faster algorithms, and the differential advantage becomes even more pronounced as n increases.

Feasibility as polynomial-time computation

What do we mean by "feasible" computation? One natural consideration might be that if two functions, f and g, are feasibly computed, then we would probably want to say that the composition $f(g(x))$ also is feasibly computed, since on any input x, we could feasibly compute $g(x)$, and then feasibly compute $f(g(x))$. Similarly, we might want always to be able to employ feasible procedures as subroutines within other feasible algorithms—for example, to do something feasible separately with every component of a given input. And we would seem to want to allow as definitely feasible the computations that take a number of steps proportional to the length of the input; these are the *linear-time* computations. But notice that if you insert a linear-time subroutine nested into a linear-time algorithm, you get an n^2 time algorithm; and if you do it again, you get a n^3 time algorithm, and so on. So we are led naturally to the class of *polynomial-time* computations, that is, computational algorithms for which there is a polynomial bound on the running time of the computation as a function of the length of the input. This class is closed under composition and subroutine insertion, and ultimately provides an extremely robust conception of feasibility, fulfilling many of our prereflective ideas about feasibility. A rich complexity theory has emerged built upon it.[8]

Worst-case versus average-case complexity

The time complexity we have discussed thus far can be described as *worst-case complexity*, because we insist that there is a polynomial function bounding the running time of the algorithm on all input of a given size. But one can easily imagine a situation where an algorithm runs very fast on almost all input and has difficulty with just a few rare cases. A competing conception of feasibility, therefore, would use *average-case complexity*, where we want to bound merely the average running time for inputs of a certain size. For example, perhaps a program takes time n^2 on all inputs of length n except for the single input $111 \cdots 11$, where it takes time 2^n. This would mean an average-time complexity of less than $n^2 + 3$ for input of size n, but a worst-case complexity of 2^n, which is much larger.

One important consideration undermining the robustness of average-case complexity as a notion of feasibility is that it does not behave well under substitution and composition. It could be that functions f and g have good average-case complexity, but $f(g(x))$ does not, simply because the output of $g(x)$ might happen nearly always to feed into the difficult cases of f. For this reason, the class of all average-case feasible algorithms is less robust in important respects than the class of worst-case feasible algorithms.

---

[8] In complexity theory, we generally expand the language beyond just 0 and 1 (using an end-of-input marker, for example) to enable the familiar compact number representations, such as binary notation. In the bloated unary input, the meaning of "polynomial time in the length of the input" would in effect mean exponential time for binary input.

The black-hole phenomenon

The *black-hole* phenomenon arises when a decision problem is difficult, but the difficulty of the problem is concentrated in a very tiny region, outside of which the problem is easy. One would not want to base an encryption scheme, for example, on a difficult decision problem if this problem had a black hole, because it would be intolerable if the hackers could rob the bank 95 percent of the time, or indeed even if they could rob the bank only 5 percent of the time. The black-hole phenomenon, therefore, concerns extreme examples where worst-case complexity is high, but average-case complexity is low. Such a problem might seem difficult because of the worst-case difficulty, yet most instances of it are easily solved.

Alexei Miasnikov had found instances of the black-hole phenomenon in decision problems arising in computational group theory, and he had inquired whether the halting problem itself might harbor a black hole. In Hamkins and Miasnikov (2006), we proved that indeed it did. What we proved is that for the standard model of Turing machine, with a one-way infinite tape writing 0s and 1s, with finitely many states and one halt state, the halting problem is decidable on a set of asymptotic density one. That is, there is a decidable set of Turing-machine programs that includes almost every program in the sense of asymptotic density—the proportion of all n-state programs that are in the set steadily converges to 100 percent as n increases—such that the halting problem is decidable for programs in this set.

One can start to get an idea of this set if you consider the set of programs that have no transition to the *halt* state. Since such a program can never achieve the *halt* state, the halting problem is decided for such programs by the algorithm saying "no." And the proportion of such programs has density $1/e^2$ as the number of states increases; this is about 13.5 percent. This means that we can easily decide the halting problem in about 13.5 percent of cases.

Miasnikov and I had initially sought to pile together such trivial or even silly reasons for nonhalting, in the hope that it would eventually exceed 50 percent, with our goal being able to say that we could decide "most" instances of the halting problem. But instead, what we found is one gigantic silly reason for being able to decide instances of the halting problem. What we proved is that with the one-way infinite tape model, almost all computations end up with the head falling off the edge of the tape before a state has been repeated. The behavior of a random program, before states have repeated (and therefore before there can be any information feedback) is essentially a random walk. It then follows from Pólya's recurrence theorem that the head will eventually return to the origin and fall off with high probability. Since it can be easily checked whether this behavior indeed occurs before a state is repeated, it follows that we can decide the halting problem with asymptotic probability one.

Decidability versus verifiability

Complexity theory makes a key distinction between decidability and verifiability. A decision problem is said to be polynomial-time decidable, as we have discussed, if there is a polynomial-time algorithm that will correctly answer instances of it. A decision problem A is polynomial-time *verifiable*, in contrast, if there is a polynomial-time algorithm p, such that $x \in A$ if and only if there is a companion witness w, of polynomial-size in x, such that p accepts (x, w). In effect, the right witness w can provide the extra information enabling positive instances of $x \in A$ to be verified as correct. (This concept of verifiability differs from the verifiability property of a proof system mentioned in chapter 5.)

Consider as an illustration the *satisfiability* problem of propositional logic. Given a Boolean assertion of propositional logic, such as $(p \vee \neg q) \wedge (q \vee r) \wedge (\neg p \vee \neg r) \wedge (\neg q \vee \neg r)$, determine if there is a satisfying instance, an assignment of truth values to the propositional variables that makes the whole assertion true. In general, there is no polynomial-time algorithm known to solve this problem; the straightforward approach, computing the truth table and inspecting it for a true outcome, takes exponential time. But the satisfiability problem is polynomial-time verifiable, since if we are also provided with the satisfying row w, then we can check correctly that it is indeed a satisfying row in polynomial time. Basically, someone claiming to have a satisfiable assertion can prove that to us easily by providing the row of the truth table where the true outcome arises.

Nondeterministic computation

These verifiable problems can also be characterized in terms of nondeterministic computation. Namely, a computational algorithm is said to be *nondeterministic* if the computational steps of it are not necessarily completely determined by the program, but at some steps, the algorithm may choose nondeterministically from a permissible list of instructions. A decision problem A is *nondeterministically* polynomial-time decidable if there is such a nondeterministic algorithm that correctly decides the decision problem. That is, an input is in A exactly when there is *some* way of making the choices so as to have the algorithm accept that input.

The class of nondeterministically polynomial-time decidable problems, it turns out, is identical to the class of polynomial-time verifiable problems, and this is the class known as NP. If a problem is polynomial-time verifiable, for example, then we can nondeterministically guess the witness w and then proceed with the verification algorithm. If the input is in the class, then there will be a way to produce a suitable witness, leading to accept; and if the input is not in the class, there will be no computation leading to accept, since no witness will work. Conversely, if a problem is nondeterministically polynomial-time decidable, then we can think of the witness w as telling us which choice to make each time we need to make a choice in the nondeterministic procedure, and then the right witness will lead us to accept. In this way, we have a dual conception of the class NP, either as the result of nondeterminism or by means of the concept of verification.

P versus NP

There are an enormous number of natural and fundamentally important problems that turn out to be in the class NP, and for many of these problems, we have not found deterministic polynomial time decision algorithms for them. Furthermore, in almost every such case, such as the satisfiability problem for Boolean expressions, certain circuit-design problems, the protein-folding problem, problems in cryptographic security, and many other problems of scientific, financial, or practical importance, the problem turns out to be NP-complete, which means that all other NP problems reduce to this problem in polynomial time.

Generally, NP problems always have at worst exponential time algorithms, since we can try out all possible companion witnesses in turn. But exponential algorithms are infeasible in practice; they simply take too long. Meanwhile, we do not know for certain whether these NP-complete problems have polynomial time algorithms.

Open Question. *Is P = NP?*

The question asks, in other words, whether any of the NP complete problems mentioned here have polynomial-time solutions. This is the single most important open question in computer science. It would be of astounding significance, with far-reaching practical advantages that hugely affect our engineering capabilities, and therefore also the state of human wealth and welfare, if the answer should turn out to be positive, with an algorithm that is feasible in practice. It would mean that finding solutions would not in principle be more difficult in verifying that they work. We would immediately gain a feasible capability for solving an enormous range of problems that currently lie out of reach. As Scott Aaronson puts it:

> If P = NP, then the world would be a profoundly different place than we usually assume it to be. There would be no special value in "creative leaps," no fundamental gap between solving a problem and recognizing the solution once it's found. Everyone who could appreciate a symphony would be Mozart; everyone who could follow a step-by-step argument would be Gauss; everyone who could recognize a good investment strategy would be Warren Buffett. (Aaronson, 2006, #9)

Most researchers, however, believe strongly that the complexity classes P and NP are not equal, and indeed, Aaronson continues, making his "philosophical argument," one of ten arguments he outlines for P ≠ NP:

> It's possible to put the point in Darwinian terms: if this is the sort of universe we inhabited, why wouldn't we already have evolved to take advantage of it? (Indeed, this is an argument not only for P ≠ NP, but for NP-complete problems not being efficiently solvable in the physical world.)

Because of its far-reaching consequences for internet security and many other issues, the P versus NP problem stands in the background of nearly every serious issue in theoretic computer science.

There is an entire zoo of hundreds of complexity classes, corresponding to various other resource-limited models of computability. For example, we have PSPACE, which is the collection of decision problems that can be solved using only a polynomial amount of space, even if they take a long time. The nondeterministic analogue of this concept is NPSPACE, but it is not difficult to see that PSPACE=NPSPACE because one can simply try out all the possible guesses in turn. Similarly, we have EXPTIME and EXPSPACE and their nondeterministic analogues, corresponding to exponential time, and on the lower end, linear time and linear space problems, as well as many other complexity classes.

Ultimately, the subject emerges in such a way that complexity theory, seen as the study of resource-limited computation, is fulfilling the hierarchy vision, while computability theory realizes the threshold vision that I had mentioned earlier in this chapter.

Computational resiliency

In the midst of the 2020 Covid-19 pandemic, which brought to light the need for resilience in our economic, medical, and political structures and organizations, Moshe Y. Vardi called for an analogous refocusing of complexity theory from computational efficiency to resilience and fault tolerance.

> I believe that computer science has yet to internalize the idea that resilience, which to me includes fault tolerance, security, and more, must be pushed down to the algorithmic level. Case in point is search-result ranking. Google's original ranking algorithm was PageRank, which works by counting the number and quality of links to a page to determine how important the website is. But PageRank is not resilient to manipulation, hence "search-engine optimization."...
>
> Computing today is the "operating system" of human civilization. As computing professionals we have the awesome responsibility as the developers and maintainers of this operating system. We must recognize the trade-off between efficiency and resilience. It is time to develop the discipline of resilient algorithms. Vardi (2020)

Computability theory reified the notion of computability via Turing machines and the other computational models, all giving rise to the same class of computable functions. Complexity theory reified the notions of feasibility and computational efficiency via polynomial-time computability and the zoo of complexity classes. What are the formal concepts of computational resilience upon which the new discipline proposed by Vardi might be founded? I find this to be a fundamentally philosophical question—a logical challenge at the heart of this proposal.

Questions for further thought

6.1 Show that the functions m^n, 2^n, and $n!$ are all primitive recursive by finding recursive definitions of them.

6.2 Show that the primitive recursive relations are closed under union, intersection, and complement.

6.3 Show that every primitive recursive function can be generated in more than one way, and thus has infinitely many construction templates. Conclude that the construction history of a function is not a property of the function itself, but of the way that one has described it. Discuss whether this means that computability is an intensional rather than an extensional feature.

6.4 Explain exactly why the defining equations given earlier in this chapter for the Ackermann function (page 198) do indeed define a function $A : \mathbb{N} \times \mathbb{N} \to \mathbb{N}$.

6.5 Describe the number $A(5, 5)$.

6.6 Design a Turing machine to compute the function

$$p(n) = \begin{cases} 0, & \text{if } n \text{ is even} \\ 1, & \text{if } n \text{ is odd.} \end{cases}$$

6.7 Provide a Turing-machine program that computes the truncated subtraction function $a \doteq 1$, which is $a - 1$ when $a \geq 1$, and otherwise is 0.

6.8 Explain precisely how the class of Turing-computable functions escapes Gödel's diagonalization argument. That is, although Turing proved that there is a universal computable function $f(n, m)$, whose individual slices $f_n(m)$ exhaust all the computable functions, what happens with the diagonalization function $d(n) = f_n(n) + 1$, as identified by Gödel? Does the diagonal argument succeed in showing that there are intuitively computable functions that are not Turing computable?

6.9 Is partiality (in the sense of partial functions) inherent to computation? Can one give a coherent account of the computable total functions on the natural numbers without any analysis or consideration of partial functions?

6.10 Criticize the following argument: List all possible Turing-machine programs: p_0, p_1, p_2, and so on. Let $f(n) = 0$, unless the output of p_n on input n is 0, in which case let $f(n) = 1$. The function f is intuitively computable, but it is not computable by any Turing machine.

6.11 Do the yellow and blue tiles pictured in the chapter (page 211) tile the plane? If you have answered, how do you reconcile this with the claim that the tiling problem is undecidable?

6.12 With the assistance of your instructor or tutor, find a suitable decision problem and prove that it is not computably decidable.

6.13 Is a computable real number a kind of real number? Or is it a program of some kind?

6.14 Discuss the success of Turing's definition of a computable real number.

6.15 What philosophical significance, if any, is caused for the concept of computable real number by the fact that the questions whether a given program computes the digits of a real number or whether two programs compute the same real number are not computably decidable? Is the identity relation $x = y$ computably decidable?

6.16 Argue that if we use an oracle A that is itself decidable, then the A-computable functions are the same as the computable functions.

6.17 Are there enough atoms in the universe to write out the value of the Ackermann function $A(5, 5)$ on a paper tape? If not, is this a problem for Turing computability as a notion of idealized computability in principle? How does your answer bear on your analysis of black-hole computation or the other physically fantastic computational methods?

6.18 Discuss the inherent tension, if any, between the ideas that feasible computation should be closed under composition and recursion and that linear-time computation is feasible, whereas computation of time complexity n^8 is not feasible in practice. Does this undermine the conception of polynomial-time computation as a notion of feasibility?

6.19 In *worst-case* complexity, we measure the speed of a computable algorithm on input of size n by looking at the longest-possible running time for inputs of that size; in *average-case* complexity, we measure the speed of the algorithm by looking at the average running time for those inputs. Discuss the advantages and disadvantages of these two conceptions.

6.20 Can you prove that worst-case polynomial-time computation is not the same as average-case polynomial time computation? If possible, identify a particular function in one class and not in the other. Can you do this with a decision problem?

6.21 Are the average-case polynomial-time computable functions closed under composition? What is the significance of your answer for the idea of using average-case time complexity as a notion of feasibility?

6.22 Invent a nondeterministic analogue of general computation and decidability to be used instead of just polynomial-time computation and polynomial-time decidability. Exactly which decision problems are nondeterministically decidable in your sense?

6.23 Explore more fully the idea that whereas computability theory may realize the Church-Turing thesis via the threshold phenomenon, complexity theory realizes the hierarchy vision. Does this speak against a Church-Turing thesis for feasible computation?

6.24 Explain the meaning of Scott Aaronson's "creative leaps" quotation on page 220. Is his Darwinian argument successful? Don't we in fact solve some NP-complete problems in the physical world, such as the protein-folding problem? Assess the other arguments of Aaronson (2006).

6.25 Learn about analog computation—as opposed to the discrete computation considered in this chapter—and discuss the distinction in connection with the Church-Turing thesis.

6.26 Learn about some of the infinitary models of computation, such as Büchi automata, the Hamkins-Kidder-Lewis model of infinite-time Turing machines or the Blum-Shub-Smale machines. Is an analogue of the Church-Turing thesis true for infinitary computability? That is, are these models equally powerful in terms of which functions they can compute?

Further reading

A. M. Turing (1936). An incredible achievement, this is Turing's seminal paper, "On Computable Numbers, with an Application to the Entscheidungsproblem," in which he (1) defines Turing machines, deriving them out of his philosophical analysis of the nature of computation; (2) proves the existence of universal computation; (3) proves the undecidability of the halting problem; (4) introduces oracle computation; and (5) analyzes the nature of computable numbers. This work is still very readable today.

Oron Shagrir (2006). An interesting account of changes in Gödel's views of Turing's work over the decades.

Robert I. Soare (1987). The standard graduate mathematics introduction to Turing degrees.

Hamkins and Miasnikov (2006). An accessible account of the black-hole phenomenon proving that the halting problem, although undecidable, is nevertheless decidable on almost all instances, with probability one, for certain models of computation.

Scott Aaronson (2006). A brief accessible account of ten arguments for P ≠ NP.

7 Incompleteness

Abstract. David Hilbert sought to secure the consistency of higher mathematics by finitary reasoning about the formalism underlying it, but his program was dashed by Gödel's incompleteness theorems, which show that no consistent formal system can prove even its own consistency, let alone the consistency of a higher system. We shall describe several proofs of the first incompleteness theorem, via the halting problem, self-reference, and definability, showing senses in which we cannot complete mathematics. After this, we shall discuss the second incompleteness theorem, the Rosser variation, and Tarski's theorem on the nondefinability of truth. Ultimately, one is led to the inherent hierarchy of consistency strength rising above every foundational mathematical theory.

Mathematical logic, as a subject, truly comes of age with Kurt Gödel's incompleteness theorems, which show that for every sufficiently strong formal system in mathematics, there will be true statements that are not provable in that system, and furthermore, in particular, no such system can prove its own consistency. The theorems are technically sophisticated while also engaged simultaneously with deeply philosophical issues concerning the fundamental nature and limitations of mathematical reasoning. Such a fusion of mathematical sophistication with philosophical concerns has become characteristic of the subject of mathematical logic—I find it one of the great pleasures of the subject. The incompleteness phenomenon identified by Gödel is now a core consideration in essentially all serious contemporary understanding of mathematical foundations.

In order to appreciate the significance of his achievement, let us try to imagine mathematical life and the philosophy of mathematics prior to Gödel. Placing ourselves in that time, what would have been our hopes and goals in the foundations of mathematics? By the early part of the twentieth century, the rigorous axiomatic method in mathematics had found enormous success, helping to clarify mathematical ideas in diverse mathematical subjects, from geometry to analysis to algebra. We might naturally have had the goal (or at least the hope) of completing this process, to find a complete axiomatization of the most fundamental truths of mathematics. Perhaps we would have hoped to discover the ultimate foundational axioms—the bedrock principles—that were themselves manifestly true and also encapsulated such deductive power that with them, we could in principle settle every question within their arena. What a mathematical dream that would be.

Meanwhile, troubling antinomies—contradictions, to be blunt—had arisen on the mathematical frontiers in some of the newly proposed mathematical realms, especially in the naive account of set theory, which exhibited enormous promise as a unifying foundational theory. Set theory had just begun to provide a unified foundation for mathematics, a way to view all mathematics as taking place in a single arena under a single theory. Such a unification allowed us to view mathematics as a coherent whole, enabling us sensibly, for example, to apply theorems from one part of mathematics when working in another; but the antinomies were alarming. How intolerable it would be if our most fundamental axiomatic systems of mathematics turned out to be inconsistent; we would have been completely mistaken about some fundamental mathematical ideas. Even after the antinomies were addressed and the initially naive set-theoretic ideas matured into a robust formal theory, uncertainty lingered. We had no proof that the revised theories were safe from new contradictions, and concern remained about the safety of some fundamental principles, such as the axiom of choice, while other principles, such as the continuum hypothesis, remained totally open. Apart from the initial goal of a complete account of mathematics, therefore, we might have sought at least a measure of safety, an axiomatization of mathematics that we could truly rely on. We would have wanted at the very least to know by some reliable finitary means that our axioms were not simply inconsistent.

7.1 The Hilbert program

Such were the hopes and goals of the *Hilbert program*, proposed in the early twentieth century by David Hilbert, one of the world's leading mathematical minds. To my way of thinking, these hopes and goals are extremely natural in the mathematical and historical context prior to Gödel. Hilbert expected, reflexively, that mathematical questions have answers that we can come to know. At his retirement address, Hilbert (1930) proclaimed:

> *Wir müssen wissen. Wir werden wissen.* (We must know. We will know.)

Thus, Hilbert expressed completeness as a mathematical goal. We want our best mathematical theories ultimately to answer all the troubling questions. Hilbert wanted to use the unifying foundational theories, including set theory, but he also wanted to use these higher systems with the knowledge that it is safe to do so. In light of the antinomies, Hilbert proposed that we should place our mathematical reasoning on a more secure foundation, by providing specific axiomatizations and then proving, by completely transparent finitary means, that those axiomatizations are consistent and will not lead us to contradiction.

Formalism

Hilbert outlined his vision of how we might do this. He proposed that we should view the process of proving theorems, making deductions from axioms, as a kind of formal mathematical game—a game in which mathematical practice consists ultimately of manipulating strings of mathematical symbols in conformance with the rules of the system. We need not encumber our mathematical ontology with uncountable (or even infinite) sets, just because our mathematical assertions speak of them; rather, let us instead consider mathematics merely as the process of formulating and working with those assertions as finite sequences of symbols. Inherent in the Hilbert program, and one of its most important contributions, is the idea that the entire mathematical enterprise, viewed as a formal game in a formal axiomatic system, may itself become the focus of metamathematical investigation, which he had hoped to undertake by entirely finitary means.

According to the philosophical position known as *formalism*, this game is indeed all that there is to mathematics. From this view, mathematical assertions have no meaning; there are no mathematical objects, no uncountable sets, and no infinite functions. According to the formalist, the mathematical assertions we make are not *about* anything. Rather, they are meaningless strings of symbols, manipulated according to the rules of our formal system. Our mathematical theorems are deductions that we have generated from the axioms by following the inference rules of our system. We are playing the game of mathematics.

One need not be a formalist, of course, to analyze a formal system. One can fruitfully study a formal system and its deductions, even when one also thinks that those mathematical assertions have meaning—a semantics that connects assertions with the corresponding properties of a mathematical structure that the assertion is about. Indeed, Hilbert applies his formalist conception principally only to the infinitary theory, finding questions of existence for infinitary objects to be essentially about the provability of the existence of those objects in the infinitary theory. Meanwhile, with a hybrid view, Hilbert regards the finitary theory as having a realist character with a real mathematical meaning.

The Hilbert program has two goals, seeking both a *complete* axiomatization of mathematics, one which will settle every question in mathematics, and a proof using strictly finitary means to analyze the formal aspects of the theory that the axiomatization is reliable. Hilbert proposed that we consider our possibly infinitary foundation theory T, perhaps set theory, but we should hold it momentarily at arm's length, with a measure of distrust; we should proceed to analyze it from the perspective of a completely reliable finitary theory F, a theory concerned only with finite mathematics, which is sufficient to analyze the formal assertions of T as finite strings of symbols. Hilbert hoped that we might regain trust in the infinitary theory by proving, within the finitary theory F, that the larger theory T will never lead to contradiction. In other words, we hope to prove in F that T is consistent. Craig Smoryński (1977) argues that Hilbert sought more, to prove in F not only that T is consistent, but that T is conservative over F for finitary assertions, meaning that any finitary assertion provable in T should be already provable in F. That would be a robust sense in which we might freely use the larger theory T, while remaining confident that our finitary conclusions could have been established by purely finitary means in the theory F.

Life in the world imagined by Hilbert

Let us suppose for a moment that Hilbert is right—that we are able to succeed with Hilbert's program by finding a complete axiomatization of the fundamental truths of mathematics; we would write down a list of true fundamental axioms, resulting in a complete theory T, which decides every mathematical statement in its realm. Having such a complete theory, let us next imagine that we systematically generate all possible proofs in our formal system, using all the axioms and applying all the rules of inference, in all possible combinations. By this rote procedure, we will begin systematically to enumerate the theorems of T in a list, $\varphi_0, \varphi_1, \ldots$, a list which includes all and only the theorems of our theory T. Since this theory is true and complete, we are therefore enumerating by this procedure all and only the true statements of mathematics. Churning out theorems by rote, we could learn whether a given statement φ was true or not, simply by waiting for φ or $\neg\varphi$ to appear, complete and reliable. In the world imagined by the Hilbert program, therefore, mathematical activity could ultimately consist of turning the crank of this theorem-enumeration machine.

The alternative

However, if Hilbert is wrong, if there is no such complete axiomatization, then every potential axiomatization of mathematics that we can describe would either be incomplete or include false statements. In this scenario, therefore, the mathematical process leads inevitably to the essentially creative or philosophical activity of deciding on additional axioms. At the same time, in this situation, we must inevitably admit a degree of uncertainty, or even doubt, concerning the legitimacy of the axiomatic choices we had made, precisely because our systems will be incomplete and we will be unsure about how to extend them, and furthermore, because we will be unable to prove even their consistency in our finitary base theory. In the non-Hilbert world, therefore, mathematics appears to be an inherently unfinished project, perhaps beset with creative choices, but also with debate and uncertainty concerning those choices.

Which vision is correct?

Gödel's incompleteness theorems are bombs exploding at the very center of the Hilbert program, decisively and entirely refuting it. The incompleteness theorems show, first, that we cannot in principle enumerate a complete axiomatization of the truths of elementary mathematics, even in the context of arithmetic, and second, no sufficient axiomatization can prove its own consistency, let alone the consistency of a much stronger system. Hilbert's world is a mirage.

Meanwhile, in certain restricted mathematical contexts, a reduced version of Hilbert's vision survives, since some naturally occurring and important mathematical theories are decidable. Tarski proved, for example, that the theory of real-closed fields is decidable, and from this (as discussed in chapter 4), it follows that the elementary theory of geometry is decidable. Additionally, several other mathematical theories, such as the theory of abelian groups, the theory of endless, dense linear orders, the theory of Boolean algebras, and many others, are decidable. For each of these theories, we have a theorem-enumeration algorithm; by turning the crank of the mathematical machine, we can in principle come to know all and only the truths in each of these mathematical realms.

But these decidable realms necessarily exclude huge parts of mathematics, and one cannot accommodate even a modest theory of arithmetic into a decidable theory. To prove a theory decidable is to prove the essential weakness of the theory, for a decidable theory is necessarily incapable of expressing elementary arithmetic concepts. In particular, a decidable theory cannot serve as a foundation of mathematics; there will be entire parts of mathematics that one will not be able to express within it.

7.2 The first incompleteness theorem

For most of the rest of this chapter, I should like to explain the proof of the incompleteness theorems in various formulations.

The first incompleteness theorem, via computability

Let us begin with a simple version of the first incompleteness theorem. To emphasize the arc of the overall argument, I shall mention "elementary" mathematics, by which I mean simply a mathematical theory capable of formalizing finite combinatorial reasoning, such as arithmetic or the operation of Turing machines. But for the moment, let me be vague about exactly what counts as part of elementary mathematics. Later on, we shall discuss the concept of arithmetization, which shows that in fact arithmetic itself is powerful enough to serve as a foundation for all such finite combinatorial mathematics.

Theorem 11 (First incompleteness theorem, variation, Gödel). *There is no computably enumerable list of axioms that prove all and only the true statements of elementary mathematics.*

By the end of the chapter, we shall have given several distinct and qualitatively different proofs of Gödel's theorem, including a version of Gödel's original proof, a strengthening due to John Barkley Rosser Sr., and another due to Alfred Tarski. This first soft argument, however, is due essentially to Turing, amounting to his solution of the Entscheidungsproblem in Turing (1936) based on the undecidability of the halting problem.

To prove the theorem, suppose toward contradiction that we could enumerate a list of axioms for elementary mathematics, which prove all and only the true statements of elementary mathematics. By systematically applying all rules of the proof system, we can computably enumerate all the theorems of our axioms, which by assumption will include all and only the truths of elementary mathematics. Given any Turing-machine program p and input n, let us turn the crank of the theorem-enumeration machine and simply wait for the assertion "p halts on n" or the assertion "p does not halt on n" to appear. Since our system is sound and complete, exactly one of these statements will appear on the list of theorems—the true one—and we will thereby come to know the answer to this instance of the halting problem. So if there were a complete axiomatization of the truths of elementary mathematics, then there would be a computable procedure to solve the halting problem. But there is no such computable procedure to solve the halting problem. Therefore, there can be no complete axiomatization of the truths of elementary mathematics. QED

The proof that we just gave actually proves a slightly stronger, more particular result. Namely, for any computably axiomatized true theory T, there must be a particular Turing-machine program p and particular input n such that p does not halt on n, but the theory T is unable to prove this. The reason for this is that we do not need the theorem-enumerating machine to observe positive instances of halting; these we can discover simply by running the programs and observing when they halt. We need the theorem-enumeration method only to discover instances of nonhalting. So, given any Turing-machine program p with input n and given any computably axiomatized true theory T, let us follow the procedure, by day, of simulating the program p on input n, waiting for it to halt, and by night, searching for a proof in the theory T that p does not halt on input n. If all true instances of nonhalting were provable in T, then this procedure would solve the halting problem. Since the halting problem is not computably decidable, therefore, there must be some nonhalting instances that are not provably nonhalting. Thus, every computably axiomatizable true theory must admit true but unprovable statements.

The Entscheidungsproblem

Before Gödel proved his incompleteness theorems, Hilbert and Ackermann (2008 [1928]) had considered the problem of determining whether a given assertion φ is logically *valid*, that is, whether φ is true in all models. The specific challenge was to find a computable procedure that would correctly answer yes or no to all such validity inquiries. This problem has become known as the Entscheidungsproblem (German for "decision problem"), although of course we now see thousands of decision problems arising in every area of mathematics.

Hilbert had expected a positive result, a computable solution to his Entscheidungsproblem, and perhaps we can imagine what such a positive solution might look like. Namely, someone would produce a specific computational algorithm and then prove that it correctly determined validity. But what would a negative resolution of the Entscheidungsproblem look like? To show that there is no computational procedure solving a certain decision problem, one would seem to need to know what counts as a computational procedure. But this was lacking at the time of the Entscheidungsproblem, which was posed prior to the development of rigorous notions of computability. Indeed, the Entscheidungsproblem is part of the historical context from which the central ideas of computability arose. The problem was ultimately solved, negatively, by Turing (1936) and Church (1936a, b), who both specifically invented notions of computability in order to do so.

I find it natural to compare the validity problem to several other related decision problems. Namely, an assertion φ is *provable* in a specific proof system if there is a proof of φ in that system; in any sound and complete proof system, of course, this is exactly the same as the validity problem; an assertion φ is *satisfiable*, in contrast, if it is true in some model; a finite set of sentences is *consistent* if they can prove no contradiction; and two assertions φ and ψ are *logically equivalent* if they have the same truth value in all models. How is the validity problem related to the satisfiability problem, the consistency problem, and the logical-equivalence problem? Are they all computably decidable? Are they computably equivalent to each other?

We have already mentioned that the validity problem is the same as the provability problem, since $T \models \varphi$ if and only if $T \vdash \varphi$. Similarly, by the completeness theorem, the consistency of a sentence is equivalent to its satisfiability, and the satisfiability of a finite list of sentences $\sigma_0, \ldots, \sigma_n$ is equivalent to the satisfiability of the conjunction $\sigma_0 \wedge \cdots \wedge \sigma_n$. Meanwhile, we may easily reduce the validity problem to the logical equivalence problem, since φ is logically valid if and only if it is logically equivalent to a statement, such as $\forall x \, x = x$, which we know is logically valid. And conversely, we may reduce the logical equivalence problem to the validity problem, since φ and ψ are logically equivalent exactly when $\varphi \leftrightarrow \psi$ is a logical validity. Similarly, we may reduce the validity problem to the satisfiability problem, or rather, to the nonsatisfiability problem, since φ is valid exactly when $\neg\varphi$ is not satisfiable. So if we had an oracle for satisfiability, we could correctly answer all inquiries about validity. And we may conversely reduce the satisfiability problem to the validity problem, since ψ is satisfiable if and only if $\neg\psi$ is not valid. Thus, these decision problems are all computably reducible to one another, and in this sense, they are equally difficult computationally; they are Turing equivalent, with the same Turing degree.

Furthermore, it turns out that all these decision problems are computably undecidable, and this is precisely the negative resolution of the Entscheidungsproblem that we have mentioned. To see this result, let us argue, as Turing did, that the halting problem reduces to the validity problem. Given a Turing-machine program p, we want to determine whether or not p halts when run on input 0. We can develop a formal language concerned with the operation of Turing machines. For example, we could introduce predicates that describe the contents of the cells of the machine and the location and state of the head at any given time; and then we could formulate axioms expressing the idea that time proceeds in discrete steps, that the machine started in the input configuration, and that it follows the program instructions from each moment of time to the next, stopping only when a *halt* state is achieved. All this is expressible in any language capable of expressing elementary combinatorial mathematics. Let φ be the assertion that if all of this holds, then "p halts," by which we mean the assertion expressed in our formal language that a halting output configuration is attained. If the program p really does halt, then indeed φ will be valid, since in any model in which the operation of Turing machines is properly encoded, the discrete sequence of computational configurations of program p will inevitably lead to the halting configuration, making φ true. Conversely, if p does not halt, then one may build a model consisting of the unending sequence of actual machine configurations, and this will make φ false. So p halts if and only if φ is valid, and we have thereby reduced the halting problem to the validity problem. Since the halting problem is not computably decidable, it follows that the validity problem also is not computably decidable, and thus, Turing has showed the undecidability of the Entscheidungsproblem.

One can also argue that the halting problem reduces to the validity problem, not by inventing a special language describing the operation of Turing machines, but rather by encoding the operation of Turing machines into arithmetic, using Gödel's idea of arithmetization (discussed later in this chapter). This will reduce the halting problem to questions of validity, provability, satisfiability, and logical equivalence, specifically for arithmetic assertions.

Theorem 12. *There are no computable procedures to determine with certainty yes-or-no whether a given arithmetic assertion is valid, whether it is provable, whether it is satisfiable, or whether two arithmetic assertions are logically equivalent.*

One should see this resolution of the Entscheidungsproblem as very closely related to the computability proof given earlier for the first incompleteness theorem. That proof had proceeded essentially by showing that the *true* arithmetic statements are not computably decidable by reducing the halting problem to this; this proof does the same for the validity decision problem, with essentially the same proof method.

Incompleteness, via diophantine equations

Here is another striking version of the first incompleteness theorem:

Theorem 13. *For any computably axiomatized true theory of arithmetic T, there is a polynomial $p(x, y, z, \dots)$ in several variables, with integer coefficients, such that the equation*

$$p(x, y, z, \dots) = 0$$

has no solution in the integers, but this is not provable in T.

This is a consequence of what has come to be known as the *MRDP theorem*, proved by (and named for) Yuri Matiyasevich, Julia Robinson, Martin Davis, and Hilary Putnam, who solved one of the outstanding open problems announced by Hilbert at the turn of the twentieth century—Hilbert's 10th problem—on his list of now-famous problems intended to guide research into the rest of the century. Hilbert had challenged the mathematical community to describe an algorithm that correctly determines whether a given polynomial equation $p(x, y, z, \dots) = 0$, in several variables with integer coefficients, has a solution in the integers or not. I find it interesting to note that in his way of phrasing the question—he asks us to present the algorithm—he seems to presume that indeed there is such an algorithm. In his way of asking the question, Hilbert reveals his expectation that mathematical problems have answers that we can come to know.

However, the shocking answer provided by the MRDP theorem is that there is no such algorithm. To belabor the point, the situation is not that we do not know of any such algorithm; rather, we can prove that it is logically contradictory for there to be such an algorithm. The MRDP theorem shows that for every existential assertion of arithmetic, there is a certain polynomial $p(e, n, x_1, \ldots, x_k)$, such that $p(e, n, x_1, \ldots, x_k) = 0$, if and only if $(x_1, \ldots, x_k)$ codes a witness for the truth of the existential assertion. In particular, since the halting problem is an existential assertion—a program e halts on input n if and only if there is a number encoding the entire halting computation—it follows that they will have reduced the halting problem to the diophantine equation problem. So the question of whether e halts on input n is equivalent to the question of whether $p(e, n, x_1, \ldots, x_k) = 0$ has an integer solution.

Arithmetization

Let us now make the move to a more precise context by considering the standard structure of arithmetic $\langle \mathbb{N}, +, \cdot, 0, 1, < \rangle$. One of Gödel's key insights was the concept of *arithmetization*, an idea underlying the entire incompleteness result. To my way of thinking, arithmetization is a truly profound idea—one of Gödel's deep insights—but it is also an instance of the phenomenon whereby a profound idea becomes mundane through familiarity. The idea of arithmetization has permeated essentially all our thinking on the topic, embedding itself in the mathematical and computer science culture and even penetrating the popular culture, to such an extent that it now seems routine.

Let me explain. *Arithmetization* consists of the idea that essentially, any finite mathematical structure can be encoded by numbers in such a way that the combinatorial reasoning and processes that we might like to employ with that structure become arithmetic manipulations on the corresponding numbers encoding those structures. Each of us is familiar, for example, with the fact that when we type our documents into a computer, the information is stored ultimately in the computer's memory. While composing our document, we naturally conceive of the document in high-level terms; it consists of chapters and paragraphs, perhaps written in certain fonts and augmented with certain spacing information. These components in turn consist of words, and ultimately individual characters, supplemented with control codes for the fonts and spacing; ultimately, each piece of information is represented simply by a number, such as in the extended ASCII system, which has 256 distinct symbols and control characters, that is, 2^8, so that each character can be represented in binary by exactly eight digits of 0 or 1. (Current systems tend to be based on Unicode, which allows for a far larger symbol set.) Thus, the entire document is represented as a gigantic sequence of 0s and 1s, with each bit represented via the presence of low or high voltage, respectively, in certain transistors on the circuit board. In this way, the entire document can be thought of as a single enormous binary number, which might be called the *Gödel code* of the entire document. The idea of using numbers to encode finite assemblages of various kinds of general information is the essence of arithmetization. Writing a document on your computer is thus an instance of arithmetization.

In the context of first-order logic, we can give each symbol in the language of arithmetic an ASCII-style code, and then view each assertion φ in this language as having a Gödel code $\ulcorner\varphi\urcorner$, a number that represents that formula with respect to this encoding. All the syntactical features of first-order logic, such as well formed formulas, free variables, terms, sentences, tautologies, proofs, and so on, deal ultimately with finite sequences of symbols, which can all be coded into arithmetic by means of this Gödel coding.

In this way, the statements of arithmetic, which talk about numbers, can be viewed as talking about the statements of arithmetic themselves and the syntactic features of those statements, including provability. An arithmetic statement might be able to make assertions even about itself, and thus, a capacity for self-reference sneaks into the language of arithmetic. In particular, we can construct formulas in the language of arithmetic that express the following concepts:

> x is the Gödel code of a well formed formula in the language of arithmetic.
>
> x is the Gödel code of an axiom of PA.
>
> x is the Gödel code of an axiom in the set of axioms whose Gödel code is enumerated by the Turing-machine program having Gödel code e.

And so on. If T is a computably enumerable theory in the language of arithmetic, then $\mathrm{Pr}_T(x)$ is the assertion, "x is the Gödel code of a theorem of T." And $\mathrm{Con}(T)$ is the assertion "T is consistent," or in other words, $\neg\,\mathrm{Pr}_T(\ulcorner 0 \neq 0\urcorner)$. The technique of arithmetization shows that all these concepts are expressible in the language of arithmetic.

One may undertake such a Gödel-coding process for essentially any finite combinatorial process. For example, we may represent a Turing-machine program e, for example, by the Gödel code number $\ulcorner e\urcorner$, which encodes all the information that we need to run this program. We may similarly find Gödel codes for "snapshots" of Turing-machine computations, which encode the tape content, head position, program, and current state of a Turing-machine computation. The computation itself is simply a sequence of such snapshots, with the property that each snapshot is related to the next in a way that accords with the operation of the program, and whose initial snapshot correctly shows the starting configuration of a computation. Arithmetization thus shows that all the basic operations and features of Turing-machine computation can be expressed in the language of arithmetic, using Gödel codes to represent the computations, and these can be manipulated via the basic arithmetic. For example, there is a formula $\varphi(e, n)$ in the language of arithmetic whose meaning asserts, "e is the code of a Turing-machine program that halts on input n." The point is that although $\varphi(e, n)$ is on its face an assertion in the language of arithmetic about the numbers e and n, nevertheless the meaning of the sentence concerns the operation of a certain Turing machine. Thus, arithmetization shows the power of arithmetic to express essentially arbitrary finite combinatorial mathematical concepts.

We face difficult philosophical issues concerning the veracity of the translation. To what extent does the arithmetic encoding of a mathematical assertion preserve the meaning of that assertion? We do not object, after all, to an author writing his or her great novel on a computer, by pointing out that it is not a novel at all, but merely a long, boring sequence of 0s and 1s, with no discernible plot or characters or story line. Feferman (1960) considers the faithfulness of an arithmetic encoding to our metamathematical intension, and Moschovakis (2006) discusses the concept of "faithful representation" in mathematical foundations, particularly set theory. Similar issues arise regularly in reverse mathematics, where one interprets mathematical structure in second-order arithmetic. The worry is that the encodings of a mathematical idea miss out on important features of those ideas.

A complete proof of Gödel's incompleteness theorem, of course, would include checking that the details of these arithmetization representations are in fact sound, and traditional accounts of Gödel's theorem would typically spend considerable time on that. But let us not dwell inordinately on those details, which by themselves offer very little mathematical insight, typically involving numerous arbitrary choices about how to implement the encoding. The fact of arithmetization is that essentially, any precise finite combinatorial process can be coded with numbers in a way that the essential features of that process are expressible in the language of arithmetic, just as essentially, any written document can in principle be saved on our computer as a sequence of 0s and 1s. This is simply no longer surprising—the details of it are unimportant; the profound idea of arithmetization has thus become mundane.

First incompleteness theorem, via Gödel sentence

Let me now give a second proof of the first incompleteness theorem, which is a little more in line with the original proof given by Gödel. We shall take as our base theory the first-order theory of Peano arithmetic (PA), discussed in chapter 1. Roman Kossak defines the theory PA with the slogan, "All the arithmetic principles you can think of in ten minutes, plus induction." Nearly everyone will think of associativity, commutativity, distributivity, additive identity, multiplicative identity, and so on, but the point of the slogan is that minor differences in these axioms are of no consequence—they will all be equivalent—provided that one has the all-important induction axiom, which is the scheme asserting that if an arithmetic statement φ is true at 0, and its truth is preserved from every number n to the next number $n + 1$, then it is true of all numbers:

$$\left[\varphi(0) \wedge \forall n\, (\varphi(n) \rightarrow \varphi(n + 1))\right] \rightarrow \forall n\, \varphi(n).$$

The theory PA is quite strong, and one can undertake essentially all the classical development of number theory within PA. Indeed, to all appearances, PA might seem to be a complete axiomatization of arithmetic, and this would have been a natural candidate with which one might have hoped to realize the Hilbert program. But it is not complete, of course, precisely because of the incompleteness theorem.

Let us begin with an enigma—the fixed-point lemma, a mathematical mystery, a logical labyrinth that shows how self-reference, the stuff of nonsense and confusion, sneaks explicitly into our beautiful number theory. It continually amazes me.

Lemma 14 (The fixed-point lemma). *For any formula $\varphi(x)$ with one free variable, there is a sentence ψ such that*

$$PA \vdash \psi \leftrightarrow \varphi(\ulcorner \psi \urcorner).$$

Thus, ψ asserts that "φ holds of my Gödel code."

Proof. Let sub be the substitution function, defined such that

$$\text{sub}(\ulcorner \varphi(x) \urcorner, m) = \ulcorner \varphi(\underline{m}) \urcorner,$$

where $\underline{m}$ is the syntactic term $1 + \cdots + 1$, with m many 1s. The sub function is a primitive recursive function, and it is representable in the language of arithmetic in accordance with the principle of arithmetization. Consider now any formula $\varphi(x)$ with one free variable x. Let $\theta(x) = \varphi(\text{sub}(x, x))$, and let $n = \ulcorner \theta(x) \urcorner$. Finally, let $\psi = \theta(\underline{n})$, which is a sentence in the language of arithmetic. Putting all this together, we observe in PA the following equivalences:

$$
\begin{aligned}
\psi \quad &\leftrightarrow \quad \theta(\underline{n}) \\
&\leftrightarrow \quad \varphi(\text{sub}(n, n)) \\
&\leftrightarrow \quad \varphi(\text{sub}(\ulcorner \theta(x) \urcorner, n)) \\
&\leftrightarrow \quad \varphi(\ulcorner \theta(\underline{n}) \urcorner) \\
&\leftrightarrow \quad \varphi(\ulcorner \psi \urcorner).
\end{aligned}
$$

Thus, the sentence ψ has the desired fixed-point property. $\qquad\square$

Let me point out a subtle notational infelicity in the notation $\varphi(\ulcorner \psi \urcorner)$ appearing in the lemma. What is intended is $\varphi(\underline{k})$, where $k = \ulcorner \psi \urcorner$, of course, so that we substitute the syntactic numeral representation of the Gödel number, rather than the number itself.

With the fixed-point lemma, we can readily construct sentences that seem to refer to themselves. For example, if $P(x)$ asserts that x is the Gödel code of an axiom of PA, then a fixed point ϕ, where $\phi \leftrightarrow P(\ulcorner \phi \urcorner)$, can be read as the assertion, "This sentence is an axiom of PA." Is it true? No, since PA has no axioms of exactly that form. If $S(x)$ asserts that x is the Gödel code of a sentence, then a fixed point σ, where $\sigma \leftrightarrow S(\ulcorner \sigma \urcorner)$, can be read as the assertion, "This is a sentence." And indeed it is.

Similar ideas lie at the heart of the fixed-point proof of the incompleteness theorem. Suppose that T is a theory in the language of arithmetic, containing and possibly extending

PA, and that the axioms of T can be enumerated by a computable procedure. It follows by the arithmetization of syntax that the predicate $\mathrm{Pr}_T(x)$, asserting that x is the Gödel code of a sentence that is provable from T, is expressible in the language of arithmetic. By the fixed-point lemma, therefore, there is a sentence ψ, known as the *Gödel sentence*, such that

$$\mathrm{PA} \vdash \psi \leftrightarrow \neg\mathrm{Pr}_T(\ulcorner \psi \urcorner).$$

If you reflect upon this situation carefully, you will come to the amazing conclusion that this sentence ψ asserts exactly its own unprovability. The Gödel sentence ψ asserts, "This sentence is not provable in T."

Perhaps you have heard of the Liar paradox, the sentence "This sentence is false," which asserts its own falsity. The Liar sentence, it seems, cannot be true, for then it would also be false; and it cannot be false, for then it would also be true. So which is it? It is the Liar paradox. The Gödel sentence ψ is not the Liar, but a cousin of it, using provability in place of truth—or, more accurately, using unprovability in place of falsity. This is a difference that matters, since although the Liar sentence can be formulated easily in natural language, the semantic notions of truth and falsity are not subject to arithmetization, and we seem unable meaningfully to express the Liar sentence in the language of arithmetic. Indeed, it must be impossible to do so, since the paradox would turn into outright contradiction, as all arithmetic assertions have truth values.

The Gödel sentence, in contrast, is expressible in arithmetic because it replaces the semantic truth concept with its syntactic analogue, provability. Since provability is amenable to arithmetization, Gödel is able to construct his sentence, "This sentence is not provable in T." This is a sentence in the formal language of arithmetic, a mathematical statement like any other in the realm of arithmetic. Thus, Gödel taps the strange logic of the Liar. Whereas the Liar sentence leads to a paradox or outright contradiction, the Gödel sentence leads instead exactly to the conclusions of the incompleteness theorem.

Namely, we prove the incompleteness theorem as follows: Assume that T is a consistent, computably axiomatizable theory of arithmetic extending PA. If the Gödel sentence ψ were provable in T, then by inspecting the proof, we could also prove that ψ was provable. That is, T would prove $\mathrm{Pr}_T(\ulcorner \psi \urcorner)$. But this is provably equivalent to $\neg\psi$, and so T will have proved both ψ and $\neg\psi$, thereby revealing inconsistency, contrary to assumption. So T does not prove ψ. But this is precisely what ψ itself asserts. So ψ is a true sentence that is not provable in T. We have therefore established the following version of the first incompleteness theorem.

Theorem 15 (First incompleteness theorem, Gödel, 1931). *Every consistent, computably axiomatizable theory of arithmetic admits true but unprovable statements.*

This refutes the first portion of the Hilbert program, for it shows that there is no way we can describe a complete axiomatization of true arithmetic by any concrete computable procedure.

7.3 Second incompleteness theorem

A 1930 conference in Königsberg was the setting of dramatic developments in the foundations of mathematics. Gödel presented his completeness theorem, providing a key component of the Hilbert program. At a round-table discussion the next day, however, Gödel also announced his first incompleteness theorem, which amounted to a decisive refutation of the Hilbert program. Nevertheless, the following day, Hilbert gave his retirement address, concluding with triumphant optimism for his program, "We must know. We will know." John von Neumann, in attendance at Gödel's announcement of the first incompleteness theorem, spoke with Gödel afterward and inquired whether the theory would also fail to prove its own consistency. Gödel had only the first incompleteness theorem at that time. Returning home, von Neumann worked out the second incompleteness theorem on his own and wrote to Gödel about it, but by that time, Gödel also had proved it, and it appears in the same paper as his first incompleteness theorem.

Theorem 16 (Second incompleteness theorem, Gödel, 1931). *If T is a consistent, computably axiomatizable theory extending PA, then T does not prove its own consistency.*

$$T \nvdash Con(T).$$

The main idea behind the proof of the second incompleteness theorem is to formalize within PA the proof of the first incompleteness theorem. Namely, we had proved the implication: if T is consistent, then T does not prove the Gödel sentence ψ. Formalizing this within PA, we establish $PA \vdash Con(T) \rightarrow \neg Pr_T(\ulcorner\psi\urcorner)$. Since PA is included within T and ψ is PA-provably equivalent to $\neg Pr_T(\ulcorner\psi\urcorner)$, this amounts to $T \vdash Con(T) \rightarrow \psi$. Consequently, T does not prove $Con(T)$, for if it did, then it would prove ψ, which it does not. Thus, we have proved the second incompleteness theorem.

Löb proof conditions

However, let us flesh out in somewhat greater detail the key initial step of that argument, where we formalize the proof of the first incompleteness theorem within PA. One can appeal to what are called the *Löb proof conditions*, named after Martin Hugo Löb, and also known as the *Hilbert-Bernays-Löb derivability conditions*:

D1: If $T \vdash \varphi$, then $PA \vdash Pr_T(\ulcorner\varphi\urcorner)$.

> If the theory T proves a sentence, then PA proves that the theory proves it.

D2: $PA \vdash Pr_T(\ulcorner\varphi\urcorner) \rightarrow Pr_T(\ulcorner Pr_T(\ulcorner\varphi\urcorner)\urcorner)$.

> The theory PA proves, in each instance, that a provable sentence is provably provable.

D3: $PA \vdash Pr_T(\ulcorner\varphi\urcorner) \wedge Pr_T(\ulcorner\varphi \rightarrow \sigma\urcorner) \rightarrow Pr_T(\ulcorner\sigma\urcorner)$.

> The theory PA proves, in each instance, that the provable sentences are closed under modus ponens.

Thus, we distinguish sharply between the metatheoretic provability relation $T \vdash \varphi$ and its arithmetical representation $\mathrm{Pr}_T(\ulcorner\varphi\urcorner)$, and the derivability conditions express certain features of how these concepts of provability are related. Each of the properties is quite reasonable and not difficult to establish. In the case of D1, for example, if $T \vdash \varphi$, then there is an actual proof of φ from T, and that proof has a Gödel code, which can be verified as the Gödel code of a proof of φ in T; the existence of such a code can then be used to prove that $\mathrm{Pr}_T(\ulcorner\varphi\urcorner)$. Similarly, D2 is established by proving in PA that if you have a proof of φ, then you can prove that it is indeed a proof. For D3, it is a simple matter of proving that if you can prove φ and you can prove $\varphi \to \sigma$, then you can prove σ simply by concatenating the proofs and adding one more instance of modus ponens.

The point is that the derivability conditions express all that we happen to need to know about the provability predicate in order to establish the second incompleteness theorem. Specifically, by the fixed-point lemma, we have the sentence ψ for which $T \vdash \psi \leftrightarrow \neg\,\mathrm{Pr}_T(\ulcorner\psi\urcorner)$. From this, it follows by D1 that $T \vdash \mathrm{Pr}_T(\ulcorner\psi \to \neg\,\mathrm{Pr}_T(\ulcorner\psi\urcorner)\urcorner)$. Thus, by D3, we know $T \vdash \mathrm{Pr}_T(\ulcorner\psi\urcorner) \to \mathrm{Pr}_T(\ulcorner\neg\,\mathrm{Pr}_T(\ulcorner\psi\urcorner)\urcorner)$. But by D2, we get $T \vdash \mathrm{Pr}_T(\ulcorner\psi\urcorner) \to \mathrm{Pr}_T(\ulcorner\mathrm{Pr}_T(\ulcorner\psi\urcorner)\urcorner)$. A few more applications of D3 and some elementary logic leads to the conclusion that $T \vdash \mathrm{Pr}_T(\ulcorner\psi\urcorner) \to \mathrm{Pr}_T(\ulcorner\mathrm{Pr}_T(\ulcorner\psi\urcorner) \wedge \neg\,\mathrm{Pr}_T(\ulcorner\psi\urcorner)\urcorner)$, which implies $T \vdash \mathrm{Pr}_T(\ulcorner\psi\urcorner) \to \mathrm{Pr}_T(\ulcorner\bot\urcorner)$. So $T \vdash \mathrm{Con}(T) \to \psi$, and therefore $T \nvdash \mathrm{Con}(T)$, as desired.

Actually, the Gödel sentence ψ is provably equivalent to $\mathrm{Con}(T)$, since ψ asserts an instance of nonprovability (its own) and any instance of nonprovability implies consistency. Thus, $T \vdash \psi \leftrightarrow \mathrm{Con}(T)$.

I once heard logician Jack Silver object to the significance of the derivability conditions, on the grounds that the predicate "is a formula" also satisfies the derivability conditions. That is, if we were to interpret $\mathrm{Pr}_T(x)$ as "x is the Gödel code of a well formed formula," then indeed we will satisfy all the derivability conditions. In the case of D1, for example, PA proves that $\ulcorner\varphi\urcorner$ is the Gödel code of a formula, and we do not even need to use the hypothesis that $T \vdash \varphi$. A similar analysis works for the other conditions.

What exactly is the criticism? Notice that the conclusion of the second incompleteness theorem does hold if we were to reinterpret the provability predicate instead by the predicate "is a formula." After all, the theory T does not prove that a contradiction is not a formula. So Silver's criticism is not aimed against the use of the derivability conditions in the proof of the second incompleteness theorem, which seems completely fine. Rather, Silver is criticizing the idea that the derivability conditions somehow sum up what it means to be a proof predicate and that those conditions fully capture the fundamental properties that one wants to see in a proof predicate. Silver's counterexample shows that they are much too weak for this. So the significance of the derivability conditions is not that they axiomatize and fully express what it means to be a proof predicate, but rather that they express certain weak properties that we expect to hold of any provability predicate—properties that are nevertheless sufficient for proving the second incompleteness theorem.

Provability logic

It is quite natural to study provability as a modal operator, for provability is surely a form of necessity. Thus, we introduce the notation $\Box \varphi$ to mean that φ is provable within some fixed axiomatic framework such as PA. One sometimes sees subscripts to denote the particular system, such as $\Box_{PA} \varphi$, and this can be seen simply as an alternative modal notation for the provability predicate $\mathrm{Pr}_{PA}(\ulcorner\varphi\urcorner)$ that we considered earlier. The modal perspective, however, focuses our attention on important aspects of how provability relates to truth, and it allows us to express features of the provability predicate in a particularly clear manner. For example, the derivability conditions can be expressed like this:

D1: If $\vdash \varphi$, then $\vdash \Box \varphi$.

D2: $\vdash \Box \varphi \to \Box\Box \varphi$.

D3: $\vdash \Box(\varphi \to \psi) \to (\Box \varphi \to \Box \psi)$.

The content of Löb's theorem (section 7.10) is then expressed by the Gödel-Löb axiom:

$$\Box(\Box \varphi \to \varphi) \to \Box \varphi.$$

The resulting provability logic, GL, is intensely studied.

7.4   Gödel-Rosser incompleteness theorem

Another way to state Gödel's first incompleteness theorem is as the assertion: Every computably axiomatizable true theory of arithmetic is incomplete. We know that a consistent, computably axiomatizable theory T does not prove the Gödel sentence ψ, which is a true sentence. But this falls short of incompleteness, since if we do not insist that T is a *true* theory, then perhaps T proves $\neg\psi$. Gödel had proved a slightly stronger version, where one assumes not that T is true, but only that it is *ω-consistent*, a technical condition on the theory that is stronger than consistency, but weaker than truth.

Meanwhile, Rosser had observed that one can drop these extra hypotheses on T completely by using a slightly different sentence. Rosser's idea is to use a fixed point not of the unprovability predicate, but rather of a more idiosyncratic predicate. Specifically, Rosser proposed to use a sentence ρ, asserting that "For every proof of ρ, there is a smaller proof of $\neg\rho$," that is, one with a smaller Gödel code. Such a sentence exists by an application of the fixed-point lemma. By means of this simple trick, Rosser avoids the need to assume in the incompleteness theorem that the theory T is true or that it is ω-consistent.

Theorem 17 (Gödel, Rosser). *Every consistent, computably axiomatizable T, containing PA, is incomplete.*

Proof. Consider the Rosser sentence ρ. If $T \vdash \rho$, then this proof has some finite Gödel code. And so, because of what ρ asserts, T must prove that there is a shorter proof of $\neg\rho$.

So there must really be such a proof, and so T is inconsistent, contrary to our assumption. Therefore, T does not prove ρ.

If, alternatively, $T \vdash \neg\rho$, then in particular, because of what $\neg\rho$ asserts, T proves that there is a proof of ρ with no smaller proof of $\neg\rho$. But since we have a proof of $\neg\rho$ of some definite finite size, it means that there must be an even smaller proof of ρ. So T is inconsistent again. Thus, T proves neither ρ nor proves $\neg\rho$. So it is incomplete. $\square$

The main point of this argument is that the extra hypotheses on the theory T that it is true or that it is ω-consistent completely fall away; all we need to know is that T is consistent. We do seem to lose the analogue of the second incompleteness theorem, however, in the Rosser context, because it is not generally true that $\mathrm{Con}(T)$ is independent of T. To see this, consider the theory $\mathrm{PA} + \neg\,\mathrm{Con(PA)}$. This theory is consistent, by the second incompleteness theorem, but it proves itself inconsistent. In other words: it is consistent with PA to assume that PA is inconsistent. Indeed, one can establish a very strange thing. Suppose that T is a computably axiomatizable theory extending PA. Then I claim that

$$T \text{ is inconsistent} \quad \text{if and only if} \quad T \vdash \mathrm{Con}(T).$$

That is, T is inconsistent if and only if T proves that T is consistent. (And no, there is no typo here.) The reason is that if T is inconsistent, then it proves everything, including the assertion $\mathrm{Con}(T)$; and conversely, if T is consistent, then the incompleteness theorem shows that it does not prove $\mathrm{Con}(T)$. So T is inconsistent if and only if T proves $\mathrm{Con}(T)$.

7.5 Tarski's theorem on the nondefinability of truth

Let us move on now to another generalization of the incompleteness theorem, due to Tarski.

Theorem 18 (Tarski). *There is no definable predicate $T(x)$ in the language of arithmetic, such that $\mathbb{N} \models \psi \leftrightarrow T(\ulcorner\psi\urcorner)$.*

In other words, arithmetic truth is not arithmetically definable.

Proof. This follows immediately from the fixed-point lemma, since there must be a sentence ψ such that $\mathrm{PA} \vdash \psi \leftrightarrow \neg T(\ulcorner\psi\urcorner)$. $\square$

Tarski actually proved a somewhat more subtle and sophisticated version of this theorem, establishing that there can be no formula $\varphi(x)$ whose satisfying instances provably fulfill the Tarskian recursive definition of truth, whether or not the predicate agrees with actual truth (although it will follow inductively that it does). It follows from Tarski's theorem that we may extend the first incompleteness theorem to the case where the theory T is not merely computably axiomatizable, but any arithmetically definably axiomatizable theory. This is truly a vast generalization of the theorem. Not only can we not computably enumerate a complete axiomatization of arithmetic, but we cannot describe in any arithmetical manner the complete theory of arithmetic.

Raymond Smullyan (1992) has argued that much of the attention and fascination accumulating around Gödel's theorems are more aptly directed to Tarski's theorem. Tarski's theorem does not rely on detailed arithmetization or coding of the proof system, but rather only on the fixed-point lemma, and it proves the much stronger result, identifying an inherent limitation on any formal language to express the concept of truth in that language.

In light of the fact that Tarski is credited with the disquotational theory of truth, including the recursive definition of truth in first-order logic for first-order structures, we find ourselves in the awkward position of asserting both that Tarski defined truth and that he proved that truth is not definable.

7.6   Feferman theories

Meanwhile, the second incompleteness theorem does not extend from computably axiomatized theories to arbitrary arithmetically defined theories, for there are arithmetically definable theories that do prove their own consistency.

To see this, consider the following theory, following an idea of Solomon Feferman (1960). Let us enumerate the axioms of PA, but with the proviso that we allow the next axiom onto the list only when the resulting theory remains consistent. Let us call this theory PA*. We have not given a computable axiomatization of PA*, since we showed in the previous chapter that there is no computable procedure to test a finite set of sentences for consistency. But we have given an arithmetically definable axiomatization of PA*, since consistency is arithmetically expressible. Notice that by its very nature, we can see that PA* is a consistent theory, since at no stage did we allow an axiom that would support an inconsistency. So we can prove in a very weak theory that PA* is consistent. In particular, this theory proves its own consistency.

Indeed, since it is known that PA proves that every finite subtheory of PA is consistent, it follows that PA* will include any given finite fragment of PA. Thus, provided that PA is consistent, it follows that PA and PA* are precisely the same theory, as sets of sentences. But the theory PA* is defined in such a way that PA* is an arithmetically definable theory that can prove its own consistency, even though it cannot prove the consistency of PA.

A similar idea, attributed to Steven Orey (see Feferman, 1960, theorem 7.6), shows that the consistency strength of a computably axiomatizable theory depends on the manner in which it is enumerated. To see this, consider any two computably axiomatizable consistent theories S and T. Fix any enumeration of the axioms of T, and consider the computable enumeration of the axioms of S by the following procedure: at stage n, enumerate the next axiom of S unless one has found a contradiction in T with a Gödel code below n, in which case one enumerates a contradiction into S. Since T is actually consistent, no such contradiction will actually be found, and so this is indeed a computable enumeration of S. But for this enumeration, we have PA $\vdash$ Con$(S) \rightarrow$ Con(T), because if no contradiction is found in S, there must not have been one in T, provably so. Thus, we were able to prove a consistency relation between two arbitrary theories. In light of this, one should not speak of the "consistency strength" of a *theory*—it is not well defined with respect to the theory—but only of the consistency strength of a specific enumeration of the theory. For example, it would be sensible to discuss ZFC plus large cardinals, if one understood that the enumeration procedure simply enumerated the axioms according to the schemata by which this theory is usually described.

7.7 Ubiquity of independence

The incompleteness theorems establish the ubiquity of independence in mathematics: every fundamental theory that we propose will admit statements that are not settled by the theory. If we start with the Peano theory PA, we are not able to prove Con(PA). What if we add this assertion to the theory? Let us simply adopt

$$PA + Con(PA)$$

as our fundamental theory. That would enable us to prove Con(PA), of course, but since the incompleteness theorem will also apply to this new theory, we will not be able to prove Con(PA + Con(PA)). But this is an assertion that we believe is true, and so we will want also to add it as an axiom. So we form the new theory

$$PA + Con(PA) + Con(PA + Con(PA)).$$

We will now want similarly to extend this theory with its consistency assertion, forming the theory

$$PA + Con(PA) + Con(PA + Con(PA)) + Con(PA + Con(PA) + Con(PA + Con(PA))).$$

 And so on, continuing the hierarchy by adding the consistency assertion of the current theory at each step. [Note that these theories can each be simplified to remove redundancy, since Con(PA + Con(PA)) implies Con(PA) and similarly in the higher cases.] Yet, no matter how long we iterate this, we will never arrive at a complete theory; the consistency assertion for the current theory will not be provable in that theory itself. Incompleteness is inevitable. Furthermore, we can iterate the process transfinitely, for example, making the following assertion: "The union of the theories obtained at all the finite stages is consistent." But still this theory will not prove its own consistency, and we can iterate still further by adding the consistency assertion for *that* theory, and so on, continuing far beyond ω. Ultimately, this line of reasoning leads one to consider the various ordinal denotation systems and the possibilities of arithmetic representations of ordinals in order to express the theories at those ordinal stages.

Tower of consistency strength

Thus, Gödel reveals a fundamental mathematical phenomenon, the underlying inherent transfinite tower of consistency strength looming above every sufficient mathematical theory. We move higher in the tower of consistency strength by adopting a stronger theory, which can prove the consistency of our previous weaker theory. The iterated consistency assertions, of course, rise in consistency strength in just this way; they are increasingly powerful true yet unprovable mathematical statements, proving the consistency of the previous theories.

But furthermore, we know of many other natural examples of statements rising in consistency strength, instantiating the transfinite tower of consistency strength that we knew must be there because of the incompleteness theorems. Perhaps the best examples of this arise with the large cardinal hierarchy (described in the next chapter), a vast tower of precise mathematical assertions concerning infinite cardinals of unimaginable size; these axioms are natural statements of infinite combinatorics, having nothing directly to do with consistency, or even arithmetic. Yet we now know that each small step up in the large cardinal hierarchy makes an enormous leap in the tower of consistency strength; a larger large cardinal generally implies not only the consistency of the smaller cardinals, but also vast transfinite iterations of consistency assertions over the smaller cardinals.

7.8   Reverse mathematics

The ubiquity of incompleteness and independence, of course, means that we shall always be faced with a hierarchy of mathematical theories; we shall inevitably need to strengthen our axioms to reach further mathematical truths. The subject of *reverse mathematics*, introduced by Harvey Friedman and greatly developed by Stephen Simpson (2009), seeks to analyze the hierarchy of theories needed to do what we have done so far, to prove the classical theorems of mathematics. Reverse mathematics begins with the observation that a huge part of mathematics can be carried out in a formal system of second-order number theory, a framework capable of interpreting essentially arbitrary countable mathematical objects and structure, including real numbers, sequences of real numbers, projectively definable functions on the real numbers, countable graphs, countable linear orders, countable groups, countable fields, and so on.

The various theorems of classical mathematics often require different axioms for their proofs, and reverse mathematics seeks to discover and classify exactly the axiomatic structure necessary in each case, finding for each theorem the optimal theory necessary to prove it. Reverse mathematics does this by undertaking mathematics in reverse: rather than merely proving the theorem from the axioms, in reverse mathematics, we seek to prove the axioms from the theorem as well, thereby finding the axiomatic system to which the theorem is equivalent over a very weak base theory. This theory is optimal, in that it is the weakest extension of the base theory capable of proving the theorem in question.

One might have expected the result of reverse mathematics to be a complicated mess of axiomatic systems, perhaps with a slightly different axiomatic framework for nearly every theorem. But on the contrary, the remarkable discovery of this subject is that the central theorems of classical mathematics fit, for the most part, into five large equivalence classes. Over a very weak base theory, it turns out that most well known theorems are equivalent to one of exactly five natural set-existence theories. Within each class, therefore, the theorems are not only equivalent to the axioms, but also to each other. To put it in slogan form, there are essentially only five core mathematical theorems up to provable equivalence over a weak base theory:[9]

RCA_0 The base theory, which makes set-existence assertions only for computable sets. This theory can prove basic arithmetic facts about the natural numbers, the integers, the rational number field, and the real number field, as well as the Baire category theorem, the intermediate value theorem, the Banach-Steinhaus theorem, the existence of algebraic closures of countable fields (but not uniqueness), and the existence and uniqueness of the real closure of a countable ordered field.

WKL_0 Weak König's lemma, which asserts that every infinite subtree of the full binary tree has an infinite path. Over the base theory, this is equivalent to the Heine-Borel theorem for the closed unit interval, the boundedness of continuous real functions on a closed interval, the uniform continuity of continuous real functions on a closed interval, the Brouwer fixed-point theorem, the separable Hahn-Banach theorem, the Jordan curve theorem, the Gödel completeness theorem for countable languages, and the uniqueness of algebraic closures for a countable field, among many other theorems.

ACA_0 Arithmetic comprehension, asserting set existence for any arithmetically definable properties of numbers, is equivalent to the sequential completeness of the real numbers, the Bolzano-Weierstrass theorem, Ascoli's theorem, the theorem that every countable vector space over $\mathbb{Q}$ has a basis, the theorem that every countable field has a transcendence base, the full König's lemma, and Ramsey's theorem, among many others.

ATR_0 Arithmetic transfinite recursion, asserting that arithmetically definable transfinite recursive definitions have solutions, is equivalent to well-order comparability, the perfect-set theorem, Lusin's separation theorem, and the principle of open determinacy for games in Baire space, among many others.

$\Pi_1^1\text{-}CA_0$ The comprehension axiom for Π_1^1-definable sets is equivalent over the base theory to the Cantor-Bendixson theorem, as well as many others.

Why choose these particular axiom systems? Are they natural? Simpson emphasizes that we may finesse this issue: the axioms are natural precisely because the results of reverse mathematics show that they are mathematically equivalent to the corresponding classical theorems, which are certainly natural.

In Gitman et al. (2020), my coauthors and I mounted a set-theoretic analogue of reverse mathematics in the context of second-order set theory, working over Gödel-Bernays

---

[9] Although the big-five phenomenon is firmly established for the bulk of core mathematics, we now also know about theorems that do not fit exactly into one of the big five, and a complex hierarchy has emerged.

set theory (but without the power-set axiom) as a base theory, a considerably stronger axiomatic framework. A rich hierarchy of natural axiomatizations and theorem equivalents has emerged. For example, the class forcing theorem (the assertion that every class forcing notion admits a forcing relation) is equivalent to the principle of elementary transfinite recursion, to the existence of Ord-iterated truth predicates, to the existence of clopen determinacy for proper class games of rank Ord+1, and to the existence of Boolean set-completions of class Boolean algebras, among others.

What is it that we learn about a mathematical theory when we place it into the reverse mathematics hierarchy? By adopting such a weak base theory, was it not inevitable that we would find that we could not prove some theorems? What does it matter that our mathematical theorems separate in strength over a very weak base theory if, over stronger theories that we believe are true, the hierarchy collapses and all of them become equally provable? Is the subject of reverse mathematics, therefore, ultimately based in skepticism about these stronger theories? Do we have such a weak base theory because we think the strengthenings of it might not be correct? As I see it, this is the central philosophical question of reverse mathematics. Part of the lesson of the incompleteness theorem is that we cannot seem ultimately to be certain about our mathematical foundations. One rational response to this, of course, is simply to be less certain of our mathematical foundations as the theory is strengthened.

Many set theorists, however, would find uncertainty only at a much higher level, preferring to work over much stronger theories. W. Hugh Woodin, for example, draws the line of uncertainty well beyond ZFC set theory and quite high into the large cardinal hierarchy; these are *much* stronger theories than any of the big five theories appearing in reverse mathematics. He points to the regularity consequences of large cardinals for the real numbers and sets of real numbers, an astonishing connection between the existence of enormous infinities and mathematical principles at the level of the real numbers, outlining his vision that they provide the natural strengthening of our mathematical theories. Large-cardinal set theorists have found numerous instances of natural mathematical statements that are equiconsistent with large cardinal hypotheses—a rich vein of mathematics. Those hypotheses are not part of the classical or core mathematical developments considered in reverse mathematics, simply because one would have had no tools to prove them without the large cardinal assumptions that arrived only much later. In this sense, the fact that most classical mathematics is captured by the big five can be taken as a reason to move beyond the big five to the mathematical riches that might await us in stronger axiomatic frameworks. One can also find the set-theoretic axioms and large cardinal hypotheses themselves to be classical principles that are not classified by any of the systems considered in reverse mathematics.

Penelope Maddy (2011, p. 83) explains how her thin realist defeats skepticism by finding sets to be "repositories of mathematical depth" that "mark off a mathematically rich vein within the indiscriminate network of logical possibilities." She continues:

> This is what defeats an Evil Demon-style concern: the Demon might somehow induce in me all the experiences I'd have if there were an external world without there actually being such a world, but he can't present a set-theoretic posit that does a maximally-efficient job of tracking mathematical fruitfulness and yet doesn't exist—because the posit just is the sort of thing that does this sort of job.

Stephen Simpson (2014 [2009]) describes Maddy as setting up an analogy between large cardinal skepticism and Cartesian skepticism as follows:

$$\frac{\text{Large cardinals}}{\text{Set theory skepticism}} = \frac{\text{Tables and chairs}}{\text{Evil daemon theories.}}$$

In opposition, Simpson sets up an alternative analogy:

$$\frac{\text{Large cardinals}}{\text{Set theory skepticism}} = \frac{\text{Gods and devils}}{\text{Religious skepticism.}}$$

Thus, he finds room for skepticism about the stronger theories:

> A series of reverse mathematics case studies has shown that the bulk of core mathematical theorems falls at the lowest levels of the hierarchy: WKL_0 and below. The full strength of first-order arithmetic appears often but not nearly so often as WKL_0. The higher levels up to $\Pi_2^1\text{-}CA_0$ appear sometimes but rarely. For details see Simpson (2009) and Simpson (2010). To me this strongly suggests that higher set theory is, in a sense, largely irrelevant to core mathematical practice. Thus the program of set-theoretic foundations is once again called into question. Simpson (2014 [2009])

We shall take this issue up further in section 8.13 of the next chapter.

7.9 Goodstein's theorem

The Gödel sentence and the Gödel-Rosser sentence are true but unprovable, and that is truly amazing. But meanwhile, perhaps one is left dissatisfied by these sentences, since what they assert in each case is some bizarre, self-referential provability assertion. Do we really care that, technically, they are true? Would it not make a more compelling incompleteness theorem for us to identify more natural examples of the independence phenomenon? Can we find more attractive true but unprovable statements in arithmetic?

Yes—in fact, there are many fascinating examples. To exhibit one of them, consider the case of Goodstein sequences. Take any positive number, such as the number 73, and write it in complete base 2, which means write it as a sum of powers of 2, but write the exponents in this way as well. Denote it by a_2:

$$a_2 = 73 = 64 + 8 + 1 = 2^{2^2+2} + 2^{2+1} + 1.$$

Now, obtain a_3 by replacing all the 2s with 3s and subtracting 1. In this case,

$$a_3 = 3^{3^3+3} + 3^{3+1} + 1 - 1 = 3^{3^3+3} + 3^{3+1}.$$

Similarly, write this in complete base 3, replace 3s with 4s, and subtract 1, to get

$$a_4 = 4^{4^4+4} + 4^{4+1} - 1 = 4^{4^4+4} + 3 \cdot 4^4 + 3 \cdot 4^3 + 3 \cdot 4^2 + 3 \cdot 4 + 3.$$

And so on. The surprising conclusion follows.

Theorem 19 (Goodstein). *For any initial positive integer a_2, there is $n \geq 2$, for which $a_n = 0$.*

That is, although it seems that the sequence is always growing larger, eventually it hits zero. So our initial impression that this process should proceed to ever larger numbers is simply not correct. The proof of Goodstein's theorem uses transfinite ordinals to measure the complexity of the numbers that arise, proving that this complexity strictly descends with each step. At each stage, the associated ordinal replaces the ns appearing in the complete base n expression of a_n with the ordinal ω. For example, the ordinals associated with the Goodstein sequence given previously begin like this:

$$a_2 = 2^{2^2+2} + 2^{2+1} + 1 \quad \mapsto \quad \omega^{\omega^\omega+\omega} + \omega^{\omega+1} + 1$$
$$a_3 = 3^{3^3+3} + 3^{3+1} \quad \mapsto \quad \omega^{\omega^\omega+\omega} + \omega^{\omega+1}.$$

At the next step, comparing with the complete base 4 representation of a_4 above, and noting the noncommutative nature of ordinal multiplication, we have

$$a_4 \mapsto \omega^{\omega^\omega+\omega} + \omega^\omega \cdot 3 + \omega^3 \cdot 3 + \omega^2 \cdot 3 + \omega \cdot 3 + 3.$$

Notice how the associated ordinals are descending. Since there can be no infinite descending sequence of ordinals, they must eventually hit zero, and the only way this happens is if the number a_n itself is zero. One can see that we had to split up the complexity of the number somewhat in moving from a_3 to a_4, although even in this case, the number did get larger. Eventually, the proof goes, the complexity drops low enough that the base exceeds the number, and from this point on, one is just subtracting 1 endlessly.

That conclusion is surprising. But this theorem truly packs a one-two punch because not only is the theorem itself surprising, but there is an amazing follow-up theorem.

Theorem 20 (Kirby, Paris, 1982). *Goodstein's theorem is not provable in the usual first-order Peano axioms of arithmetic PA.*

That is, the statement of Goodstein's theorem is independent of PA. It was a statement about finite numbers that is provable in set theory, using the axiom of infinity provided by ZFC set theory, but it is not provable in PA.

Kirby and Paris (1982) introduced an attractive variant of their theorem—namely, the *Killing the Hydra* game. One begins with a finite tree, which we think of as a multiheaded serpent called the Hydra. The game is played according to the Hydra rule: you can chop off any head of the Hydra, but when you do so, you should follow the neck down one node (unless already at the bottom), and then replicate the part of the Hydra remaining through that neck or above, replacing with n copies on the nth move. Can you win? Is there a strategy that will enable us to cut off all the heads? Here are a few moves in a sample instance:

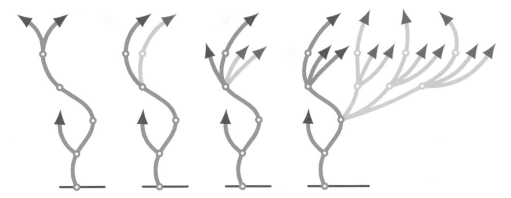

The first, amazing theorem is that every strategy is a winning strategy! No matter how you proceed to cut off the heads, eventually, all of them will be cut off. One can begin to see how this could be true by noticing that although the Hydra does initially get larger, nevertheless the complexity of branching in the Hydra is simplifying. For example, the first Hydra here has a branching node at level 4, but after the first cut, this will never arise again; similarly, the third Hydra has a level-five head growing out of a degree four node, but no subsequent Hydra will have such a node. Ultimately, one associates each Hydra with a countable ordinal below ϵ_0, as a measure of complexity, and then proves that whenever you cut off a head and grow the resulting new heads, the newly associated ordinal is strictly smaller. Since there is no infinite descending sequence of ordinals, it must be that you eventually get to zero, having cut off all the heads. Meanwhile, the second amazing theorem is that the assertion that every strategy is a winning strategy is not provable in PA.

Although Gödel established the necessary ubiquity of independence in arithmetic, it is in set theory where the independence phenomenon has really taken hold. Set theory overflows with hundreds (perhaps thousands) of natural examples of fundamental assertions that are now known to be independent of the fundamental axioms of set theory. This includes prominent questions, such as the continuum hypothesis and essentially every nontrivial assertion in infinite combinatorics, as well as the axiom of choice, assertions about cardinal characteristics of the continuum, and, if consistent, the existence of any of the various large cardinals. The independence phenomenon is pervasive throughout set theory.

7.10 Löb's theorem

If the Liar sentence asserts its own falsity, let us consider the truth-teller sentence, which asserts its own truth. And just as the Gödel sentence asserts its own unprovability, let us consider the *Löb* sentence, which asserts its own provability. So we have four sentences:

Truth-teller: **Liar:**

This sentence is true.   This sentence is false.

Löb: **Gödel:**

This sentence is provable. This sentence is not provable.

Many people undertake the analysis of the Liar sentence by arguing that if the Liar sentence is true, then it must be false; and if it is false, then it must be true. And so, they conclude, the sentence is paradoxical. Meanwhile, the corresponding analysis of the truth-teller sentence seems to go nowhere: if it is true, then it is true; and if it is false, then it is false. Sometimes people conclude from this that the truth-teller sentence could be either true or false; it is somehow not determined. But that analysis is not actually a proof of this claim; just because one does not immediately find a contradiction does not mean that there is not a deeper conclusion to be made. It is like saying that you searched for the keys and did not find them, therefore, they could be anywhere! But they are actually in a specific place, and they probably could not be on the Moon or inside Grant's Tomb. Can the truth-teller really be either true or false? If you think so, then perhaps you might be inclined to expect a similar situation for the Löb sentence. But that conclusion would be wrong in light of Löb's theorem.

Theorem 21 (Löb). *The Löb sentence is actually provable, and hence true. More generally, for any arithmetic sentence ψ, if $PA \vdash \Pr_{PA}(\ulcorner\psi\urcorner) \to \psi$, then $PA \vdash \psi$.*

Before giving the proof of the theorem, let us consider a somewhat more fanciful version of it. Let me prove to you that Santa Claus exists. Consider the statement, "if this sentence is true, then Santa Claus exists." Without taking a stand yet on whether it is true, let us call this sentence *s*. Since it has the form of a conditional statement, we might try to prove it by assuming the antecedent and arguing for the consequent. Thus, we assume the hypothesis of *s* is true. But the hypothesis of *s* is simply *s* itself. So we assume *s* is true. In that case, by *s*, we may deduce the conclusion of *s*, which is that Santa Claus exists. So we have proved that if the hypothesis of *s* is true, then so is the conclusion. So we have proved that *s* is true. But in that case, we may now use *s*, by observing that its hypothesis is true (with no assumption now) and therefore also its conclusion. So Santa Claus exists.

The logic of the argument appears to show that any statement *s* of the form "*s* implies *t*" must be true, and this informal version of the argument is known as *Curry's paradox*, named after Haskell Curry. It is also now known as *Löb's paradox*, since the proof of Löb's theorem follows very similar reasoning, except that it now becomes legitimate arithmetical reasoning instead of fanciful nonsense.

Proof of Löb's theorem. Assume that PA $\vdash \mathrm{Pr}_{\mathrm{PA}}(\ulcorner\psi\urcorner) \to \psi$. By the fixed-point lemma, let σ be a sentence for which PA $\vdash \sigma \leftrightarrow (\mathrm{Pr}_{\mathrm{PA}}(\ulcorner\sigma\urcorner) \to \psi)$. Since the left-to-right implication is provable, it is also provably provable, and from this and the derivability conditions, it follows in PA that $\mathrm{Pr}_{\mathrm{PA}}(\ulcorner\sigma\urcorner)$ implies $\mathrm{Pr}_{\mathrm{PA}}(\ulcorner\mathrm{Pr}_{\mathrm{PA}}(\ulcorner\sigma\urcorner) \to \psi\urcorner)$, which in turn implies $\mathrm{Pr}_{\mathrm{PA}}(\ulcorner\mathrm{Pr}_{\mathrm{PA}}(\ulcorner\sigma\urcorner)\urcorner) \to \mathrm{Pr}_{\mathrm{PA}}(\ulcorner\psi\urcorner)$. But provability implies being provably provable, so we can prove that $\mathrm{Pr}_{\mathrm{PA}}(\ulcorner\sigma\urcorner) \to \mathrm{Pr}_{\mathrm{PA}}(\ulcorner\mathrm{Pr}_{\mathrm{PA}}(\ulcorner\sigma\urcorner)\urcorner)$. By putting this together, we have proved that $\mathrm{Pr}_{\mathrm{PA}}(\ulcorner\sigma\urcorner) \to \mathrm{Pr}_{\mathrm{PA}}(\ulcorner\psi\urcorner)$. By assumption, this latter conclusion implies ψ itself, and so we have proved that $\mathrm{Pr}_{\mathrm{PA}}(\ulcorner\sigma\urcorner) \to \psi$. But this is equivalent to σ, and so we have proved σ in PA. From this, it follows that $\mathrm{Pr}_{\mathrm{PA}}(\ulcorner\sigma\urcorner)$ also is provable, and so by the choice of σ, we deduce that PA proves ψ, as desired. □

7.11 Two kinds of undecidability

In mathematical practice, we find at least two notions of undecidability. On the one hand, a decision problem can be computably undecidable, meaning that there is no computable procedure that correctly solves all instances of the problem. For example, the halting problem is undecidable, as is the tiling problem and the diophantine equation problem. On the other hand, mathematicians say that a statement is undecidable in a theory T, when the theory neither proves nor refutes the statement. For example, the continuum hypothesis is undecidable in ZFC set theory; and the Gödel sentence is undecidable in PA. This usage of the term "undecidable" does not directly involve computation, only logic.

Although the two notions are distinct, people are sometimes sloppy about distinguishing them, which can lead to confusion. For example, the undecidability of the continuum hypothesis in ZFC is not about our ability or inability to write a computer program that will determine whether it is true, but rather is the claim that ZFC proves neither it nor its negation.

Meanwhile, the two notions of undecidability are nevertheless deeply connected. What I claim is that every computably undecidable decision problem is saturated with provably undecidable instances. To explain what I mean, suppose that $A \subseteq \mathbb{N}$ is a decision problem that is computably undecidable, in the sense that there is no computable procedure to determine on input n whether $n \in A$ or $n \notin A$. For any computably axiomatizable true theory T, in which the decision problem is expressible, there must be infinitely many instances n such that the theory T does not settle the question of whether $n \in A$; more specifically, I claim that there must be infinitely many instances for which $n \notin A$, but this is not provable. The reason? Otherwise, we could build a computable decision procedure for A by searching either for membership in the enumeration of A or for proofs of nonmembership. Since the decision problem was undecidable, this procedure cannot succeed in all cases, or even in all but finitely many cases. So there must be infinitely many numbers for which

$n \notin A$, but this is not provable. No matter how strong your theory is, there will always be instances of the halting problem, for example, that the theory simply does not settle. Thus, computable undecidability leads to instances of provable undecidability, connecting the two notions.

Questions for further thought

7.1   Suppose that T is a computably axiomatizable theory extending PA. Is it correct to say that T is consistent if and only if T does not prove $\mathrm{Con}(T)$?

7.2   Can you supply a consistent theory extending PA that proves its own inconsistency? (Yes, *in*consistency; this is not a typographical error.)

7.3   Discuss the principle that a decidable theory cannot serve as a foundation of mathematics.

7.4   Turing and Church showed that the Entscheidungsproblem is not computably decidable, but is it computably enumerable? That is, is there a computable procedure to enumerate all the valid assertions? Is there a computable procedure to enumerate all provable sentences? All satisfiable sentences? All consistent finite sets of sentences? All pairs of logically equivalent assertions?

7.5   Turing showed that the Entscheidungsproblem is not computably decidable by reducing the halting problem to the validity problem. Is there a converse reduction? That is, can you reduce validity questions to questions of halting?

7.6   Show that the validity decision problem, the satisfiability problem, the consistency problem (for a finite list of sentences), and the logical-equivalence problem (for two sentences) are all computably reducible to one another and to the halting problem.

7.7   Suppose that you have an undecidable decision problem A. Prove that there must be some particular numbers n, such that the question of whether $n \in A$ or not is independent of PA. Would it help for us to strengthen PA to a stronger theory?

7.8   A statement is said to be *decidable* with respect to a theory when the theory either proves or refutes it; a decision problem is said to be *decidable* when there is a computational procedure that will correctly compute the answer for any given instance. How are these two uses of the word "decidable" related?

7.9   What is the difference, if any, between asserting $T \vdash \varphi$ and asserting the arithmetical representation $\mathrm{Pr}_T(\ulcorner \varphi \urcorner)$? Are these two assertions substitutable for one another in all circumstances?

7.10   Suppose that T is an arithmetically definable theory T, and consider the arithmetic assertion $\mathrm{Con}(T)$, asserting that T is consistent. To what extent does the assertion $\mathrm{Con}(T)$ depend on intensional aspects of T, that is, on the way that we described the theory T, rather than extensional aspects, concerning only the set of sentences that are axioms of T? If we had described exactly the same theory but in a different manner, would the resulting assertion $\mathrm{Con}(T)$ have the same truth value? Would it have the same deductive power as an arithmetic hypothesis?

7.11 Suppose that a colleague provides you with a list of several newly formulated general axioms of arithmetic and proceeds to prove from those axioms that they are consistent. What is your response?

7.12 Can we prove, as a general principle or scheme, that whenever a statement is provable, then it is true? In light of Löb's theorem, exactly which instances of this principle are provable?

7.13 Explain how the fixed-point lemma shows that there is a sentence fulfilling Rosser's property, a sentence ρ that is provably equivalent to the assertion that for every proof of ρ, there is a smaller proof of $\neg\rho$.

7.14 Assuming PA is consistent, can you identify a specific sentence ψ for which we cannot prove the implication, "If ψ is provable in PA, then it is true." What is the philosophical significance of this for provability as a certification of truth?

7.15 Consider the driving motivations of the Hilbert program, apart from Gödelian considerations. Would it actually be reassuring to have a theory T that proved its own consistency? Would that be a reason to trust in T?

7.16 Argue that for every computably enumerable true theory of arithmetic T, there is a stronger such theory T' that proves the consistency of T. Is this a reason to trust in T? Similarly, prove that for every consistent theory T, there is another stronger consistent theory T' that proves that T is inconsistent. Is this a reason to distrust T?

7.17 In what sense is it correct to say that Tarski defined truth and also proved that truth is not definable?

7.18 Is there a consistent arithmetic theory, with an arithmetically definable set of axioms, that proves its own consistency? Can such a theory have a computably decidable set of axioms?

7.19 Show that every theory with a computably enumerable set of axioms can be axiomatized by a computably decidable set of axioms. [Hint: Replace each assertion φ in the computably enumerable axiomatization with a logically equivalent assertion $\varphi \wedge \varphi \wedge \cdots \wedge \varphi$, in such a way that the new set of resulting axioms is computably decidable. This method is known as *Craig's trick*, named for the result of William Craig (1953).]

7.20 There seems to be something at least a little sensible about the principle that if you are willing to make an assertion ψ, then you should also be willing to assert that ψ is consistent. Suppose we make a formal proof system by adding to our deduction system a rule of inference: from ψ, deduce $\text{Con}(\psi)$. (Note, the Con operator here refers to consistency in the old system.) What can you say about this proof system? Will it be sound and complete?

7.21 How are the Gödel sentence and the Löb sentence and our analysis of them fundamentally similar to or different from the Liar sentence, "This sentence is false" and the truth-teller sentence, "This sentence is true," respectively?

Further reading

Raymond Smullyan (1992). An delightful introduction to Gödel's incompleteness theorems in the author's inimitable style.

Richard Zach (2019). An excellent survey of Hilbert's program at the *Stanford Encyclopedia of Philosophy*.

David Hilbert (1930). This is Hilbert's 1930 radio-broadcast retirement address, in which he famously says, "We must know. We will know."

Joel David Hamkins (2018). This is my article on the universal algorithm and arithmetic potentialism, dealing with some consequences of the incompleteness phenomenon that I find fascinating.

8 Set Theory

Abstract. We shall discuss the emergence of set theory as a foundation of mathematics. Cantor founded the subject with key set-theoretic insights, but Frege's formal theory was naive, refuted by the Russell paradox. Zermelo's set theory, in contrast, grew ultimately into the successful contemporary theory, founded upon a cumulative conception of the set-theoretic universe. Set theory was simultaneously a new mathematical subject, with its own motivating questions and tools, but it also was a new foundational theory with a capacity to represent essentially arbitrary abstract mathematical structure. Sophisticated technical developments, including in particular, the forcing method and discoveries in the large cardinal hierarchy, led to a necessary engagement with deep philosophical concerns, such as the criteria by which one adopts new mathematical axioms and set-theoretic pluralism.

Set theory appeared in earnest in the late nineteenth century with Dedekind's treatment of arbitrary sets of natural numbers and Cantor's ideas on cardinality, the ordinals, transfinite arithmetic, and transfinite recursion. Set theory is a wellspring of mathematical thought, an independent mathematical subject with its own deep questions, problems, and methods, many with important applications in other parts of mathematics. Set theory also provides tools and a conceptual framework for mathematicians in general to deal with arbitrary mathematical structure. Thus, set theory plays two roles: it is simultaneously its own mathematical subject and an ontological foundation for mathematics—a common forum in which one can find realizations of structures and ideas from any part of mathematics. On both sides of this dichotomy, set theory engages with fundamentally philosophical issues, perhaps far more than many other mathematical subjects do. The mathematical core of set theory, for example, concerns the nature of infinity, a topic perhaps formerly thought of primarily as philosophical. The set-theoretic analysis of infinity has been so successful and clarifying that one cannot now seriously undertake a philosophical discussion of infinity without touching upon the set-theoretic conceptions. Meanwhile, in its role as a foundational theory, set theory is even more explicitly engaged with philosophical issues. What role do axioms play in a foundational theory? By what criteria are we to add or remove axioms from a foundational theory? May we judge the truth of strong axioms by the fruitfulness of their consequences? Are the axioms of set theory describing a unique intended set-theoretic universe? Or is there a multiverse of set-theoretic conceptions and corresponding set-theoretic worlds? Is the set-theoretic universe a completed totality, or does it have ultimately a character of potentiality?

8.1   Cantor-Bendixson theorem

Several key set-theoretic ideas germinate in the Cantor-Bendixson theorem, for whose proof and analysis Cantor had developed his set-theoretic ideas on arbitrary sets of real numbers, the ordinals, well-orders, and transfinite recursion. The theorem grows out of Cantor's 1872 investigation of the process of iteratively casting out isolated points from a closed set of real numbers. A point is *isolated* in a set if it stands alone in some small neighborhood as the only element of the set; or equivalently, if it is not a limit of other points from the set. We form a new set, the *derived* set, by casting out all isolated points from the original set. The derived set of a closed set is also closed because every limit point of the derived set must be in the original set, but it cannot have been isolated there. The curious phenomenon to observe with derived sets is that, although we cast out all isolated points from the original set, the derived set can exhibit newly isolated points—points that were not isolated in the original set but become isolated in the derived set. And if we cast these out in a next step, the resulting set could again exhibit freshly isolated points; again and again we might cast out isolated points, iterating the process many times.

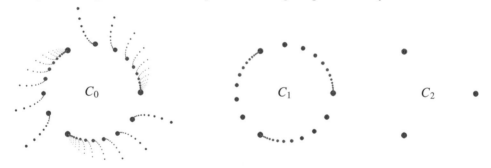

Consider the set C_0 at left. This set has many isolated points, on the spiral fringes waving out from the central circle, but the points on the main circle of C_0, being limits of those fringe points, are not isolated. The derived set C_1, therefore, has cast out all the fringe points, but keeps the central circle points. Most of these points, however, have become isolated in C_1, and so in the next step of the Cantor-Bendixson process, therefore, they are cast out, leaving us with just three points in the derived set C_2. In the general case, with another more complex set, we might continue similarly to cast out isolated points again and again, forming the further derivatives C_3, C_4, and so on. The process stops only when we reach a *perfect* set, which is a closed set having no isolated points (perhaps empty).

The *rank* of the original closed set is the number of steps it takes for the process to terminate. The set shown above has rank 3, because after three steps, there are no remaining isolated points. In general, however, sets can have much higher ranks. Perhaps the reader can construct a set of rank 7, or rank n for any finite number n. Cantor realized that a set can have infinite rank. Imagine an initial set C_0 with a sequence of points x_0, x_1, x_2, and so on, converging to a point x, but where x_n has rank n, and all the points converging to x disappear by some finite stage; this would leave x isolated at the limit stage $C_\omega = \bigcap_n C_n$, the set of points that have survived through all the finite stages.

How confusing and strange this might seem. We cast out isolated points again and again, but even after doing so infinitely many times, we still have isolated points with what is left in the limit set C_ω. Since this limit set is closed, however, it makes sense to try yet again, to form the next Cantor-Bendixson derivative $C_{\omega+1}$, and then perhaps $C_{\omega+2}$, and so on. The Cantor-Bendixson process appears to be inherently transfinite. Will we ever complete the process? What is the nature of this iteration? What are the stages of this construction? It proceeds by a recursive process, defining the later derived sets from the earlier ones, but the recursion is transfinite, proceeding beyond the finite stages.

Cantor realized that in order to describe the construction, he needed a way to talk independently about the stages of the construction. The stages themselves form a certain kind of order structure. He needed a number system transcending the finite numbers to serve as the names or indices for the stages in this transfinite recursive process. This is exactly what the ordinals provide. Cantor invented the ordinals specifically in order to make sense of the iterative Cantor-Bendixson process. He came to the concepts of well-orders, ordinals, and transfinite recursion—concepts at the very center of set theory as it is known today. Ultimately, for any closed set C_0, one defines all the Cantor-Bendixson derivatives C_α, for every ordinal α, in a transfinite recursive process: given C_α, the next derivative $C_{\alpha+1}$ is obtained by casting out isolated points, and at limit ordinals, C_λ is the intersection of all earlier C_α.

The Cantor-Bendixson theorem is the claim that for any closed set of real numbers (and in any finite or indeed countable dimension), there is some countable ordinal stage α at which the derived set C_α has no isolated points; that is, the Cantor-Bendixson process terminates, after countably many steps, in a perfect set. To prove this, simply imagine that every cast-out isolated point x is associated with a rational isolating neighborhood, a reason it was isolated, that is, a rational interval (p, q) containing x and no other points of the derived set at that stage; we shall never use the same isolating neighborhood again because it can no longer isolate another point at some later stage, as there are none left in it. Since there are only countably many rational intervals, it follows that there will also be only countably many points cast out altogether. Therefore, there must be some countable ordinal stage of the construction by which we will have achieved a perfect set.

Cantor made a remarkable conclusion from this observation. Namely, the theorem shows that every closed set of real numbers is the union of a countable set (the isolated points cast out during the construction) and a perfect set (the points remaining at the end, if any); and since Cantor had already proved that every nonempty perfect set of real numbers is equinumerous with the whole real number line, he concluded that every closed set of real numbers is either countable or equinumerous with the whole real number line. Thus, he proved that the continuum hypothesis holds for closed sets of real numbers.

8.2   Set theory as a foundation of mathematics

At first, set theory was used as a mathematical tool brought into other mathematical domains. Dedekind had imagined arbitrary sets of numbers when giving his categorical axiomatization of the natural numbers, and Cantor worked with arbitrary sets of real numbers. Gradually, however, the capacity of set theory to express essentially any kind of mathematical structure was realized, and one could begin to view set theory as providing an ontological foundation for the rest of mathematics.

Penelope Maddy describes set theory as providing a "meta-mathematical corral," an arena inside of which one may introduce and analyze mathematical structures and ideas. She points to Karl von Staudt's early resolution of the use of virtual objects in geometry:

> The "scandal" was resolved in mid-century by von Staudt, using proto-set-theoretic techniques, in particular a precursor of the method of equivalence classes: for example, the point at infinity where two horizontal lines meet is identified with what we'd now see as the set of lines parallel to these two...In this and related ways, von Staudt managed to build surrogates for heretofore suspicious, possibly dangerous new items (like points at infinity) out of uncontroversial, unproblematic materials (ordinary lines), and to redefine the relevant relations so as to validate the existing, successful theory...As time went by, it became clear that the construction tools needed for this "building" process—tools von Staudt regarded as "logical"—were actually set-theoretic in character. Maddy (2017)

Set-theoretic conceptions thus provided a framework for understanding what mathematical structure could be. In light of the power and utility of set theory, Hilbert had said,

> *Aus dem Paradies, das Cantor uns geschaffen, soll uns niemand vertreiben können.* (No-one shall cast us from the paradise that Cantor has created for us.) Hilbert (1926, p. 170)

To be precise in twentieth-century mathematics often came to mean to specify one's mathematical structure in a set-theoretic fashion. What is an order? When pressed, a mathematician might say that it is a set with a certain binary relation on it that is reflexive, transitive, and antisymmetric. What is a binary relation on a set? When pressed, a mathematician might say that it is a set of ordered pairs of elements from that set. What is a group? When pressed, a mathematician might say that it is a set with a certain binary operation on it—one that is associative, with an identity element and inverses.

Yiannis Moschovakis describes the process of faithful representation that occurs in mathematics when we use one kind of mathematical structure or conceptual framework to represent mathematical objects that arise naturally in another. Descartes founded analytic geometry, for example, by conceiving of points in terms of their coordinates, revealing the hidden connection between algebra and geometry. The classical curves and forms—ancient, elegant, and distinguished geometric royalty—gave up their secret names: each is precisely described by an elementary algebraic equation. By associating points with their coordinates (x, y), Descartes made profound new mathematical insights and enabled the far-reaching mathematical developments to come. Today, we typically identify points in the plane by their coordinates (x, y), and this is an instance of faithful representation. The geometric structure and relations of the points in which we were interested, even when conceived without coordinates, carry over to their representations in analytic geometry via coordinates. We need not insist that points *are* coordinate pairs (x, y)—why should we? Mathematicians had proceeded for centuries with a purely geometric conception of point—yet we may fruitfully analyze them as if they were.

Similarly, mathematicians found in set theory the capacity to express essentially any mathematical structure. Moschovakis summarizes the attitude:

> We will discover within the universe of sets *faithful representations* of all the mathematical objects we need, and we will study set theory on the bases of the lean axiomatic system of Zermelo **as if all mathematical objects were sets**. The delicate problem in specific cases is to formulate precisely the correct definition of a "faithful representation" and to prove that one such exists. (Moschovakis, 2006, p. 34, emphasis original)

Thus, set theory had become a grand unified theory of mathematics.

Naturally, one can take set theory merely as one of many possible foundations of mathematics. Other contemporary foundations might include category-theoretic foundations, such as the elementary theory of the category of sets or univalent foundations via homotopy type theory. Meanwhile, a more extreme view of set-theoretic foundations is set-theoretic reductionism, the view that the foundations of mathematics are inherently set-theoretic, that metaphysically all there is to mathematical objects are sets. Penelope Maddy specifically rejects this view:

> The other purported foundational role for set theory that seems to me spurious is what might be called the **Metaphysical Insight**. The thought here is that the set-theoretic reduction of a given mathematical object to a given set actually reveals the true metaphysical identity that object enjoyed all along. (Maddy, 2017, p. 6)

She rejects this as a goal for a successful foundational theory, but meanwhile finds set theory to fulfill the other legitimate foundational goals she identifies:

> At first glance, this may look like just one more instance of the set-theoretic reduction that underlies the **Meta-mathematical Corral**, but in fact there's something more going on. It isn't that we have an explicit mathematical item—the ordered pair, or the numbers as described by Peano—which we then "identify" with a set that can play the same role, do

the same jobs. Instead, in this case, we have a vague picture of continuity that's served us
well enough in many respects, well enough to generate and develop the calculus, but now
isn't precise enough to do what it is being called upon to do: allow for rigorous proofs of
the fundamental theorems. For that we need something more exact, more precise, which
Dedekind supplies. This isn't just a set-theoretic surrogate, designed to reflect the features
of the pre-theoretic item; it's a set-theoretic improvement, a set-theoretic replacement of
an imprecise notion with a precise one. So here's another foundational use of set theory:
Elucidation. The replacement of the imprecise notion of function with the set-theoretic
version is another well known example. (Maddy, 2017, p. 8–9)

From the logic perspective, the rise of set theory as a unifying mathematical foundation
was an important historical development. While mathematics diversified with ever-greater
specialized complexity, mathematicians would sometimes seek to apply results from one
area within another. For example, one might use notions and theorems from topology and
analysis in order to prove results in algebra, or conversely. This practice would be logically
incoherent unless the subjects were part of a common mathematical framework. By providing a unifying context, therefore—a single theory in which one can view all mathematical
arguments as taking place—set theory thereby logically facilitated this transfer.

Imagine a balkanized mathematics, in contrast, where mathematics is divided into separate realms, each with its own fundamental axioms, but without a common underlying
theory. Might not the subjects of geometry, algebra, and analysis have been totally separate
mathematical efforts? Mathematics exhibits at least a limited version of such balkanization today, where the axiom of choice, for example, is routinely assumed in many subjects,
especially algebra and analysis, but it is resisted and remarked upon by mathematicians
in others. One can easily imagine much worse conflicts. Suppose, for example, that one
subject had developed in a foundational theory with something like the power of Zermelo
set theory, but with the axiom of determinacy, and another subject proceeded in Zermelo-
Fraenkel ZFC set theory. These two foundations conflict with each other, and lead to
profoundly different conclusions about fundamental facts in topology and analysis, with
contradictory visions of the nature of Lebesgue measure, for example. In such a context,
one would not be able reliably to borrow results from one subject for use in the other.

Later in this chapter, I shall discuss set-theoretic pluralism, which emphasizes the multi-
faceted nature of contemporary set theory, with numerous incompatible versions of the
theory, some with the continuum hypothesis and some without, and some with large cardinals and some without. Does this undermine the capacity of set theory to serve as a
unifying foundation? A similar question arises in geometry: does the splintering of geometry into various incompatible forms—Euclidean geometry, spherical geometry, hyperbolic
geometry, elliptical geometry—undermine geometry as a whole to provide a foundation for
geometrical thinking and analysis? I do not believe it does; indeed, on the contrary, geometry is greatly enriched by that splintering in terms of foundations, since it can clarify
when a particular geometrical concept depends on those foundational aspects.

Another kind of balkanization arises today from the difficulty in translating between alternative foundational systems. In addition to set theory, for example, there are other foundational theories aiming to play a unifying foundational role in mathematics, such as category theory, type theory, and a newer hybrid, homotopy type theory. These foundational theories each implement the doctrine of faithful representation; truly we may find faithful representations of our mathematical ideas in diverse foundational theories. Yet a foundational conflict arises when the translation of mathematical methods and ideas between alternative foundations is difficult or unclear, as often seems to be the case, unfortunately, in current mathematical practice. Logicians report difficulties, for example, in translating constructions between set theory and category theory or homotopy type theory. The fundamental conceptions differ so greatly that even simple constructions in one realm become complicated or confusing in another. For this reason, I believe that mathematics would benefit by having more mathematicians who are expert in multiple foundational realms.

8.3   General comprehension principle

Let us discuss the specific axioms of set theory in greater detail. What are the core principles of set theory? Cantor had worked in an intuitive set theory whose fundamental principles were not made explicit; he lacked a formal theory. Later, formal set-theoretic systems were made by Frege and ultimately by Zermelo and others, including Abraham Fraenkel, Paul Bernays, Kurt Gödel, and John von Neumann.

At the very center of set theory, of course, is the idea that one can consider a collection of things as a single abstract thing, a *set* consisting of those things, and only those things. For example, we have the set of all red things, the set of all natural numbers, the set of all people in this room, and perhaps the set of all sets of zebras. Which collections can be brought together in this way? Well, all of them, of course. If we have any way of describing or conceiving of a collection, then according to the naive version of set theory, we can form the set consisting of just those objects. This naive idea leads to what is known as the *general comprehension principle*.

General comprehension principle. *For any property φ, one may form the set of all x with property $\varphi(x)$. That is,*

$$\{ x \mid \varphi(x) \}$$

is a set.

Despite its simple clarity and attraction, nevertheless this principle was the cause of a rude coming-of-age for set theory in its teenage years. It was a mathematical disaster. The simple fact is that the general comprehension principle is wrong. It is false in the generality in which it is stated; it has self-contradictory instances. Some properties simply cannot define sets at all, upon pain of contradiction.

Theorem 22 (Russell, 1901). *The general comprehension principle has false instances.*

Proof. To see this, let R be the set of all sets x that are not members of themselves; that is,

$$R \quad = \quad \{\, x \mid x \text{ is a set and } x \notin x \,\}.$$

By the general comprehension principle, this is a set. Let us ask: Is R a member of itself? The members of R are precisely the sets x with $x \notin x$, and therefore we see that

$$R \in R \quad \Longleftrightarrow \quad R \notin R.$$

But this is a contradiction, and so this instance of the general comprehension principle cannot be true. □

Consider the allegory of the barber who shaves all those in town who do not shave themselves. Does the barber shave himself? Well, if he does not, then he should; and if he does, then he should not. This is precisely the contradictory logic of the Russell paradox. Or consider the barista who happily serves up coffee each morning to all and only those who do not make coffee for themselves. Does she make coffee for herself? If so, then she should not; and if not, then she should.

You might say that the property $x \notin x$ is rather strange to consider, since we do not often expect sets to be members of themselves. After all, the set of all elephants is not an elephant (it is a *set* of elephants), and it is therefore not a member of itself; the set of all prime numbers is not itself a prime number, for it is not a number at all. Very well then, the objection shows that the property of being non-self-membered is quite ordinary—it seems to hold of most, or even all, sets—and therefore the general comprehension principle is false for some quite ordinary properties. Indeed, ZF set theory proves that every set obeys $x \notin x$, and in this theory, therefore, the Russell set would be identical to the universal set.

Russell wrote to Frege about the inconsistency he had discovered—a devastating objection, since Frege had woven the general comprehension principle deeply into his account. Russell's letter arrived just as Frege was preparing to publish. To his credit, Frege included an epilogue addressing the objection:

> Hardly anything more unwelcome can befall a scientific writer than to have one of the foundations of his edifice shaken after the work is finished. This is the position into which I was put by a letter from Mr. Bertrand Russell as the printing of this volume was nearing completion. The matter concerns my basic law V. (Frege, 1893/1903, epilogue), translation (Frege, 2013, p. 253)

Frege's Basic Law V

Frege's trouble with the Russell paradox and the general comprehension principle is commonly laid at the feet of his Basic Law V, which, in the case of concept extensions, asserts:

Basic Law V. *The extension of a concept F is the same as the extension of G if and only if* $Fx \leftrightarrow Gx$ *for every x.*

$$\varepsilon F = \varepsilon G \quad \Longleftrightarrow \quad \forall x (Fx \leftrightarrow Gx).$$

The law arises naturally from the case of functional extensions, where one might want to say that a function F is extensionally identical with another function G if and only if the two functions give rise to exactly the same course of values on objects, meaning that they give the same output for every input: $\forall x\, F(x) = G(x)$. Indeed, this functional version of Basic Law V is fully general for Frege, because in his ontology, concepts are regarded as functions mapping objects to the true/false values. In this sense, Basic Law V asserts that the extensions of two concepts are the same if and only if the true/false course of values of the two concepts agree.

Underlying the principle is the implicit idea that to every concept F, we have assigned an object εF, the *extension* of the concept F, an object that represents the concept. The law itself asserts exactly that these extensions are classification invariants for the coextension relation on concepts.

To my way of thinking, however, and despite the extensive literature on this topic (as well as Frege's own remark, quoted just previously), I find it misleading to place the blame of the Russell inconsistency squarely on Basic Law V. The source of trouble in Frege's system is not the explicit central claim of this law, which is innocuous on the natural contemporary readings of it; rather, the trouble arises from the implicit commitment underlying Frege's use of this law, namely, that we are to take the extension of a concept εF itself as an object x falling under the scope of the universal quantifier $\forall x$ appearing on the right-hand side. So I should like to separate the two claims of Basic Law V, the explicit from the implicit. The natural contemporary reading of Basic Law V, I claim, is exactly as the principle of class extensionality, stated on the next page, a principle widely viewed as harmless and indeed taken as a fundamental axiom in all the currently standard class theories, such as Gödel-Bernays set theory or Kelley-Morse set theory. Meanwhile, the implicit claim of Basic Law V amounts to the general comprehension principle.

Let me explain. In the current class theories, what we generally mean by the extension of a concept F is the class of instances:

$$\varepsilon F = \{\, x \mid Fx \,\}.$$

All the standard class theories have various class comprehension principles, which assert that for suitable predicates F, we may indeed form the class $\{\, x \mid Fx \,\}$. In Gödel-Bernays set theory, for example, we assert class comprehension for all properties F expressible in the first-order language of set theory, whereas in Kelley-Morse set theory, we assert it also for properties F assertible in the second-order language of set theory, where we can quantify over classes of sets. Note that the Russell paradox is undertaken with a mild quantifier-free first-order instance $\{\, x \mid x \notin x \,\}$.

Meanwhile, the principle of class extensionality, an axiom in all the standard class theories, is precisely the claim that every class is determined by its elements:

Class extensionality. *Two classes, A and B, are equal if and only if they have all the same elements.*

$$A = B \quad \Longleftrightarrow \quad \forall x\,(x \in A \leftrightarrow x \in B).$$

In particular, when we define classes by predicates, we have

$$\{\,x \mid Fx\,\} = \{\,x \mid Gx\,\} \quad \Longleftrightarrow \quad \forall x\,(Fx \leftrightarrow Gx).$$

With our class interpretation of extension, therefore, this amounts exactly to the central claim of Basic Law V:

$$\varepsilon F = \varepsilon G \quad \Longleftrightarrow \quad \forall x\,(Fx \leftrightarrow Gx).$$

In other words, on the contemporary reading of class extension, Basic Law V is the principle of class extensionality, which is a fundamental axiom in all the contemporary class theories and not thought to be inconsistent.

So what gives? Why do people say that Frege's Basic Law V is inconsistent, while set theorists continue to affirm the very same principle in all the standard accounts of class theory? The explanation is that it is not the explicit central equivalence asserted by Basic Law V that is problematic, but rather the implicit underlying claim that the extension εF of any concept F is itself an object x about which we may inquire whether the concept holds of it or not. This is the same critical difference as between the class comprehension and general comprehension principles. While the class comprehension axiom asserts that $\{\,x \mid Fx\,\}$ is a class, the class of all objects x with property Fx, it does not assert that this class is itself a set or an object x to which we might apply the predicate F. The general comprehension principle, in contrast, does assert that $\{\,x \mid Fx\,\}$ is a set, an object, to which we might consider whether the property F is applicable. It is this further claim that Russell uses in his paradox: if $R = \{\,x \mid x \notin x\,\}$ is a set, then we may inquire whether $R \in R$ or not, and thereby find ourselves with the contradictory conclusion $R \in R \leftrightarrow R \notin R$.

At bottom, the Russell argument shows that there can be no totality of objects, such that every class F of objects is assigned to a different extension object εF. If there were such an assignment, we could form the class R of all such extension objects $x = \varepsilon F$ for which $\neg Fx$. Consider now the case $x = \varepsilon R$, and therefore we see that Rx holds if and only if $\neg Rx$, which is a contradiction. In this way, just as Cantor showed that every set has more subsets than elements, Russell has used essentially identical logic to show that in every totality of objects, there must be more classes than objects.

Ultimately, therefore, Frege has gone wrong not in the main equivalence stated by Basic Law V, which on the contemporary reading is the innocuous principle of class extensionality, but in the associated underlying claim that extensions of concepts εF are themselves objects to which we might consider applying the concept. This implicit claim amounts to the general comprehension principle, asserting that every class extension $\{\, x \mid Fx \,\}$ is a set. So let us place the blame where it belongs: not on Basic Law V, but on the general comprehension principle.

8.4 Cumulative hierarchy

So the general comprehension principle is inconsistent. What is to be done? Does the objection destroy set theory? Perhaps one took the general comprehension principle to express a core set-theoretic idea, a fundamental set-theoretic truth—namely, that one can pick out any definable collection of objects and thereby form a set. Yet the principle is contradictory. Is set theory incoherent?

No. Let us think a little more carefully about the nature of the set theoretic universe that we had been imagining and develop an account of the *cumulative hierarchy*, a vision of how the set-theoretic universe is formed. Initially, we may have had some mathematical objects, the *urelements*, which are not sets but which are the objects that will constitute the stuff out of which our sets will be formed, upon which our set-theoretic universe will be constructed. In possession of the urelements, we form all possible sets of them. We form, for example, all the singleton sets $\{a\}$, one for each individual urelement a, and the doubleton sets $\{a, b\}$, and so on, in all possible ways, including the empty set $\varnothing$, and perhaps the set of all urelements.

If one imagines the urelements as forming the bottom layer of the set-theoretic universe, then the sets of urelements form a layer just above them. At the next level, we form all sets whose elements are on these first two layers, that is, the sets of sets of urelements and urelements. Some of these sets we already had, of course, the sets consisting of urelements only, and so the hierarchy is cumulative. Accumulating sets in this way, we thereby form the set-theoretic universe in a

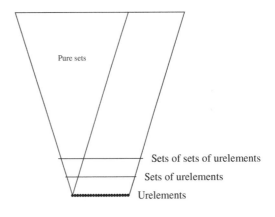

well founded hierarchy of stages. At each stage, we form all the sets whose elements were constructed at earlier stages. The fundamental picture is that a set appears in the cumulative hierarchy only at a stage after all its elements have appeared. The *pure* sets are those that have no urelements appearing hereditarily amongst their elements; they have no urelements as elements, nor as elements-of-elements nor as elements-of-elements-of-elements, and so on. The pure sets are those that arise in the cumulative hierarchy when one begins with only the empty set $\varnothing$. At the next stage, one adds the set $\{\varnothing\}$, and then $\{\varnothing, \{\varnothing\}\}$ and $\{\{\varnothing\}\}$, and so on.

Set theorists quickly realized that urelements were not needed for any mathematical purpose because every mathematical structure that arises in set theory with urelements can be constructed isomorphically in the pure sets. According to the philosophy of structuralism, therefore, the realm of pure sets serves just as well as a foundation of mathematics; and since the pure set theory is also conceptually cleaner as well as more elegant, set theory came typically to be cast in the pure form only, without urelements. The view is that the set-theoretic universe thereby consists entirely of sets—everything is a set—and when one uses this kind of set theory as a foundation of mathematics, therefore, every mathematical object is regarded as a set. The entirety of mathematics consists of sets, sets of sets, and sets of sets of sets, in a vast hierarchy founded ultimately upon the empty set (which is to say, upon nothing).

Highlighting the curiosity of this situation, Tourlakis writes:

> [I find] it extremely counterintuitive, especially when addressing undergraduate audiences, to tell them that all their familiar mathematical objects—the "stuff of mathematics" in Barwise's words—are just perverse "box-in-a-box-in-a-box..." formulations built from an infinite supply of empty boxes. (Tourlakis, 2003, p. xiii)

Nevertheless, the pure form of set theory might be seen ultimately as fulfilling the logicist program of reducing mathematics to logic. Namely, by faithfully representing mathematical structure in set theory, one has reduced mathematics to set theory, which can be seen as part of logic. Many philosophers would disagree with this, however, finding set-theoretic assertions such as the axiom of infinity and the axiom of choice to be mathematical rather than logical assertions. Other philosophers remark that there seems to be too little at stake in the logicist program and the question of whether a particular mathematical foundation counts as logic or not. What does it matter how we apply this label?

But let me come now to a critical point concerning the cumulative hierarchy understanding of how the set-theoretic universe is formed. Namely, this perspective does not provide any support at all for the general comprehension principle. In the cumulative hierarchy, a set forms at a stage only after all its elements have formed. Thus, we do not expect to find the set of all sets, for example, or other sets whose elements appear at stages unboundedly in the hierarchy. Also, we expect never to see $x \in x$, since this would require x to have appeared at a stage earlier than itself. Thus, the intuitive pull that we might had felt initially for the general comprehension principle simply falls away; it is not actually part of our more reflectively considered set-theoretic view. We may abandon the general comprehension principle as a naive mistake—one that is also very easily refuted.

8.5 Separation axiom

The cumulative hierarchy perspective, meanwhile, does seem to justify a certain weakening of the general comprehension principle, a weakening known as the *separation axiom* (also sometimes known as the *restricted comprehension axiom*), which allows you to pick out a subset of a set already given. This axiom constitutes one of the central axioms of Zermelo set theory, an axiomatization put forth by Zermelo in 1908.

Separation axiom. *For any property φ and any given set A, one may form the set of all elements x in A with property φ. That is,*

$$\{\, x \in A \mid \varphi(x)\,\}$$

is a set.

One likely finds the separation axiom to be underspecified without a more precise account of what counts as a "property." Zermelo's original formulation of the axiom was extremely broad, taking the axiom to encompass any kind of property or conception, regardless of whether it was expressible in a given formal language. Set theorists today typically understand Zermelo's original formulation of the axiom as an axiom in second-order logic. Some set theorists, however, find it incoherent to try to provide an axiomatization of set theory while using second-order logic, which itself amounts to a kind of set theory. After all, on that account, the separation axiom is basically saying: every subclass of A that there is, exists as a set. The objection is that if we already understand classes so well in the metatheory, why do we need the object set theory at all? For this reason, many philosophers (but definitely not all) regard second-order logic as a form of set theory itself, in need of clarification and setting out with set-theoretic axioms and principles. What has become known today as *Zermelo set theory*, in contrast, is the purely first-order version of his theory, with the separation axiom schema stated only for properties φ expressible in the first-order language of set theory, the language having only the $\in$ relation and =, as well as the usual logical connectives and quantifiers.

Some mathematicians and philosophers have argued that the restriction of the general comprehension principle to the separation axiom is an ad hoc response to the Russell inconsistency—a minor dodge that undercuts the particular details of the Russell argument, but which may not touch whatever fundamental underlying problem there is. From this view, therefore, we should have little confidence in the consistency of Zermelo's theory. A counterargument, however, is that the general comprehension principle was never actually part of the cumulative universe picture upon which our set-theoretic views are based; the appeal of general comprehension was simply misplaced support for the separation axiom. Thus, the fix was not ad hoc, but rather the correction of a naive error, realigning the axioms with our vision of how the set-theoretic universe unfolds.

Ill-founded hierarchies

To my way of thinking, the initial attraction (and ultimate rejection) of the general comprehension principle follows a pattern that sometimes occurs in logic and mathematics, a pattern in which a mathematical concept makes a small step up in some hierarchy of complexity or abstraction—a step up that is fine when the concept is used within a well founded hierarchy, but leads to paradox when the hierarchical nature is naively ignored.

Consider the concept of truth, for example, which also follows this pattern. It is unproblematic to add a truth predicate to any language without one, applying the disquotational theory of truth, if one seeks to apply the truth predicate only to assertions in the base language. Truth remains unproblematic even when applied in a well founded hierarchy of iterated truth predicates—truth about truth, and truth about truth about truth, and so on— a hierarchy where truth assertions are made at a level of the hierarchy about assertions referring to truth only at earlier levels of the hierarchy. One gives the disquotational interpretation of the truth predicate at each level inductively. Indeed, the existence of iterated truth predicates for the set-theoretic universe turns out to be equivalent to certain strong axioms of second-order set theory, which have been used (among other things) to clarify the power of class forcing; see Gitman et al. (2020).

If one ignores the well founded hierarchy of iterated truth, however, by treating the truth predicate as self-applicable without any well founded hierarchy, then one faces paradoxical nonsense assertions, such as the Liar paradox, "This sentence is not true." These assertions have a semblance of meaning because we are used to being able to make truth assertions sensibly, and even truth-about-truth assertions, as part of a well founded hierarchy. That is, the ill-founded self-applicable case enjoys an appearance of legitimacy, which arises from its superficial resemblance to our legitimate references to truth in the well founded case. But actually, there is no robust meaningful theory of truth here outside of a well founded hierarchy of disquotation instances.

Similarly, we are used to forming sets of objects, subject to a property, and doing so without problem or paradox, simply because in our usual mathematical practice, most such instances of set-formation happen to arise properly in the cumulative hierarchy. We freely form sets of natural numbers and sets of real numbers, or sets of topologies on a given set, and so on, and this is well founded and unproblematic. It is when we attempt to form sets by general comprehension while ignoring the cumulative hierarchy underlying set formation that we find ourselves in situations like the Russell paradox.

According to this view, the attraction of the general comprehension principle is a naive expectation that ill-founded set definitions are meaningful, an unwarranted extrapolation from the well founded case; but it is an illusion—an illusion dispelled by the Russell contradiction. Reflective contemplation about the hierarchical nature of sets removes the pull of the general comprehension principle, and one realizes that the principle was misguided.

Self-applicable truth predicates are, from this view, similarly naive; namely, while the construction of a well founded hierarchy of truth predicates is both unproblematic and useful, the self-applicable truth predicates are superficially similar, but illusory and contradictory in light of the Liar paradox. With the well founded hierarchy in mind, the ill-founded case appears ill motivated or meaningless.

Other philosophers, however, disagree with the account I have just described; they find meaning in the Liar paradox and in the general comprehension principle. Various communities of logicians are focused on the general theory of self-applicable truth predicates or on versions of set theory with the general comprehension principle. For example, Quine's New Foundations version of set theory retains a restricted form of general comprehension, allowing for a universal set (but not the Russell set), and there is work on paraconsistent set theory, which studies a version of set theory in which the general comprehension principle is true (and also false). There is a rich philosophical literature on self-applicable truth predicates, and investigators do not take the subject to be ill motivated or meaningless.

Impredicativity

Some philosophers have emphasized a different objection, laying the blame on what they call the *impredicative* nature of the general comprehension principle. A definition is impredicative when it defines an object by means of a property whose quantifiers range over a realm that includes the object itself being defined. This is seen as incoherent if the definition is meant to pick out or construct the set; how can we do so using a property that already refers to the set itself? We could define a set of natural numbers predicatively, for example, by a property whose quantifiers ranged only over the natural numbers and not over sets of natural numbers; or similarly with the real numbers or with larger collections. In the main step of the Russell paradox argument, in contrast, we define the Russell set $R = \{ x \mid x \notin x \}$, and this is impredicative, since we are attempting to define R by letting the variable x range over the collection of all sets x, of which we hope R is a member. According to this view, impredicative definitions are seen as invalid or unreliable as a means of definition, and one should allow only predicative definitions in mathematics.

From the predicative perspective, the move from general comprehension to separation does not solve the impredicativity objection because many instances of the separation axiom—those involving definitions with quantifiers ranging over the entire universe of sets—remain impredicative. When I define a set of natural numbers using a property with a universal quantifier $\forall x$, such as with $A = \{ n \in \omega \mid \forall x \varphi(x, n) \}$, then the definition is impredicative because A itself is one of the xs that is considered. For this reason, the predicative formulations of set theory allow only very restricted forms of the separation axiom, which make set-existence assertions for subsets $\{ a \in A \mid \varphi(a) \}$ only when all quantifiers appearing in φ are bounded by other sets. Such a formula is said to have complexity Δ_0, and this weak form of separation is known as Δ_0-*separation*.

Meanwhile, mathematics abounds with the routine use of elementary impredicative definitions. For example, the closure of a set A in a topological space is often defined as the intersection of all closed sets containing A; this is technically impredicative because the closure $\overline{A}$ is itself one of those closed sets. Similarly, the subgroup $\langle A \rangle$ generated by a set $A \subseteq G$ in a group G is often defined as the intersection of all subgroups of G containing A; this is again impredicative because $\langle A \rangle$ itself is one of those subgroups. Even the usual definition of the greatest common divisor of two integers is impredicative in this sense, and there are many similar cases. Nevertheless, in all these cases, we typically also have an equivalent predicative definition, often one of a much more constructive nature. For example, the closure of a set is also obtained by adding to the set all its cluster points, the points that cannot be isolated from it by a small open set; and the subgroup generated by a set can be described concretely by using the finite group-theoretic terms that may be formed using elements of that set. To my way of thinking, mathematical insight is often achieved by contemplating the interplay of equivalent impredicative and predicative constructions, and it would seem to be a loss for mathematics if all these impredicative definitions were judged illegitimate.

8.6 Extensionality

Perhaps the quintessential set-theoretic axiom is the axiom of extensionality.

Axiom of extensionality. *If sets have the same elements, then they are equal.*

$$\forall a \forall b \left[(\forall x\ x \in a \leftrightarrow x \in b) \implies a = b \right].$$

This axiom expresses the core set-theoretic idea that a set is determined by its elements. When sets are constituted by the same elements, then they are the same set.

There is surely something to that view, and yet it turns out that one may still undertake quite a rich set theory without extensionality. Without extensionality, one might have more than one set with the same elements. In this case, we might naturally want to say that these sets are equivalent in some sense, and indeed, we would want to say generally that two sets are equivalent when they have the same elements. We would then seek to treat such equivalent sets as "the same" in our set theory. But then, of course, we will want to extend the equivalence relation by saying that two sets are equivalent if their elements were previously seen to be equivalent; and then one will want to extend the equivalence relation again, and then again. Ultimately, one is pushed to a limit equivalence relation $\sim$, a *bisimulation*, such that $a \sim b$ implies that every $x \in a$ has $x \sim y$ for some $y \in b$, and every $y \in b$ has $x \sim y$ for some $x \in a$, and whenever $a \sim b$ and $a \in u$, then there is $u \sim v$ with $b \in v$. It is remarkable that by means of such a bisimilarity relation, one can interpret an extensionality-based set theory within set theory without the axiom of extensionality. To my way of thinking, this observation has a certain affinity with the univalence axiom of homotopy type theory, which also provides something like a bisimilarity criterion for identity.

Other axioms

There are several more Zermelo axioms. The axiom of *pairing* says that for every a and b, there is a pair set $\{a, b\}$ with only a and b as members. The axiom of *union* says that for any set A, there is a union set $\bigcup A = \{ x \mid x \in a \text{ for some } a \in A \}$, the union of the sets in A. The axiom of *power set* says that for every set A, there is a *power* set $P(A) = \{ B \mid B \subseteq A \}$, consisting of exactly the subsets of A.

Two additional axioms were added later, to form the Zermelo-Fraenkel ZFC axiomatization of set theory. First, the axiom of *regularity* or *foundation* expresses a certain consequence of our expectation that the sets of the cumulative hierarchy are formed in a well founded sequence of stages. If the hierarchy is well founded, then every nonempty set of sets will have one or more elements that appear earliest among all the elements of the set. In particular, such a set will have no elements in common with the original set. So every nonempty set A has an element $a \in A$ such that $a \cap A = \varnothing$, and this assertion is precisely the regularity axiom.

8.7 Replacement axiom

Next, the *replacement axiom* addressed a certain weakness of Zermelo's original axiomatization, which was noticed and corrected by Abraham Fraenkel. To appreciate this issue, suppose that for each natural number $n \in \omega$, we have defined a set a_n. We would like to form the set

$$\{ a_n \mid n \in \mathbb{N} \}.$$

But actually, it is not always possible to prove that this is a set in Zermelo's original set theory. The replacement axiom redresses this by asserting that this set does in fact exist.

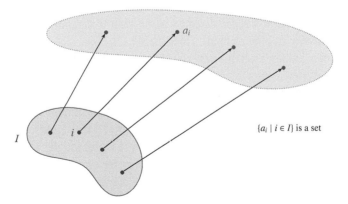

More generally, if I is a set, and for each $i \in I$ there is a unique object a_i with a certain property, then the replacement axiom asserts that the collection $\{ a_i \mid i \in I \}$ is a set. The axiom is called *replacement* because we are in effect replacing the elements of the set I

with the defined objects a_i. Zermelo-Fraenkel (ZF) set theory is the theory resulting from all the axioms mentioned so far.

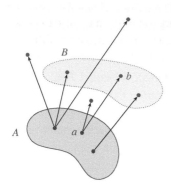

$\exists B \,\forall a \in A \,\exists b \in B \,\varphi(a, b)$

The replacement axiom turns out to admit diverse equivalent characterizations. For example, consider the axiom of *collection*, which asserts that if A is a set, and for every $a \in A$, there is an object b (and possibly many such objects) such that $\varphi(a, b)$, then there is a set B such that every $a \in A$ has some $b \in B$ with $\varphi(a, b)$. Thus, we have collected into a set B at least one witness b for each a in A. One can equivalently axiomatize ZF using collection and separation instead of replacement.

The replacement axiom is also equivalent over the other Zermelo axioms to the principle of transfinite recursion, asserting that definitions by well ordered transfinite recursion are legitimate (and this equivalence holds even without the axiom of choice). It is also equivalent to the reflection principle, asserting that for any formula φ, there is a level V_κ of the cumulative hierarchy for which φ is absolute between V_κ and the entire set-theoretic universe V. The replacement axiom is also equivalent over the Zermelo axioms to the well ordered replacement axiom, which asserts the instances of replacement for which the index set I is well ordered. It turns out that well ordered replacement implies full replacement, even without the axiom of choice.

The number of infinities

As an instance of replacement, let us revisit a claim we had discussed in chapter 3 concerning the number of infinite cardinalities. Let us denote the nth infinite cardinal by $\aleph_n$, taking these cardinals to be naturally represented by *initial* von Neumann ordinals (ordinals not equinumerous with any smaller ordinal). Does the set $\{\aleph_n \mid n \in \omega\}$ exist? Well, yes, if one has the replacement axiom, then this set exists by an instance of replacement, since ω is a set and we are replacing each $n \in \omega$ with $\aleph_n$.

This fact was important in the argument we had given in chapter 3 that there are uncountably many different infinite cardinalities. Namely, using the axiom of replacement and the other Zermelo-Fraenkel axioms, we may form the set $\{\aleph_n \mid n \in \omega\}$ and let $\aleph_\omega = \sup_{n \in \omega} \aleph_n$, since the supremum of any set of ordinals exists simply by taking the union. More generally, we can prove in Zermelo-Fraenkel set theory that for any ordinal α, the cardinal $\aleph_\alpha$ exists, the αth infinite cardinal, and since there are uncountably many ordinals, we have thus proved that there are uncountably many infinite cardinals. Further, the class of infinite cardinals cannot be put in a one-to-one correspondence with any set, since if κ_i is a cardinal for each $i \in I$, then by the replacement axiom, we can form $\{\kappa_i \mid i \in I\}$ and therefore let $\kappa = \sup_{i \in I} \kappa_i$ be the supremum, so 2^κ will be larger than any κ_i. So there are more infinite cardinalities than any given one of them.

One might have said that the number of infinite cardinalities exceeds any one of them, but it is problematic to refer to the *number* of infinite cardinalities, precisely because they do not form a set; the cardinals form a proper class, like the class of all ordinals, the class of all sets, or the Russell class, classes which in each case are too large in a sense to be sets, upon pain of contradiction. In this sense, there is no "number" of infinite cardinalities.

Some class theories, such as Gödel-Bernays set theory or Kelley-Morse set theory, include the global axiom of choice, which implies that all proper classes are equinumerous. In this sense, these theories prove that all proper classes have the same number of elements. Meanwhile, set theorists are investigating the range of possibilities when one drops the global axiom of choice, and it is known to be consistent with the other axioms of Gödel-Bernays set theory, for example, that the class of all sets is not linearly ordered by any class relation, and from this, it follows that it is not equinumerous with the class of all ordinals, which are canonically well ordered.

The argument given here that there are more infinities than any one of them relies fundamentally on the replacement axiom, and it turns out that without the replacement axiom, we cannot prove even that $\aleph_\omega$ exists; in Zermelo set theory, there might not be any set of this size. That is a strong claim, a *metamathematical* claim, since it is a claim that we cannot establish a certain fact in a certain theory. How does one show such a metamathematical claim—that is, that something is *not* provable? We are not claiming that the theory refutes the claim, since our stronger theories with the replacement axiom affirm the claim. So we want to prove that something is not provable, but not by refuting it.

What we need to do is to find a model of the Zermelo axioms, without replacement, in which the only infinite cardinals that exist are the $\aleph_n$ for finite n. And indeed this is possible. In any model of ZFC, consider the set $V_{\omega+\omega}$, the rank-initial segment of the cumulative hierarchy up to rank $\omega + \omega$. Let us view $\langle V_{\omega+\omega}, \in \rangle$ as a possible model of set theory. It is not difficult to see that this model satisfies all the Zermelo axioms of set theory, with the axiom of foundation. But it does not satisfy the axiom of replacement because ω is a set in this model and the map $n \mapsto V_{\omega+n}$ is definable, but the range of this map is not a set, since the union of it is the whole model, which is not an element of itself. Since the cardinality of $V_{\omega+n}$ is $\beth_n$, it follows that there is no set in this model of cardinality $\beth_\omega$. If the original ZFC model satisfied the generalized continuum hypothesis, then the only infinite cardinals of this model would be the $\aleph_n$ for finite n; every infinite set would have size $\aleph_n$ for some finite n.

Note that the model $V_{\omega+\omega}$ has comparatively few von Neumann ordinals—only those up to $\omega+\omega$—and so even though every set in this model may be well orderable there, this does not mean that every set is bijective with an ordinal, since you need the replacement axiom to perform the Mostowski collapse, the isomorphism of any well-order with an ordinal (named for Andrzej Mostowski). So a further point of this argument is that Zermelo set theory, as opposed to Zermelo-Fraenkel set theory, lacks a robust theory of von Neumann ordinals.

8.8   The axiom of choice and the well-order theorem

Zermelo formulated his axioms in 1908 in order to defend his 1904 theorem showing that the real numbers (and indeed any set) can be well ordered. The *well-order theorem* asserts that any set A admits a well-order, a linear order relation $<$ on A such that every nonempty subset of A has a least element with respect to the order. Thus, there will be a least element of A with respect to this order, then a next least element, then a next least element, and so on, and then a next least element after all those, and so on. At any stage, one will either have exhausted all the elements of A or else there will be a next element. Any countable set admits a well-ordering, of course, since the countability of the set itself induces such an order by enumeration. In the uncountable case, however, even with the set of real numbers, it is not easy to imagine what a well-ordering would look like. Can you define a well-ordering of the real numbers?

Well, the question of whether you can *define* a well-ordering of the real numbers is not quite the same as the question of whether there is one. Perhaps there is a strange, undefinable well-order? Perhaps there is an order that does not obey any nice rule or definition? Such was the kind of order that Zermelo's argument might have provided. He based his argument on the principle that one can freely "choose" elements from the sets in any family of nonempty sets. That is, for any set of nonempty sets, there is a way of picking an element from each of them.

Axiom of choice (AC). *For any set U consisting of nonempty sets, there is a choice function, a function f with domain U, such that $f(u) \in u$ for every $u \in U$.*

The function f is in effect a way of choosing elements from the sets u in U. This axiom is fundamentally different from the other axioms of set theory, which each make set-existence assertions, but in a way that explicitly describes the nature and elements of the object asserted to exist. Namely, the union set consists of the elements of elements of the given set; the pair set of a and b is the set whose only elements were a and b; the power set of a set consists of all the subsets of that set. With the axiom of choice, however, we are told that there is a choice function, but we are not told how to construct it or anything about which choices it might make. The axiom tells us that there is a function with a certain property, but indeed in nontrivial instances, we may expect that there could be many such functions. The axiom does not specify a particular such function; rather, it simply asserts that there is at least one function with that property.

To illustrate this definability issue more concretely, consider that we have a collection of pairs of shoes, and we want to choose one shoe from each pair. This is quite easy, for we could decide always to choose the left shoe from each pair, for example, and this does not require the axiom of choice, because we can define this function explicitly, without appealing to the axiom of choice. But if, in contrast, we had had an infinite collection of pairs of socks, indistinguishable within each pair, then we do not seem necessarily to have a way of choosing one sock from each pair. The axiom of choice, nevertheless, ensures that indeed there is a way of choosing, and in fact once we have one choice function, we can easily construct many more.

Zermelo used the axiom of choice in order to prove the well-order theorem; he constructed a well-ordering for any given set A. To see how, fix the set A and let U be the collection of nonempty subsets of A. By the axiom of choice, there is a choice function for U. This choice function provides a way of choosing an arbitrary element from any given nonempty subset of A. Zermelo's keen idea was simply to iteratively choose elements from A, choosing from amongst the elements not yet chosen. To begin, we choose an element of A. Among the elements remaining, choose the next element, and so on. At any stage, if there are any elements remaining, choose one of the remaining elements to be the next element in the order. In this way, one defines recursively a well ordered enumeration of A.

A bit more formally, one should fix a particular choice function on the set of nonempty subsets of A, and then consider all well ordered sequences from A that conform with that choice function in the manner I just described, so that the next point on the sequence is always the element of A that the choice function picks from the elements not yet on the sequence. All such well ordered sequences cohere with one another, since they can have no least place of disagreement. The union of all the sequences, therefore, is a maximal such well ordered sequence from A, which cannot leave any elements of A remaining, for if there were points remaining, then we could make a longer sequence by using f to choose an element from those remaining elements. So A is well orderable.

Zermelo thus proved that the axiom of choice implies the well-order principle, that every set can be well ordered. Many mathematicians find this quite remarkable—a bit of mathematical magic. Starting with the axiom of choice, a principle that many find initially to be totally unobjectionable (or even banal), one concludes with the mysterious, and frankly suspicious, conclusion that every set can be well ordered, a conclusion that most mathematicians cannot readily imagine even in such concrete cases as the real numbers. Mathematicians tell it as a joke: the axiom of choice is obviously true, the well-order theorem is obviously false, and about Zorn's lemma, who can tell? But of course, they are all equivalent.

The axiom of choice turns out to be equivalent to dozens of various fundamental principles. The converse of Zermelo's implication is true, because if every set has a well-order, then one can easily build choice functions: for any set A of nonempty sets, simply well order the union of those sets and then pick the least element of any set with respect to that order. The axiom of choice is equivalent to Zorn's lemma, which asserts that if a partial order has the property that every linear chain is bounded above, then it has a maximal element. The axiom of choice is also equivalent to the assertion that every vector space has a basis; it is equivalent to Tychonoff's theorem, which asserts that the product of compact spaces is compact; it is equivalent to the assertion that any two cardinalities are comparable: given two sets A and B, one of them is at least as large as the other.

Paradoxical consequences of AC

The axiom of choice is important in many parts of mathematics and routinely assumed. Yet the axiom remains controversial, first because of its nonconstructive nature, but second because of some counterintuitive consequences that it has, particularly with regard to measure-related issues.

For example, Giuseppe Vitali proved that the axiom of choice implies that there are nonmeasurable sets of real numbers with respect to the Lebesgue measure or indeed with respect to any countably additive translation-invariant measure. To do this, he first defined that two real numbers are equivalent $x \sim y$ if they differ by a rational number. Let V be a *Vitali* set, one that chooses exactly one element from each equivalence class. The translation set $V + q$ is disjoint from $V + r$ for distinct rational numbers, since otherwise $u + q = v + r$, for some $u, v \in V$, and hence $u - v = r - q$ is rational; but V has only one member of each equivalence class, so $u = v$, and thus $q = r$, contrary to assumption. Since every number is equivalent to some number in V, it follows that the rational translations $V + q$ are disjoint and cover $\mathbb{R}$. We may assume that V is contained in the unit interval and so V will have infinitely many disjoint translations appearing in the interval $[0, 2]$. Since they all have the same measure, this measure cannot be nonzero. But it cannot be zero either, since then the whole real number line $\mathbb{R}$ would be a countable union of measure zero sets, and hence measure zero itself. So the Vitali set cannot have a measure.

More extreme examples of nonmeasurable sets arise in the Banach-Tarski paradox, which asserts that the unit ball in three-space $\mathbb{R}^3$ can be partitioned into seven disjoint pieces, such that by applying certain rigid motions to these pieces, one can form fully two copies of the unit ball, of the same size as the original, with no gaps or holes or missing points. The nature of these sets, of course, is quite bizarre—one should not imagine carving the sets with a knife—they are each dense in the unit ball, mixed amongst each other. The sets are necessarily nonmeasurable, since rigid motion preserves measure.

Paradox without AC

Meanwhile, it would be a form of confirmation bias to discuss only counterintuitive consequences of the axiom of choice, without also discussing the counterintuitive situations that can occur when the axiom of choice fails. Although mathematicians often point to what are perceived as strange consequences of the axiom of choice, a fuller picture is revealed by also mentioning that many of the situations that can arise when one drops the axiom are perhaps even more bizarre.

For example, it is relatively consistent with the axioms of set theory without the axiom of choice that there can be a nonempty tree T, with no leaves, but which has no infinite path. That is, every finite path in the tree can be extended to further steps, but there is no path that goes forever. This situation can arise even when countable choice holds (so countable families of nonempty sets have choice functions), and this highlights the difference between the countable choice principle and the principle of *dependent choice*, where one makes countably many choices in succession. Finding a branch through a tree is an instance of dependent choice, since the later choices depend on which choices were made earlier.

Without the axiom of choice, a real number can be in the closure of a set of real numbers $X \subset \mathbb{R}$, but not the limit of any sequence from X. Without the axiom of choice, a function $f : \mathbb{R} \to \mathbb{R}$ can be continuous at a point x in the sense that every convergent sequence $x_n \to x$ has a convergent image $f(x_n) \to f(x)$, but not continuous in the ϵ, δ sense. Without the axiom of choice, a set can be infinite, but have no countably infinite subset. Indeed, without the axiom of choice, there can be an infinite set, with all subsets either finite or the complement of a finite set. Thus, it can be incorrect to say that $\aleph_0$ is the smallest infinite cardinality, since these sets would have an infinite size that is incomparable with $\aleph_0$.

Without the axiom of choice, there can be an equivalence relation on $\mathbb{R}$, such that the number of equivalence classes is strictly greater than the size of $\mathbb{R}$. That is, you can partition $\mathbb{R}$ into disjoint sets, such that the number of these sets is greater than the number of real numbers. Bizarre! This situation is a consequence of the axiom of determinacy and is relatively consistent with the principle of dependent choice and the countable axiom of choice.

Without the axiom of choice, there can be a field with no algebraic closure. Without the axiom of choice, the rational field $\mathbb{Q}$ can have different nonisomorphic algebraic closures. Indeed, $\mathbb{Q}$ can have an uncountable algebraic closure as well as a countable one. Without the axiom of choice, there can be a vector space with no basis, and there can be a vector space with bases of different cardinalities. Without the axiom of choice, the real numbers can be a countable union of countable sets, yet still uncountable. In such a case, the theory of Lebesgue measure is a complete failure.

To my way of thinking, these examples support a call for balance in the usual conversation about the axiom of choice regarding counterintuitive or surprising mathematical facts. Namely, the typical way of having this conversation is to point out the Banach-Tarski result and other counterintuitive consequences of the axiom of choice, heaping doubt on the axiom of choice; but a more satisfactory conversation would also mention that the axiom of choice rules out some downright bizarre phenomena—in many cases, more bizarre than the Banach-Tarski-type results.

Solovay's dream world for analysis

In a major foundational development, a celebrated theorem of Robert Solovay shows that if the existence of an inaccessible cardinal is consistent with ZFC, then it is consistent with ZF plus the principle of dependent choice (a weakening of the axiom of choice) that every set of real numbers is Lebesgue measurable. Some mathematicians have described Solovay's theory as providing a dream world for analysis—a world without pain and contradiction, where the nonmeasurable monsters are banished, all sets of real numbers are measurable, all functions are measurable, and mathematicians have π for breakfast every day and transform coffee into theorems.

Solovay told me once that when he proved his theorem, he expected that the analysts would immediately adopt this theory as the new foundational theory for analysis. Gone would be the worries about nonmeasurability; mathematics would be free and easy. One would have expected him to be carried on the shoulders of the analysts, hailed as a hero. But that did not happen. Analysts continued to work essentially in the usual ZFC set theory, keeping the axiom of choice, and paying attention to their various σ-algebras of measurable sets. Why was Solovay's theory not adopted wholesale?

My explanation is that the pull of the axiom of choice as a fundamental principle is simply too strong—stronger than critics of the axiom admit. Perhaps analysts, like most mathematicians, see the axiom of choice as expressing a fundamental mathematical truth. To live in Solovay's dream world would be to live in a mathematical fantasy rather than in the real mathematical world, where the axiom of choice holds. Theorems proved in that dream world, from this view, would concern only an imaginary mathematics, and therefore would not provide genuine mathematical insight into the actual mathematical questions of the real mathematical world.

But what is this "real" mathematical world? What are we talking about with these various mathematical worlds? An alternative, more pragmatic explanation might be simply that analysts preferred to worry about whether their sets were measurable rather than about whether they were inadvertently making an unwarranted use of the axiom of choice.

8.9   Large cardinals

Let us turn now to large cardinals, the higher infinite. The existence assertions for these sublime concepts of infinity are the strongest known axioms in mathematics. Large cardinals are infinities so vast that we cannot prove that they exist using the usual axioms of set theory; instead, we postulate their existence with axioms, the large cardinal axioms, which form a sweeping hierarchy of strength. The larger large cardinals generally imply not only the existence, but also the consistency of the smaller cardinals. The smaller cardinals, if consistent, are consequently too weak to prove the existence, or even the consistency, of the larger cardinals.

The large cardinal hierarchy is thereby a natural instantiation of the transfinite tower of iterated consistency strength that we know must exist in light of Gödel's incompleteness theorem. Namely, by the incompleteness theorem, we know that every computable axiomatization of mathematics is inherently transcended by an iterated tower of consistency assertions: no such theory T proves its own consistency $\mathrm{Con}(T)$, and $T + \mathrm{Con}(T)$ does not prove the further consistency $\mathrm{Con}(T + \mathrm{Con}(T))$, and so on. Thus, a tower of consistency strength rises above the theory T. How remarkable it is that in set theory, such a tower of consistency strength is realized not only by self-referential syntactic consistency assertions, but also by fundamental principles of infinite combinatorics, which are natural assertions of independent mathematical interest.

Large cardinals were studied from the earliest days of set theory. Hausdorff had introduced the inaccessible cardinals; Zermelo used them in his categoricity result (theorem 24) for the models of second-order ZF. In time, we had the hyperinaccessible cardinals, the Mahlo cardinals, measurable cardinals, supercompact cardinals, huge cardinals, and much more. A rich, fascinating hierarchy of infinity emerged—a core discovery of twentieth-century set theory. Here is a small fragment of the large cardinal hierarchy:

Superhuge cardinals
Huge cardinals
Almost-huge cardinals
Supercompact cardinals
Superstrong cardinals
Strong cardinals
Tall cardinals
Measurable cardinals
Subtle cardinals
Weakly superstrong cardinals
Unfoldable cardinals
Totally indescribable cardinals
Weakly compact cardinals
Greatly Mahlo cardinals
Mahlo cardinals
Uplifting cardinals
Hyperinaccessible cardinals
Inaccessible cardinals
Worldly cardinals

Evidently, words have failed us. The large cardinal terminology merely hints at the staggering enormity of these infinities. Each large cardinal concept, of course, has a mathematically precise definition, and some of these are quite technical. Let me try to convey in an elementary manner some of the flavor of large cardinal analysis.

Strong limit cardinals

Consider the countable infinity of the natural numbers, the smallest infinity, which is denoted ω when considered as an ordinal and $\aleph_0$ when considered as a cardinal, although in ZFC, one regards cardinals as initial ordinals—those not equinumerous with any smaller ordinal—and so these notations refer to the same object. This smallest infinity $\aleph_0$ exhibits certain mathematical properties and features not found amongst the finite sets. For example, since the power set of a finite set is finite, the cardinal $\aleph_0$ is closed under the power set operation: every set smaller than $\aleph_0$ has a power set that also is smaller than $\aleph_0$. Are there uncountable cardinals κ like that, such that every set of size less than κ has its power set also less than κ?

Yes, these are precisely the *strong limit* cardinals. Let us discover the first uncountable strong limit cardinal. In chapter 3, we considered iterating the power set operation on the natural numbers:

$$\mathbb{N} < P(\mathbb{N}) < P(P(\mathbb{N})) < P\left(P(P(\mathbb{N}))\right) < \cdots$$

The cardinalities of these sets are precisely the beth numbers $\beth_n$, with supremum $\beth_\omega$. This cardinal $\beth_\omega$ is uncountable because after the first step, all the sets on the sequence are uncountable; and it is a strong limit because if X has a size of less than $\beth_\omega$, then it will be bounded in size by some $\beth_n$, and so the power set $P(X)$ will be bounded by the next cardinal $\beth_{n+1}$, which is still smaller than $\beth_\omega$. So $\beth_\omega$ is an uncountable, strong limit cardinal.

Regular cardinals

Another property of infinity is that every partition of an infinite set into finitely many pieces must have an infinite piece. With regard to the countable infinity $\aleph_0$, we can express this principle as the assertion: every partition of a set of size $\aleph_0$ into fewer than $\aleph_0$ pieces has at least one piece of size $\aleph_0$. Are there uncountable cardinals κ like that, such that every partition of κ into fewer than κ pieces has a piece of size κ? Yes, these are exactly the *regular* cardinals. One of the consequences of the axiom of choice in ZFC is that every infinite successor cardinal κ^+ is regular.

Aleph-fixed-point cardinals

In chapter 3, we discussed the infinite cardinals

$$\aleph_0 < \aleph_1 < \aleph_2 < \cdots < \aleph_\omega < \aleph_{\omega+1} < \cdots < \aleph_\alpha < \cdots$$

Consider the cardinal $\aleph_\omega$, the supremum of $\aleph_n$ for finite n. This cardinal is not regular, because it is the supremum of the cardinals $\aleph_n$ for $n \in \omega$; that is, unlike ω, it is a *singular* cardinal, the supremum of a short sequence of smaller cardinals, meaning a sequence of length less than $\aleph_\omega$ using cardinals of size less than $\aleph_\omega$. In this sense, $\aleph_\omega$ is small and accessible; one can climb to $\aleph_\omega$ from below on a ladder having rungs at each $\aleph_n$. Regular cardinals κ, in contrast, are never the supremum of any smaller-than-κ set of smaller-than-κ cardinals.

After $\aleph_\omega$, we can form the cardinals $\aleph_{\omega+1}$, and so on, $\aleph_\alpha$ for every ordinal α. The cardinal $\aleph_\alpha$ is the αth infinite cardinal. One might typically expect $\alpha < \aleph_\alpha$, and this is certainly true initially, with $1 < \aleph_1$ and $\omega < \aleph_\omega$ and $\omega^2 + 5 < \aleph_{\omega^2+5}$. But could there be an ordinal λ with $\lambda = \aleph_\lambda$? This would be an $\aleph$-*fixed-point* cardinal—a cardinal λ with λ many cardinals below λ. Thus, the cardinal λ would be the λth cardinal. It follows that such a λ would have to be very large—indeed, bigger than $\aleph_\alpha$ for every $\alpha < \lambda$. Can this happen?

Yes, it can. Let us discover the first $\aleph$-fixed-point cardinal. Start with the cardinal $\aleph_0$, which is the same as the ordinal ω, the smallest infinity. Next, count out that many infinite cardinals, ω many, arriving at the cardinal $\aleph_\omega$, a much larger cardinal. And then count out *that* many additional cardinals, arriving at $\aleph_{\aleph_\omega}$. And now count out *that* many cardinals, and so on. Repeating this process infinitely many times, let λ be the supremum of the cardinals that arise:

$$\omega \qquad \aleph_\omega \qquad \aleph_{\aleph_\omega} \qquad \aleph_{\aleph_{\aleph_\omega}} \qquad \cdots$$

Because of the nature of the construction, if $\alpha < \lambda$, then α will be less than one of those cardinals on the list, and so we will have counted out α many cardinals by the next cardinal on the list. So for every $\alpha < \lambda$, there are at least α many cardinals below λ. Thus, λ must be the λth cardinal, and so $\aleph_\lambda = \lambda$, an $\aleph$-fixed point, as desired.

Inaccessible and hyperinaccessible cardinals

We have observed so far both that the countable infinity $\aleph_0$ is regular and that it is a strong limit cardinal. Can we find an uncountable cardinal with both of these properties? These cardinals, the uncountable regular strong limit cardinals, are precisely the *inaccessible* cardinals. In some sense, an inaccessible cardinal κ relates to the cardinals below κ just as $\aleph_0$ relates to the finite cardinals. Every inaccessible cardinal must be an $\aleph$-fixed point because if κ is inaccessible, then it is a limit cardinal, and hence $\aleph_\lambda$ for some limit ordinal λ; but since $\aleph_\lambda$ is the supremum of the λ sequence of smaller $\aleph_\alpha$, by regularity, it must be that $\lambda = \kappa$. And so $\kappa = \aleph_\kappa$ is an $\aleph$-fixed point.

Let us climb higher into the large cardinal hierarchy. An inaccessible cardinal κ that is itself a limit of inaccessible cardinals is said to be 1-*inaccessible*; such a cardinal κ is the κth inaccessible cardinal; it is a fixed point in the enumeration of the inaccessible cardinals, transcending mere inaccessibility. But similarly transcending *this* notion, a 2-inaccessible cardinal is an inaccessible limit of 1-inaccessible cardinals; more generally, an α-inaccessible cardinal is a limit of β-inaccessible cardinals for every $\beta < \alpha$. Transcending that entire hierarchy, a cardinal κ is *hyperinaccessible* when κ is κ-inaccessible. Climbing still higher, a cardinal is 1-hyperinaccessible when it is a limit of hyperinaccessible cardinals, and α-hyperinaccessible when it is a limit of β-hyperinaccessible cardinals for every $\beta < \alpha$. And what of fixed points in this hierarchy? You guessed it—the *hyperhyperinaccessible* cardinals are those inaccessible cardinals κ that are κ-hyperinaccessible. Erin Carmody (2017) describes a vast generalization of hyperinaccessiblity, including the *richly* inaccessible cardinals, the *utterly* inaccessible cardinals, and others.

So we have described a hierarchy of inaccessible and hyperinaccessible cardinals. But are there any inaccessible cardinals? The answer is extremely interesting. Suppose that κ is an inaccessible cardinal, and consider the set V_κ, the rank-initial segment of the cumulative universe up to stage κ. It turns out that this is a very rich collection of sets. One may consider it as a miniature set-theoretic universe on its own. Indeed, if κ is an inaccessible cardinal, then all the ZFC axioms are true in V_κ if one considers it as a model of set theory. To see this, notice first that the axioms of extensionality and foundation are true in any transitive set; the axioms of pairing, union, and power set are true in V_λ for any limit ordinal λ; the axiom of infinity is true in V_λ whenever $\lambda > \omega$; the axiom of separation is true in every V_α, since any subset of a set in V_α is also in V_α. Indeed, V_α is a model of the second-order separation axiom. The axiom of choice is true in every V_λ, provided that it is true in the ambient universe V, since the choice functions for a set have rank just slightly more than the set. Finally, if κ is inaccessible, then one can show that every set $x \in V_\kappa$ has size less than κ, and so by the regularity of κ, every function $x \to V_\kappa$ will be bounded in rank by some V_α for $\alpha < \kappa$. Thus, the image of the function will be an element of V_κ, and so we achieve ZFC in V_κ, as desired.

Zermelo (1930) had characterized the sets V_κ for κ inaccessible, the rank-initial segments of the cumulative universe up to an inaccessible cardinal, as exactly the models of his second-order set theory ZF_2, a relative categoricity result on a par with Dedekind's analysis, for it shows the extent to which second-order set theory determines the set-theoretic universe. These set-theoretic worlds V_κ, for κ inaccessible, are now known as the Zermelo-Grothendieck universes; they were rediscovered by Alexander Grothendieck many years later and have become a central universe concept in category theory. The axiom of universes is the assertion that every set is an element of some Zermelo-Grothendieck universe; this is equivalent to the large-cardinal axiom that there are unboundedly many inaccessible cardinals.

Notice that if κ is the smallest inaccessible cardinal, then V_κ is a model of set theory with no inaccessible cardinals inside it. It is, therefore, relatively consistent with the axioms of set theory that there are no inaccessible cardinals. In other words, if the ZFC axioms of set theory are consistent, then we cannot prove from them that there is an inaccessible cardinal. But not only can we not prove in ZFC that inaccessible cardinals exist, we cannot even prove that their existence is relatively consistent. That is, we cannot prove that $\text{Con}(\text{ZFC}) \to \text{Con}(\text{ZFC} + \exists \text{ inaccessible})$. The reason is that, as we have observed, the existence of an inaccessible cardinal implies the consistency of ZFC, and so if $\text{Con}(\text{ZFC})$ implied the consistency of an inaccessible cardinal, then it would imply its own consistency, contrary to Gödel's second incompleteness theorem, applied with the theory $\text{ZFC} + \text{Con}(\text{ZFC})$. Thus, the existence of an inaccessible cardinal transcends ZFC in consistency strength.

That observation exemplifies a principal feature of the large cardinal hierarchy—namely, that each level of the hierarchy completely transcends those below it. From the existence of a smaller large cardinal, we cannot prove the existence of the larger large cardinals; and indeed, from a smaller large cardinal, we cannot prove even the consistency of the larger cardinals with the axioms of set theory. From one inaccessible cardinal, we cannot prove the consistency of a second one. From 100 inaccessible cardinals, we cannot prove that there is another one, because if κ is the 101st inaccessible cardinal, then V_κ will be a model of the axioms of set theory and have precisely 100 inaccessible cardinals.

Robin Solberg and I have investigated the inaccessible categoricity phenomenon, by which inaccessible cardinals can be categorical in various senses, Hamkins and Solberg

(2020). For example, an inaccessible cardinal κ is first-order *sententially categorical* if V_κ is characterized by augmenting ZFC_2 with a single first-order sentence, and similarly for a second-order sentence or a theory. The smallest inaccessible cardinal κ, for example, is the only one for which V_κ satisfies the statement "There are no inaccessible cardinals." And similarly, the fifth one thinks, "there are exactly four." Since categoricity is a feature of the smallest large cardinal instances, noncategoricity is the largeness notion here. Meanwhile, we have separated the various categoricity notions, and amongst other interesting facts, we have proved that if there are sufficiently many inaccessible cardinals, then there must be gaps in the sententially categorical inaccessible cardinals.

Linearity of the large cardinal hierarchy

Set theorists have observed that the large cardinal hierarchy seems to have a roughly linear nature with respect to consistency strength. Given two large cardinal notions, we can usually prove that the consistency of one of them implies the consistency of the other. Some set theorists imbue this observation with deep significance. Since one does not ordinarily expect mathematical axioms to line up in this linear fashion, the fact that the large cardinal hypotheses do may be evidence of the naturality of the hierarchy, perhaps indicating the truth or consistency of the axioms. To my way of thinking, the significance of the linearity phenomenon is not yet adequately explored philosophically.

One obvious counterpoint to the argument is that in many instances, the large cardinal hypotheses were formulated by strengthening an already known axiom—we had strengthened the strongly compact cardinal concept to supercompact, and then again to almost huge, and to huge and superhuge—and so in this respect, it is unsurprising that they get stronger in a linear fashion; that's how we made them. Yet it would be wrong to suggest that the entire hierarchy was generated in that manner, and there is a genuine linearity phenomenon. The fact that inaccessible cardinals, weakly compact cardinals, Ramsey cardinals, measurable cardinals, and strongly compact cardinals are linearly related by strength is indeed surprising, especially when considering the original definitions of these notions, which generalized diverse aspects of the countable infinity $\aleph_0$ to the uncountable. Why should those properties have been linearly ordered by consistency strength?

The large cardinal hierarchy admits several natural orders, including direct implication, consistency implication, and the size of the least instance, and these do not always agree. The least superstrong cardinal, for example, is always smaller than the least strong cardinal when they both exist, even though the former has a higher consistency strength. Similarly, the least strongly compact cardinal can be consistently either larger than or smaller than the least strong cardinal. Because of these kinds of examples, the linearity phenomenon is concentrated in the consistency-strength relation. In that context, there are no known natural counterexamples to linearity, but there are open cases that can be taken as counterexample candidates. For example, we do not know the consistency-strength relation between a supercompact cardinal and two strongly compact cardinals. Meanwhile, using the Gödel fixed-point lemma, one can manufacture artificial counterexamples to linearity, statements ϕ and ψ, whose consistency strengths are not linearly related, so neither Con(ZFC + φ) nor Con(ZFC + ψ) provably implies the other. These statements, however, do not correspond to any naturally occurring large cardinal hypotheses.

Another more subtle counterpoint to the linearity phenomenon consists of the observation that we lack general methods for proving instances of nonlinearity in consistency strength. We usually prove a failure of consistency implication by proving the converse implication, but this method, of course, can never establish nonlinearity. In this sense, the observed linearity phenomenon might be an instance of confirmation bias: we see no instances of nonlinearity because we have no tools for observing nonlinearity. In this case, it would not be significant that we do not see it.

Large cardinals consequences down low

The existence of sufficiently large large cardinals turns out to have some surprising consequences for sets down low. For example, if there is a measurable cardinal, then every Σ_2^1-definable set of real numbers is Lebesgue measurable, and if there is a proper class of Woodin cardinals, then every projectively definable set of real numbers is Lebesgue measurable and has the property of Baire. This is an astonishing connection between the existence of extremely large infinities and very welcome regularity properties for definable sets of real numbers and definable functions on the real numbers.

Why should the existence of these enormous infinities have such attractive regularity consequences for sets of real numbers, which are comparatively low in the cumulative hierarchy? Some set theorists have taken these regularity consequences to be evidence for the truth of the large cardinal axioms, on the grounds of fruitful consequences. The large cardinal theory seems to be providing a rich and coherent unifying foundation for mathematics.

8.10   Continuum hypothesis

Let us turn again in more detail to the continuum hypothesis, which we introduced and discussed in chapter 3. Cantor had noticed from the earliest days that essentially all naturally arising sets of real numbers are either countable or have the same size as $\mathbb{R}$. The set of integers, the set of rational numbers, and the set of algebraic numbers are each countable. The set of transcendental numbers, the Cantor set, and many other naturally defined sets of real numbers have size continuum. Cantor never found a set of real numbers $A \subseteq \mathbb{R}$ of intermediate size; and so he asked the following question:

Question 23 (Cantor). Is every set of real numbers $A \subseteq \mathbb{R}$ either countable or equinumerous with $\mathbb{R}$?

The continuum hypothesis, as we mentioned in section 3.9, is the assertion that yes, indeed, every set $A \subseteq \mathbb{R}$ is either countable or has size continuum; in other words, it asserts that there are no infinities between the integers $\mathbb{Z}$ and the real numbers $\mathbb{R}$. Cantor struggled to settle the continuum hypothesis for the rest of his life. Hilbert gave a stirring speech at the dawn of the twentieth century, identifying the mathematical problems that would guide mathematics for the next century, and the very first problem on Hilbert's list was Cantor's continuum hypothesis. Yet, the question remained open for decades.

In 1938, Kurt Gödel constructed an artificial mathematical universe, called the *constructible universe L*, and proved that all the ZFC axioms of set theory are true in L, including the axiom of choice, and furthermore that the continuum hypothesis also is true there. Therefore, it is consistent with ZFC that the continuum hypothesis holds. In other words, if ZFC is consistent, then we cannot refute the continuum hypothesis. The constructible universe is ingenious, containing only the bare minimum of sets that absolutely must be present on the basis of the axioms of set theory, given that one will include all the ordinals. Meanwhile, Paul Cohen in 1962 provided a complementary answer. Cohen constructed another kind of artificial mathematical universe, inventing the method of *forcing* to do so, in which the axioms of set theory are all true but the continuum hypothesis fails. In other words, it is consistent with the axioms of set theory that this principle is false. The conclusion is that if the ZFC axioms of set theory are consistent, then we cannot settle the continuum hypothesis question on the basis of those axioms. We cannot prove it, since it fails in Cohen's universe. We cannot refute it, since it holds in Gödel's universe. Thus, the continuum hypothesis is *independent* of the axioms of set theory—neither provable nor refutable. Cohen had provided not only a universe where the continuum hypothesis fails, but also an extremely powerful tool, the model-construction method of forcing.

Pervasive independence phenomenon

To my way of thinking, the pervasive ubiquity of independence for the fundamental principles of set theory is the central discovery of set-theoretic research in the late twentieth century. The resolution of the continuum hypothesis was merely the first of hundreds (or even thousands) of such independence results. Forcing allows us to construct diverse extensions of any given set-theoretic universe. With forcing, we construct models of set theory exhibiting specific truths or having specific relations to other models of set theory. With forcing, we have discovered a chaotic jumble of set-theoretic possibilities. Whole parts of set theory exhaustively explore the combinations of statements realized in models of set theory, and also study the methods supporting this exploration. Would you like the set-theoretic principles CH + $\neg\diamondsuit$ to be true? Do you want $2^{\aleph_n} = \aleph_{n+2}$ for all n? Suslin trees? Would you like to separate the bounding number from the dominating number? Do you want an indestructible strongly compact cardinal? Would you prefer that it is also the least measurable cardinal? We build set-theoretic models to order. The pervasive independence phenomenon, largely or even completely unexpected by early-twentieth-century experts, is now everywhere to be found and fundamental to our understanding of the subject.

8.11 Universe view

Fine—so we can neither prove nor refute the continuum hypothesis; it is independent of the ZFC axioms. But still, mathematicians want to know: is it true or not? Although this question may seem innocent at first, interpreting it sensibly leads directly to extremely difficult philosophical issues. The main trouble arises from the fact that we do not actually have an adequate account of what it means to say absolutely that a mathematical statement is "true." We can express truth in a structure, for a given structure, but when it comes to set-theoretic truth, which structure are we talking about?

Many set theorists have the idea that there is a definite set-theoretic reality—a deterministic set-theoretic world of all the actual sets, in which all the ZFC axioms are true and more. The view is realist—the set-theoretic objects enjoy a real existence—and also singularist, asserting the uniqueness of this set-theoretic realm. According to this view, every set-theoretic assertion has a definite truth value, whether or not it can be proved on the basis of the current axioms. The set-theoretic universist describes the set-theoretic universe unfolding via the cumulative hierarchy. At each ordinal stage α, we have gathered in the set V_α all the sets that have been accumulated so far. The next stage $V_{\alpha+1}$ includes all subsets of V_α. Which subsets do we include? *All of them.* And which ordinal levels α does the hierarchy proceed through? *All of them.* It is a core commitment of the universe view that indeed there are absolute notions of "all subsets" and "all ordinals," meaningful independent of and prior to any particular axiomatization of set theory. These absolute notions are what the axiomatizations are trying to capture. According to the universe view, the independence phenomenon is explained away as being caused by the weakness of our axioms. Statements like the continuum hypothesis are independent of the ZFC axioms, simply because we had not been sophisticated enough in our initial formulations of the axioms to adopt a theory that settled it. The task before us is to search for the fundamental principles that will decide these questions.

Categoricity and rigidity of the set-theoretic universe

Let me briefly argue that the set-theoretic universe V is *rigid*—that is, that the set-theoretic universe, considered as a mathematical structure $\langle V, \in \rangle$, admits no nontrivial isomorphism with itself. To see this, suppose that π is such an isomorphism, a class bijection $\pi : V \to V$ for which $x \in y \leftrightarrow \pi(x) \in \pi(y)$. It follows by $\in$-induction, I claim, that $\pi(x) = x$ for every set x. Namely, if this is true for all elements $x \in a$, then since $x \in a$ implies that $x = \pi(x) \in \pi(a)$, we see immediately that $a \subseteq \pi(a)$; and conversely, if $y \in \pi(a)$, then $y = \pi(x)$ for some x, which must have $x \in a$ since $\pi(x) \in \pi(a)$; so $x = \pi(x) = y$, and therefore $y \in a$. So $\pi(a) = a$ for every set a, as desired.

An essentially identical argument shows that every transitive set is rigid, where a set A is *transitive* if it contains all the hereditary elements of its elements; in other words, if $b \in a \in A \implies b \in A$, a characterization showing why this property is called "transitive."

The rank-initial segments V_α of the cumulative hierarchy, for example, are transitive sets, and every set has a transitive closure, which is the smallest transitive set containing it. In particular, every rank-initial segment V_α is rigid.

Rigidity can be seen as a proto-categoricity phenomenon. After all, if the set-theoretic universe V is rigid, then every set in the set-theoretic universe plays a distinct structural role, in the sense of abstract structuralism. In this sense, set theory proves its own proto-categoricity as an internal claim about the sets. Each transitive set plays precisely its own structural role in the set-theoretic universe, and no other. Zermelo (1930), building on essentially similar ideas, proved the quasi-categoricity result for set theory.

Theorem 24 (Zermelo's quasi-categoricity theorem). *For any two models of second-order set theory* ZFC_2, *either they are isomorphic, or one is isomorphic to a rank-initial segment of the other. More specifically, the models of* ZFC_2 *are precisely the models isomorphic to a rank-initial segment* $\langle V_\kappa, \in \rangle$ *for an inaccessible cardinal* κ.

Proof sketch. The linearity claim of the first sentence is a consequence of the inaccessible characterization in the second sentence. Every model of second-order set theory ZFC_2 is well founded, since the second-order foundation axiom amounts to actual well-foundedness. By the Mostowski collapse, it follows that the model is isomorphic to a transitive set $\langle M, \in \rangle$. By the second-order separation and power set axioms, it follows that M is closed under actual power sets, and so $M = V_\kappa$ for some ordinal κ. By the second-order replacement axiom, it follows that κ must be regular. By the infinity axiom, it follows that $\kappa > \omega$. And by the power set axiom, it follows that κ is a strong limit cardinal. So κ is an inaccessible cardinal. Conversely, if κ is inaccessible, one may easily prove that $V_\kappa \models ZFC_2$ by verifying each axiom. $\square$

As I mentioned earlier, those models V_κ for κ inaccessible were later rediscovered by Grothendieck and are now known as *Zermelo-Grothendieck* universes.

Donald Martin (2001) argued, on essentially similar grounds as Zermelo in the quasi-categoricity theorem, that there is ultimately at most one concept of set: any two concepts of set can be compared level-by-level through their respective cumulative hierarchies, and the isomorphism that we have built from the first V_α to the second $\bar{V}_\alpha$ extends to the next stage $V_{\alpha+1}$, since each has *all* the sets, and ultimately extends to the two respective entire universes.

Georg Kreisel (1967) argued that because the set-theoretic universe is sufficiently determined in second-order logic by ZFC_2, it follows that the continuum hypothesis has a determinate truth value. That is, since the truth of the continuum hypothesis is revealed already in $V_{\omega+2}$, it is therefore held in common with all the models V_κ of ZFC_2. In this sense, the continuum hypothesis has a definite answer in second-order set theory, even if we do not know which answer it is. This argument has been defended at length more recently by Daniel Isaacson (2011), who analyzes the nature of definite mathematical structure, arguing that on the basis of the categoricity results, the set-theoretic universe itself is a definite mathematical structure.

Critics view this as sleight of hand, since second-order logic amounts to set theory itself, in the metatheory. That is, if the interpretation of second-order logic is seen as inherently set-theoretic, with no higher claim to absolute meaning or interpretation than set theory, then to fix an interpretation of second-order logic is precisely to fix a particular set-theoretic background in which to interpret second-order claims. And to say that the continuum hypothesis is determined by the second-order set theory is to say that, no matter which set-theoretic background we have, it either asserts the continuum hypothesis or it asserts the negation. I find this to be like saying that the exact time and location of one's death is fully determinate because whatever the future brings, the actual death will occur at a particular time and location. But is this a satisfactory proof that the future is "determinate"? No, for we might regard the future as indeterminate, even while granting that ultimately something particular must happen. Similarly, the proper set-theoretic commitments of our metatheory are open for discussion, even if any complete specification will ultimately include either the continuum hypothesis or its negation. Since different set-theoretic choices for our interpretation of second-order logic will cause different outcomes for the continuum hypothesis, the principle remains in this sense indeterminate.

Other philosophers take second-order set theory to have an absoluteness that gives meaning to the corresponding determinacy of the continuum hypothesis. Button and Walsh (2016, 2018) contrast various attitudes that one can have toward this and similar categoricity results, such as the *algebraic* attitude, which treats the object theory model-theoretically, having a rich class of different models; the *infer-to-stronger-logic* attitude, which presumes the definiteness of $\mathbb{N}$ and infers the definiteness of the logic; and the *start-with-full-logic* attitude, which takes the second-order categoricity results at face value.

8.12 Criterion for new axioms

By what criterion are we to adopt new set-theoretic axioms? Are we to adopt only axioms expressing principles in fundamental intrinsic agreement with our prior conception of the subject? Or can the fruitful consequences of a mathematical theory be evidence in favor of the truth of the theory? Do we want our mathematical foundations to be very strong or very weak?

Set theorists with the universe perspective often point to the unifying consequences of large cardinals as evidence that those axioms are on the right track toward the ultimate theory. This is a *consequentialist* appeal, in which we judge a mathematical theory by its consequences and explanatory power. Some set theorists have proposed that we should treat large cardinal axioms by scientific criteria. The theories make testable predictions about future mathematical discoveries, and the fact that they pass those tests is evidence in favor of their truth. The *overlapping consensus* phenomenon, by which different incompatible extensions of set theory sometimes have strong consequences in common, can be viewed as an instance of this, providing support for the truth of those common consequences. Other set theorists point to the unifying elegance of the large cardinal axioms. Ultimately, in light of the independence phenomenon, the justificatory grounds for new axioms cannot consist entirely in proof, precisely because the new truths are independent of the prior axioms; in this sense, we must search for nonmathematical (or at least nondeductive) justifications of the axioms.

Intrinsic justification

Set theorists commonly take the principal axiomatizations of set theory to have *intrinsic* justification in terms of their conformance with the concept of set. The axioms express intuitively clear principles about sets, particularly with respect to the cumulative hierarchy conception of set. In this sense, the Zermelo axioms and the Zermelo-Fraenkel axioms are often taken to express fundamental truths of what we mean by the concept of set. Many set theorists similarly find intrinsic justification for some extensions of ZFC into the large cardinal hierarchy. One finds intrinsic grounds to accept the existence of inaccessible, hyperinaccessible, and Mahlo cardinals, and sometimes much more.

Extrinsic justification

Kurt Gödel, hoping to find new fundamental principles settling the open questions, explains how we might be led to adopt these new axioms, even when they lack a character of intrinsic necessity, but rather simply in light of their unifying explanatory power:

> Furthermore, however, even disregarding the intrinsic necessity of some new axiom, and even in case it had no intrinsic necessity at all, a decision about its truth is possible also in another way, namely, inductively by studying its "success," that is, its fruitfulness in consequences and in particular in "verifiable" consequences, i.e., consequences demonstrable without the new axiom, whose proofs by means of the new axiom, however, are considerably simpler and easier to discover, and make it possible to condense into one proof many different proofs. The axioms for the system of real numbers, rejected by the intuitionists, have in this sense been verified to some extent owing to the fact that analytical number theory frequently allows us to prove number theoretical theorems which can subsequently be verified by elementary methods. A much higher degree of verification than that, however, is conceivable. There might exist axioms so abundant in their verifiable consequences, shedding so much light upon a whole discipline, and furnishing such powerful methods for solving given problems (and even solving them, as far as that is possible, in a constructivist way) that quite irrespective of their intrinsic necessity they would have to be assumed at least in the same sense as any well established physical theory. (Gödel, 1947, p. 521)

The large cardinal hypotheses fulfill this vision in many respects, in light of their strong regularity consequences for sets of real numbers. The large cardinal axioms explain why definable sets of real numbers behave so well. Yet, Gödel's specific hope that large cardinals would settle the continuum hypothesis turned out to be wrong. Using forcing, we can turn the continuum hypothesis on and off in successive forcing extensions, and the Lévy-Solovay theorem shows that we can do this while preserving all the known large cardinals. Therefore, those large cardinal axioms do not decide the continuum hypothesis question. A huge part of set theory aims at the precise interaction between large cardinals and forcing. For example, Richard Laver identified the *indestructibility* phenomenon of certain large cardinals, by which they are preserved with forcing from a certain large class. Arthur Apter and I (1999) established the *universal indestructibility* phenomenon by constructing models of set theory with numerous large cardinals, all of which are indestructible.

Meanwhile, we have mentioned the astonishing regularity consequences of large cardinals for definable sets of real numbers, which might be taken as a positive instance of the phenomenon that Gödel had in mind. In case after case, we have very welcome consequences of large cardinals, which explain and unify our understanding of what is going on at the level of the real numbers. Penelope Maddy (2011) explains how it works:

> What we've learned here is that what sets are, most fundamentally, is markers for these contours, what they are, most fundamentally, is maximally effective trackers of certain strains of mathematical fruitfulness. From this fact about what sets are, it follows that they can be learned about by set-theoretic methods, because set-theoretic methods, as we've seen, are all aimed at tracking particular instances of effective mathematics. The point isn't, for example, that "there is a measurable cardinal" really means "the existence of a measurable cardinal is mathematically fruitful in ways x, y, z (and this advantage isn't outweighed by accompanying disadvantages)"; rather, the fact of measurable cardinals being mathematically fruitful in ways x, y, z (and these advantages not being outweighed by accompanying disadvantages) is evidence for their existence. Why? Because of what sets are: repositories of mathematical depth. They mark off a mathematically rich vein within the indiscriminate network of logical possibilities. (Maddy, 2011, pp. 82–83)

The extrinsic fruitful consequences of strong axioms become evidence for their truth. The large cardinal axioms enable us to develop a robust unified theory for the projectively definable sets of real numbers.

What is an axiom?

What does it mean to adopt, or even to consider, an axiom? Since the late twentieth century, there has been a shift in meaning for this term as it is used in set theory. Formerly, in a mathematical development or foundational theory, an axiom was a self-evident truth—a fundamental, rock-solid truth, about which there was no question. In geometry, Euclid's axioms express the most fundamental geometric truths, from which many other truths flow. In arithmetic, Dedekind's axioms express the central core of our beliefs about the nature of the natural numbers with the successor operation. In set theory, Zermelo's axioms aim to express fundamental truths about the nature of the concept of set.

Contemporary set theory, however, is awash in diverse axioms, often standing in conflict with one another. We have Martin's axiom, the axiom of constructibility, the axiom of determinacy, the diamond axiom $\diamond$, the square principles $\square_\kappa$, the large cardinal axioms, the ground axiom, and so on. Cantor called his principle the continuum hypothesis, but if named today, it would be the continuum axiom. The word "axiom" in set theory has come to mean a statement expressing a concise fundamental principle, whose consequences are to be studied. Maybe the axiom is true, maybe it is not, but we want to investigate it to find out when it holds, when it cannot hold, what are its consequences, and so on.

Solomon Feferman (1999) despairs of this new usage:

> It's surprising how far the meaning of axiom has been stretched from the ideal sense in practice, both by mathematicians and logicians. Some even take it to mean an arbitrary assumption, and so refuse to take seriously what purpose axioms are supposed to serve. (p. 100)

To my way of thinking, however, the new usage arises directly from the case of geometry, where the Euclidean axioms had the status of self-evident truths until the discovery of non-Euclidean geometry tempted us to alternative incompatible axiomatizations. So we find ourselves considering various incompatible, yet concisely stated, geometric statements, which distill the point of geometric disagreement. We no longer think that there is an absolute geometric truth, about which we should make assertions of a self-evident nature. Rather, we find various kinds of geometry, whose different natures can be succinctly expressed by geometric assertions. This is the contemporary use of axiom. Because of the ubiquitous independence phenomenon, set theory overflows with these independent, concisely stated points of set-theoretic disagreement, and thus we have many incompatible axioms in set theory.

Meanwhile, the word "axiom" is also commonly used in another related way, again not in the sense of self-evident truth, but rather as a means of defining a mathematical domain of discourse. One speaks, for example, of the axioms of group theory, or the field axioms, or the axioms of a metric space, or of a vector space. These axioms are used to define what it means to be a group or a field or a metric space or a vector space. Many mathematical subjects orbit around such particular kinds of mathematical structure, which can be defined by a list of properties called axioms, and in these cases, the axioms are used essentially as a definition rather than to express a fundamental truth. Ultimately, this usage is not so different from the concise-fundamental-principle usage described above, since one can take the axioms used in that sense also to be defining a mathematical realm. The axiom of constructibility, for example, in effect defines the kind of set-theoretic universe that will be studied, and it is similar with Martin's axiom and the ground axiom.

Feferman distinguishes between foundational axioms and definitional axioms, disputing that one can have a purely definitional foundation of mathematics. We cannot only have axioms that define the realm; they must also tell us what is in it.

8.13   Does mathematics need new axioms?

So we are faced with a vast spectrum of set-theoretic possibility. And forever shall it be, in light of the incompleteness theorem, which shows that we can never provide a full account of set-theoretic truth; all our set-theoretic theories will necessarily be incomplete. Shall we search for new axioms anyway, for the new true principles of set theory, extending our current theory? This is in a sense what set theorists have been doing with the large cardinal hierarchy.

Absolutely undecidable questions

Kurt Gödel uses the incompleteness phenomenon to argue that if human mathematical reasoning is fundamentally mechanistic, then there must be absolutely undecidable questions—questions that we shall be unable in principle to solve.

> For if the human mind were equivalent to a finite machine, then objective mathematics not only would be incompletable in the sense of not being contained in any well-defined axiomatic system, but moreover there would exist *absolutely* unsolvable diophantine problems of the type described above, where the epithet "absolutely" means that they would be undecidable, not just within some particular axiomatic system, but by *any* mathematical proof the human mind can conceive. So the following disjunctive conclusion is inevitable: *Either mathematics is incompletable in this sense, that its evident axioms can never be comprised in a finite rule, that is to say, the human mind (even within the realm of pure mathematics) infinitely surpasses the powers of any finite machine, or else there exist absolutely unsolvable diophantine problems of the type specified.*
> (Gödel, 1995 [*1951], emphasis original)

Peter Koellner discusses at length the question of whether there could be such absolutely undecidable questions in set theory, questions that would remain independent of any justified axiomatization:

> There is a remarkable amount of structure and unity beyond ZFC and that a network of results in modern set theory make for a compelling case for new axioms that settle many questions undecided by ZFC. I will argue that most of the candidates proposed as instances of absolute undecidability have been settled and that there is not currently a good argument to the effect that a given sentence is absolutely undecidable. (Koellner, 2010, p. 190)

Koellner is optimistic regarding our ability to find the justifications that might lead us to the true set theory. After considering how numerous instances have played out, he concludes:

> There is at present no solid argument to the effect that a given statement is absolutely undecidable. We do not even have a clear scenario for how such an argument might go. (Koellner, 2010, p. 219)

He argues that our current best theories (ZFC plus large cardinals) may actually be complete enough for the natural mathematical assertions in which we may be interested, and there seem to be no absolutely undecidable mathematical questions.

Strong versus weak foundations

One of the central unresolved questions at the core of the foundational dispute—to my way of thinking, a question deserving considerably more attention from philosophers than it has received—is whether it is better to have a strong axiomatic foundation of mathematics or a weak one. In discussions of set theory as a foundation of mathematics, some mathematicians emphasize that the replacement axiom is not needed for ordinary mathematics and that ZFC is much stronger than needed. And therefore weaker foundational systems are proposed and defended, such as systems amounting to just Zermelo set theory or modified versions of it, with the bounded replacement axiom, but still weak in comparison with ZFC or the large cardinal hypotheses. Indeed, in section 7.8 we saw how the subject of reverse mathematics reveals that very weak axiomatic systems in second-order arithmetic suffice for the vast bulk of core mathematics. Does one want one's foundational theory to be as weak as possible? Many mathematicians seem to think so.

Meanwhile, John Steel has argued that the strongest foundations will win out in the end, not the weakest.

> What is important is just that our axioms be true, and as strong as possible. Let me expand on the strength demand, as it is a fundamental motivation in the search for new axioms.
>
> It is a familiar but remarkable fact that all mathematical language can be translated into the language of set theory, and all theorems of "ordinary" mathematics can be proved in ZFC. In extending ZFC, we are attempting to strengthen this foundation. Surely strength is better than weakness! Professor Maddy has labelled the "stronger is better" rule of thumb *maximize*, and discussed it at some length in her recent book (Maddy, 1997). I would say that what we are attempting to maximize here is the *interpretative power* of our set theory. In this view, to believe that there are measurable cardinals is to seek to naturally interpret all mathematical theories of sets, to the extent that they have natural interpretations, in extensions of ZFC + "there is a measurable cardinal." Maximizing interpretative power entails maximizing formal consistency strength, but the converse is not true, as we want our interpretations to preserve meaning. Steel, in (Feferman et al., 2000, p. 423)

From his view, we should seek the fundamental principles that will enrich our mathematical knowledge, providing enduring deep insights. At the moment, the large cardinal axioms are the strongest-known principles in mathematics, by the best measures, which include not only direct implication, but also consistency strength, and these axioms identify a rich structure theory for sets of real numbers that is not anticipated in any competing foundational system. Ultimately, Steel argues, these consequences will be the reason that large cardinal set-theoretic foundations will endure. Any successful foundation of mathematics must ultimately incorporate the large cardinal hypotheses and their consequences.

The existence of sufficient large cardinals, for example, implies the axiom of determinacy in the inner model $L(\mathbb{R})$, which itself implies numerous regularity properties for projectively definable sets of real numbers: they are measurable; they fulfill the property of Baire; and they obey the continuum hypothesis. This is a welcome continuation of Cantor's original strategy for settling the continuum hypothesis, since he had begun with the closed sets, at the bottom level of the projective hierarchy. Current work is pressing on with H_{ω_2}

in place of the real numbers, but there are many complicating issues. Getting an attractive theory up to $\aleph_2$, unfortunately, seems to mean ruining parts of the attractive theory that we had for the sets of real numbers, thus tending to undermine the earlier argument from fruitful consequences. Will the justified theories stabilize as we reach upward? To grasp the delectable higher fruit, it seems, we must trample upon the lower delicacies that had initially attracted us.

Shelah

Saharon Shelah, in contrast, despairs at the idea that we shall find compelling new fundamental principles:

> Some believe that compelling, additional axioms for set theory which settle problems of real interest will be found or even have been found. It is hard to argue with hope and problematic to consider arguments which have not yet been suggested. However, I do not agree with the pure Platonic view that the interesting problems in set theory can be decided, we just have to discover the additional axiom. My mental picture is that we have many possible set theories, all conforming to ZFC. I do not feel "a universe of ZFC" is like "the sun," it is rather like "a human being" or "a human being of some fixed nationality." (Shelah, 2003, p. 211)

He asserts:

> My feeling is that ZFC exhausts our intuition except for things like consistency statements, so a proof means a proof in ZFC. Shelah (1993)

Feferman

Meanwhile, Feferman argues that mathematics does not need new axioms:

> I am convinced that the Continuum Hypothesis is an inherently vague problem that no new axiom will settle in a convincingly definite way. Moreover, I think the Platonistic philosophy of mathematics that is currently claimed to justify set theory and mathematics more generally is thoroughly unsatisfactory and that some other philosophy grounded in inter-subjective *human* conceptions will have to be sought to explain the apparent objectivity of mathematics. Finally, I see no evidence for the practical need for new axioms to settle open arithmetical and finite combinatorial problems. The example of the solution of the Fermat problem shows that we just have to work harder with the basic axioms at hand. However, there is considerable theoretical interest for logicians to try to explain what new axioms *ought to be accepted* if one has already accepted a given body of principles, much as Gödel thought the axioms of inaccessibles and Mahlo cardinals *ought to be accepted* once one has accepted the Zermelo-Fraenkel axioms. In fact this is something I've worked on in different ways for over thirty years; during the last year I have arrived at what I think is the most satisfactory general formulation of that idea, in what I call the *unfolding of a schematic formal system*. Feferman (1999)

Feferman's paper was followed up in a panel discussion with Feferman, Friedman, Maddy, and Steel held at the ASL annual meeting in 2000, Feferman et al. (2000).

8.14 Multiverse view

The pluralist or multiverse view in set theory is the topic of an ongoing vigorous debate in the philosophy of set theory. According to set-theoretic pluralism, there is a huge variety of concepts of set, each giving rise to its own set-theoretic world. These various set-theoretic worlds exhibit diverse set-theoretic and mathematical truths—an explosion of set-theoretic possibility. We aim to discover these possibilities and how the various set-theoretic worlds are connected. From the multiverse perspective, the pervasive independence phenomenon is taken as evidence of distinct and incompatible concepts of set. The diversity of models of set theory is evidence of actual distinct set-theoretic worlds.

I have described the situation like this:

> As a result, the fundamental objects of study in set theory have become the models of set theory, and set theorists move with agility from one model to another. While group theorists study groups, ring theorists study rings and topologists study topological spaces, set theorists study the models of set theory. There is the constructible universe L and its forcing extensions $L[G]$ and nonforcing extensions $L[0^\sharp]$; there are increasingly sophisticated definable inner models with large cardinals $L[\mu]$, $L[\vec{E}]$, and so on; there are models V with much larger large cardinals and corresponding forcing extensions $V[G]$, ultrapowers M, cut-off universes L_δ, V_α, H_κ, universes $L(\mathbb{R})$, HOD, generic ultrapowers, boolean ultrapowers and on and on and on. As for forcing extensions, there are those obtained by adding one Cohen real number, or many, or by other c.c.c. or proper (or semiproper) forcing, or by long iterations of these, or by the Lévy collapse, or by the Laver preparation or by self-encoding forcing, and on and on and on. Set theory appears to have discovered an entire cosmos of set-theoretic universes, revealing a category-theoretic nature for the subject, in which the universes are connected by the forcing relation or by large cardinal embeddings in complex commutative diagrams, like constellations filling a dark night sky. (Hamkins, 2012, p. 3)

The task at hand in the foundations of mathematics is to investigate how these various alternative set-theoretic worlds are related. Some of them fulfill the continuum hypothesis and others do not; some have inaccessible cardinals or supercompact cardinals and others do not. Set-theoretic pluralism is thus an instance of plenitudinous platonism, since according to the multiverse view, all the different set theories or concepts of set that we can imagine are realized in the corresponding diverse set-theoretic universes.

Dream solution of the continuum hypothesis

After the independence of the continuum hypothesis was discovered, set theorists yearned to solve it on intrinsic grounds, by finding a new set-theoretic principle—one that everyone agreed was valid for our intended concept of set, and from which we could either prove or refute the continuum hypothesis. I argue, however, that as a result of our rich experience with our diverse set-theoretic concepts, we can no longer expect to find such a "dream solution"—a solution in which we discover a new fundamental set-theoretic truth that settles the continuum hypothesis. My view is that there will be no dream solution because our experience in the set-theoretic worlds with the contrary truth value for the continuum hypothesis will directly undermine our willingness to regard the new axiom as a fundamental set-theoretic truth.

> Our situation with CH is not merely that CH is formally independent and we have no additional knowledge about whether it is true or not. Rather, we have an informed, deep understanding of how it could be that CH is true and how it could be that CH fails. We know how to build the CH and ¬CH worlds from one another. Set theorists today grew up in these worlds, comparing them and moving from one to another while controlling other subtle features about them. Consequently, if someone were to present a new set-theoretic principle Φ and prove that it implies ¬CH, say, then we could no longer look upon Φ as manifestly true for sets. To do so would negate our experience in the CH worlds, which we found to be perfectly set-theoretic. It would be like someone proposing a principle implying that only Brooklyn really exists, whereas we already know about Manhattan and the other boroughs. Hamkins (2015)

According to the multiverse view, the continuum hypothesis question is settled by our deep understanding of *how* it is true and false in the multiverse.

> On the multiverse view, consequently, the continuum hypothesis is a settled question; it is incorrect to describe the CH as an open problem. The answer to CH consists of the expansive, detailed knowledge set theorists have gained about the extent to which it holds and fails in the multiverse, about how to achieve it or its negation in combination with other diverse set-theoretic properties. Of course, there are and will always remain questions about whether one can achieve CH or its negation with this or that hypothesis, but the point is that the most important and essential facts about CH are deeply understood, and these facts constitute the answer to the CH question. Hamkins (2012)

Analogy with geometry

There is a strong analogy between the pluralist nature of set theory and what has emerged as an established plurality in the foundations of geometry. Classically, the study of geometry from the time of Euclid until the late nineteenth century was understood to be about a single domain of knowledge—what we might describe as the one true geometry, which many (including Kant) took to be the geometry of physical space. But with the rise of non-Euclidean geometry, the singular concept of geometry splintered into distinct, separate concepts of geometry. Thus, we have spherical geometry, hyperbolic geometry, and elliptical geometry. It was no longer sensible to ask whether the parallel postulate is true, as though every geometrical question has a determinate answer for the one true geometry. Rather, one asks whether it is true in this geometry, or that geometry. Similarly, in set theory, it is not sensible to ask whether the continuum hypothesis is true or not. Rather, one asks whether it holds in this set-theoretic universe, or that one.

Pluralism as set-theoretic skepticism?

Set-theoretic pluralists are sometimes described as set-theoretic skeptics, in light of their resistance to the universe view; for instance, see Koellner (2013). But to my way of thinking, this label is misapplied, for the multiverse position is not especially or necessarily skeptical about set-theoretic realism. We do not describe a geometer who works freely sometimes in Euclidean and sometimes in non-Euclidean geometry as a geometry skeptic. Such a geometer might simply recognize that there are a variety of geometries to consider—each fully real, each offering a different geometric conception. Some may be more interesting or more useful, but she is not necessarily in denial about the reality of any of them. Similarly, the set-theoretic pluralist, who holds that there are diverse incompatible concepts of set, is not necessarily a skeptic about set-theoretic realism.

Plural platonism

In my experience, a few decades ago, it used to be that when set theorists described themselves as "platonist," part of what was usually meant was the idea that there was ultimately a unique, final platonic universe in which set-theoretic statements would achieve their absolute truth value. That is, the singular nature of the set-theoretic universe was commonly taken to be part of platonism. Today, however, this usage seems to have relaxed. I argued in Hamkins (2012) that platonism should concern itself with the real existence of mathematical and abstract objects rather than with the question of uniqueness. According to this view, therefore, platonism is not incompatible with the multiverse view; indeed, according to plenitudinous platonism, there are an abundance of real mathematical structures, the mathematical realms that our theories are about, including all the various set-theoretic universes. And so one can be a set-theoretic platonist without committing to a single and absolute set-theoretic truth, precisely because there are many concepts of set, each carrying its own set-theoretic truth. This usage in effect separates the singular-universe claim from the real-existence claim, and platonism concerns only the latter.

Theory/metatheory interaction in set theory

In its role as a foundation of mathematics, set theory provides an arena for undertaking model theory—the study of all the models of a given theory. Model theorists have observed that many fundamental model-theoretic questions are affected by aspects of the set-theoretic background in which they are considered. The existence of saturated models of a given theory and cardinality, for example, often depends on set-theoretic principles of cardinal arithmetic—things generally work out much better for saturated models, for instance, under the generalized continuum hypothesis. Many early model-theoretic questions were seen in this way to have an essentially set-theoretic character.

One of the theories that set theorists would like to analyze in this model-theoretic manner way is ZFC itself. We seek to undertake the model theory of ZFC, and so we have here an interaction of object theory with metatheory. Namely, while working in ZFC as our background theory, we want to prove theorems about the models of ZFC. Are these two versions of ZFC the same? This is precisely the theory/metatheory distinction, which can be confusing in the case of set theory, since we seem in some sense to be talking about the same theory.

Set-theoretic independence results, for example, are metatheoretic in character. We prove, in ZFC, that the theory ZFC, if consistent, neither proves nor refutes the continuum hypothesis. The proof is ultimately like any other such proof in model theory: from any given (set-sized) model of ZFC, we can construct other models of ZFC, in some of which the continuum hypothesis holds and in others of which it fails. In a strong sense, the core results establishing the independence phenomenon constitute a subject that could be called the *model theory of set theory*.

The multiverse perspective ultimately provides what I view as an enlargement of the theory/metatheory distinction. There are not merely two sides of this distinction, the object theory and the metatheory; rather, there is a vast hierarchy of metatheories. Every set-theoretic context, after all, provides in effect a metatheoretic background for the models and theories that exist in that context—a model theory for the models and theories one finds there. Every model of set theory provides an interpretation of second-order logic, for example, using the sets and predicates existing there. Yet a given model of set theory M may itself be a model inside a larger model of set theory N, and so what previously had been the absolute set-theoretic background, for the people living inside M, becomes just one of the possible models of set theory, from the perspective of the larger model N. Each metatheoretic context becomes just another model at the higher level. In this way, we have theory, metatheory, metametatheory, and so on, a vast hierarchy of possible set-theoretic backgrounds. This point of view amounts to a denial of the need for uniqueness in Maddy's metamathematical corral conception; it seems to enlarge our understanding of metamathematical issues when different metamathematical contexts offer competing metamathematical conclusions and when the independence phenomenon reaches into metamathematical contexts.

Summary

Set theorists have discovered the large cardinal hierarchy—an amazing clarification of the higher infinite. Set theorists have also discovered the pervasive independence phenomenon, by which many statements in mathematics are neither provable nor refutable. These technical developments have led to a philosophical crisis concerning the nature of mathematical truth and existence. As a result, there is an ongoing philosophical debate in set theory and the philosophy of set theory concerning how we are to overcome those issues. The debate between the singular and pluralist views in set theory is ongoing, a central dispute in the philosophy of set theory. Please join us in what I find to be a fascinating conversation.

Questions for further thought

8.1   Show that the separation axiom implies that there is no set of all sets.

8.2   Draw a closed set in the plane with Cantor-Bendixson rank $\omega + 1$, and describe how the Cantor-Bendixson process proceeds.

8.3 Argue that all the Zermelo axioms (that is, the first-order set theory without the replacement axiom) are true in $V_{\omega+\omega}$, considered as a model of set theory with the $\in$ relation. Conclude that one cannot prove in Zermelo set theory that $\aleph_\omega$ exists.

8.4 Using the result of the previous exercise question, explain why ZFC proves Con(Z), the consistency assertion for Zermelo set theory.

8.5 Assuming that the generalized continuum hypothesis is consistent with ZFC, show that it is consistent with Zermelo's axioms that there are only countably many different infinities.

8.6 If ω_1 is the first uncountable ordinal, argue that V_{ω_1} satisfies all the Zermelo axioms, plus the countable replacement axiom, asserting all instances of replacement on a countable set. Bonus question: argue that the consistency strength of ZFC strictly exceeds that of the Zermelo theory plus countable replacement, which strictly exceeds the plain Zermelo theory (assuming that these are all consistent).

8.7 Does ZFC prove that the class of all self-membered sets $\{ x \mid x \in x \}$ is a set?

8.8 Discuss the differences and interrelations between Frege's Basic Law V, the principle of class extensionality, and the class comprehension axiom, which for suitable predicates φ asserts that $\{ x \mid \varphi(x) \}$ is a class? And how do they relate to the general comprehension principle?

8.9 Find an explicit contradictory instance of Basic Law V, under the presumption that εF is an x that falls under the scope of the universal quantifier $\forall x$ on the right-hand side. [Hint: Let F be the concept "x does not fall under the concept of which x is the extension," and let G be the concept "x falls under the concept of which r is the extension," where $r = \varepsilon F$.]

8.10 What are the reasons why one would want one's foundational theory of mathematics to be as weak as possible? What reasons might one have to want a very strong foundational theory?

8.11 Give several natural examples of impredicative definitions that are widely used in mathematics, along with a predicative equivalent. Do the impredicative definitions strike you as illegitimate?

8.12 Explain why we cannot prove the relative consistency implication Con(ZFC) $\rightarrow$ Con(ZFC + $\exists$inaccessible cardinal) in ZFC, unless ZFC is inconsistent. [Hint: We can prove that if κ is inaccessible, then V_κ is a model of ZFC, and so Con(ZFC). Now consider the second incompleteness theorem with respect to the theory ZFC + Con(ZFC).]

8.13 Prove that the countable union of countable sets is countable. It is known that one cannot prove this result in ZF, without the axiom of choice, assuming that ZF is consistent. Where exactly in your proof did you use the axiom of choice?

8.14 Prove that the choice-function version of the axiom of choice is equivalent to the assertion that for every collection A of nonempty disjoint sets, there is a set x such that $x \cap a$ has exactly one element for every $a \in A$.

8.15 Elementary model-theoretic results show that if there is a model of ZFC, then there is such a model $\langle M, \in^M \rangle$ with numerous nontrivial automorphisms. How can you reconcile that with the rigidity claim made in section 8.11? [Hint: Contrast "internal" rigidity in the object theory with "external" rigidity of a model.]

8.16 Discuss whether the axioms of Euclidean geometry can be taken as self-evident. What about the axiomatizations of non-Euclidean geometry? Is there a sense in which these axioms are more like definitions than fundamental truths?

8.17 Is there any connection between the categoricity of a mathematical theory and the rigidity of its structures?

8.18 Is categoricity a property of mathematical structures or mathematical theories? In what way does it make sense to say that a mathematical structure is categorical?

8.19 By emphasizing "verifiable consequences" in his "axioms so abundant" quotation (page 289), is Gödel insisting that the new axioms are conservative over the previous axioms? If so, then how much support would this position offer to the proponents of large cardinal axioms? If not, then how would the consequences be verified?

8.20 Discuss the issue of intrinsic and extrinsic justification in the context of the axioms of geometry. Does one adopt the parallel postulate for intrinsic reasons?

8.21 Discuss intrinsic and extrinsic justification in the context of arithmetic, such as with regard to the Dedekind axioms of arithmetic. How does the issue apply to the Riemann hypothesis, an arithmetic assertion with numerous consequences? And how does it apply to other outstanding arithmetic questions, such as the abc conjecture, the twin-prime conjecture, or the Collatz conjecture, some of which have numerous consequences in number theory, while others have fewer.

8.22 Is set-theoretic pluralism a form of skepticism about set-theoretic realism?

Further reading

Penelope Maddy (2017). An excellent essay on the nature of set theory and the roles it plays in the foundations of mathematics, providing a precise elementary vocabulary that greatly assists discussion of the roles to be played by a foundation and the goals for this that one might have.

Thomas Jech (2003). The standard graduate-level introduction to set theory; it is excellent.

Akihiro Kanamori (2004). The standard graduate-level introduction to large cardinals; it also is excellent.

Tim Button and Sean Walsh, in Button and Walsh (2016). A survey of work on the significance of categoricity arguments in the philosophy of mathematics.

Joel David Hamkins (2012). An introduction to the multiverse view in set theory.

Joel David Hamkins (2015). My argument concerning the impossibility of the dream solution of the continuum hypothesis.

Solomon Feferman, Harvey M. Friedman, Penelope Maddy, and John R. Steel (2000). A back-and-forth exchange of essays on the question of whether mathematics needs new axioms.

Charles Parsons (2017). An account of major developments in the debate on impredicativity.

Credits

The Russell paradox was also known independently to Zermelo and Cantor in the late 1890s, who had written letters to Dedekind, Hilbert, and others. The barista allegory is due to Barbara Gail Montero. Solovay's theorem, the one providing a model in which every set of real numbers is Lebesgue measurable, appears in Solovay (1970). The fact that the axiom of replacement is equivalent to the principle of transfinite recursion is due to me, following some ideas of Jeremy Rin; the case without the axiom of choice, which is considerably more interesting, is due to myself and Alfredo Roque-Feirer. The fact that without the axiom of choice, the rational field can have multiple nonisomorphic algebraic closures is due to Hans Läuchli.

Bibliography

Aaronson, Scott. 2006. Reasons to believe. Shtetl-Optimized, the blog of Scott Aaronson. https://www.scottaaronson.com/blog/?p=122.

Apter, Arthur W., and Joel David Hamkins. 1999. Universal indestructibility. *Kobe Journal of Mathematics* 16 (2): 119–130. http://wp.me/p5M0LV-12.

Balaguer, Mark. 1998. *Platonism and Anti-Platonism in Mathematics*. Oxford University Press.

Balaguer, Mark. 2018. Fictionalism in the philosophy of mathematics, Fall 2018 edn. In *The Stanford Encyclopedia of Philosophy*, ed. Edward N. Zalta. Metaphysics Research Lab, Stanford University. https://plato.stanford.edu/archives/fall2018/entries/fictionalism-mathematics.

Ball, W. W. Rouse. 1905. *Mathematical Recreations and Essays*, 4th edn. Macmillan and Co. Available via The Project Gutenberg. http://www.gutenberg.org/ebooks/26839.

Beall, Jeffrey, and Greg Restall. 2000. Logical pluralism. *Australasian Journal of Philosophy* 78 (4): 475–493.

Beall, Jeffrey, and Greg Restall. 2006. *Logical Pluralism*. Oxford University Press.

Benacerraf, Paul. 1965. What numbers could not be. *The Philosophical Review* 74 (1): 47–73. http://www.jstor.org/stable/2183530.

Benacerraf, Paul. 1973. Mathematical truth. *Journal of Philosophy* 70: 661–679.

Berkeley, George. 1734. *A Discourse Addressed to an Infidel Mathematician*. The Strand. https://en.wikisource.org/wiki/The_Analyst:_a_Discourse_addressed_to_an_Infidel_Mathematician.

Bishop, Errett. 1977. Book Review: Elementary Calculus. *Bulletin of the American Mathematical Society* 83 (2): 205–208. doi:10.1090/S0002-9904-1977-14264-X.

Blackwell, David, and Persi Diaconis. 1996. A non-measurable tail set, eds. T. S. Ferguson, L. S. Shapley, and J. B. MacQueen, Vol. 30 of *Lecture Notes–Monograph Series*, 1–5. Institute of Mathematical Statistics. doi:10.1214/lnms/1215453560.

Blasjö, Viktor. 2013. Hyperbolic space for tourists. *Journal of Humanistic Mathematics* 3 (2): 88–95. doi:10.5642/jhummath.201302.06.

Borges, Jorge Luis. 1962. *Ficciones*. English translation. Grove Press. Originally published in 1944 by Sur Press.

Borges, Jorge Luis. 2004. *A Universal History of Infamy*. Penguin Classics. Translated by Andrew Hurley. Originally published in 1935 by Editorial Tor as *Historia Universal de la Infamia*.

Burgess, John P. 2015. *Rigor and Structure*. Oxford University Press.

Button, Tim, and Sean Walsh. 2016. Structure and categoricity: Determinacy of reference and truth value in the philosophy of mathematics. *Philosophia Mathematica* 24 (3): 283–307. doi:10.1093/philmat/nkw007.

Button, Tim, and Sean Walsh. 2018. *Philosophy and Model Theory*. Oxford University Press. doi:10.1093/oso/9780198790396.001.0001.

Buzzard, Kevin. 2019. Cauchy reals and Dedekind reals satisfy "the same mathematical theorems." MathOverflow question. https://mathoverflow.net/q/336191.

Camarasu, Teofil. What is the next number on the constructibility sequence? And what is the asymptotic growth? Mathematics Stack Exchange answer. https://math.stackexchange.com/q/3514953.

Cantor, Georg. 1874. Über eine Eigenschaft des Inbegriffs aller reelen algebraischen Zahlen. *Journal für die Reine und Angewandte Mathematik (Crelle's Journal)* 1874 (77): 258–262. doi:10.1515/crll.1874.77.258.

Cantor, Georg. 1878. Ein Beitrag zur Mannigfaltigkeitslehre. *Journal für die Reine und Angewandte Mathematik (Crelle's Journal)* 84: 242–258. doi:10.1515/crll.1878.84.242. http://www.digizeitschriften.de/en/dms/img/?PID=GDZPPN002156806.

Carmody, Erin Kathryn. 2017. Killing them softly: Degrees of inaccessible and Mahlo cardinals. *Mathematical Logic Quarterly* 63 (3-4): 256–264. doi:10.1002/malq.201500071.

Carroll, Lewis. 1894. *Sylvie and Bruno Concluded*. Macmillan and Co. Illustrated by Harry Furniss; available at http://www.gutenberg.org/ebooks/48795.

Cavasinni, Umberto. 2020. Dedekind defined continuity. Twitter post, @UmbertoCi. https://twitter.com/UmbertoCi/status/1271971631716618241.

Cep, Casey. 2014. The allure of the map. *The New Yorker*. https://www.newyorker.com/books/page-turner/the-allure-of-the-map.

Chao, Yuen Ren. 1919. A note on "continuous mathematical induction." *Bulletin of the American Mathematical Society* 26 (1): 17–18. https://projecteuclid.org:443/euclid.bams/1183425067.

Cheng, Eugenia. 2004. Mathematics, morally. Text of talk given to the Cambridge University Society for the Philosophy of Mathematics. http://eugeniacheng.com/wp-content/uploads/2017/02/cheng-morality.pdf.

Church, Alonzo. 1936a. A note on the Entscheidungsproblem. *Journal of Symbolic Logic* 1 (1): 40–41. doi:10.2307/2269326.

Church, Alonzo. 1936b. An unsolvable problem of elementary number theory. *American Journal of Mathematics* 58 (2): 345–363. doi:10.2307/2371045.

Clark, Pete. 2019. The instructor's guide to real induction. *Mathematics Magazine* 92: 136–150. doi:10.1080/0025570X.2019.1549902.

Clarke-Doane, Justin. 2017. What is the Benacerraf problem? In *New Perspectives on the Philosophy of Paul Benacerraf: Truth, Objects, Infinity*, ed. Fabrice Pataut, 17–43. Springer.

Clarke-Doane, Justin. 2020. *Morality and Mathematics*. Oxford University Press.

Colyvan, Mark. 2019. Indispensability arguments in the philosophy of mathematics, Spring 2019 edn. In *The Stanford Encyclopedia of Philosophy*, ed. Edward N. Zalta. Metaphysics Research Lab, Stanford University. https://plato.stanford.edu/archives/spr2019/entries/mathphil-indis.

Craig, William. 1953. On axiomatizability within a system. *Journal of Symbolic Logic* 18: 30–32. doi:10.2307/2266324.

Dauben, Joseph W. 2004. Topology: invariance of dimension. In *Companion Encyclopedia of the History and Philosophy of the Mathematical Sciences*, ed. Ivor Grattan-Guiness, Vol. 2, 939–946. Routledge. doi:10.4324/9781315541976.

Davis, Martin. 1977. Review: J. Donald Monk, Mathematical Logic. *Bull. Amer. Math. Soc.* 83 (5): 1007–1011. https://projecteuclid.org:443/euclid.bams/1183539465.

Davis, Martin, and Melvin Hausner. 1978. The joy of infinitesimals, J. Keisler's Elementary Calculus. *The Mathematical Intelligencer* 1 (3): 168–170. doi:10.1007/BF03023265.

De Morgan, Augustus. 1872. *A Budget of Paradoxes*. Volumes I and II. Project Gutenberg. https://en.wikisource.org/wiki/Budget_of_Paradoxes.

Dedekind, Richard. 1888. Was sind und was sollen die Zahlen? (What are numbers and what should they be?) Available in Ewald, William B. 1996. *From Kant to Hilbert: A Source Book in the Foundations of Mathematics*, Vol. 2, 787–832. Oxford University Press.

Dedekind, Richard. 1901. *Essays on the Theory of Numbers. I: Continuity and Irrational Numbers. II: The Nature and Meaning of Numbers*. Dover Publications. Authorized translation 1963 by Wooster Woodruff Beman.

Descartes, René. 2006 [1637]. *A Discourse on the Method of Correctly Conducting One's Reason and Seeking Truth in the Sciences*. Oxford University Press. Translated by Ian Maclean.

Dirac, Paul. 1963. The evolution of the physicist's picture of nature. *Scientific American*. Republished June 25, 2010. https://blogs.scientificamerican.com/guest-blog/the-evolution-of-the-physicists-picture-of-nature.

Dominus, Mark. 2016. In simple English, what does it mean to be transcendental? Mathematics Stack Exchange answer. https://math.stackexchange.com/q/1686299.

Ehrlich, Philip. 2012. The absolute arithmetic continuum and the unification of all numbers great and small. *Bulletin of Symbolic Logic* 18 (1): 1–45. doi:10.2178/bsl/1327328438.

Ehrlich, Philip. 2020. Who first characterized the real numbers as the unique complete ordered field? MathOverflow answer. https://mathoverflow.net/q/362999.

Enayat, Ali. 2004. Leibnizian models of set theory. *The Journal of Symbolic Logic* 69 (3): 775–789. http://www.jstor.org/stable/30041757.

Evans, C. D. A., and Joel David Hamkins. 2014. Transfinite game values in infinite chess. *Integers* 14: 2–36. http://jdh.hamkins.org/game-values-in-infinite-chess.

Ewald, William Bragg. 1996. *From Kant to Hilbert*. Vol. 2 of *A Source Book in the Foundations of Mathematics*. Oxford University Press.

Feferman, S. 1960. Arithmetization of metamathematics in a general setting. *Fundamenta Mathematicae* 49: 35–92. doi:10.4064/fm-49-1-35-92.

Feferman, Solomon. 1999. Does mathematics need new axioms? *American Mathematical Monthly* 106 (2): 99–111.

Feferman, Solomon, Harvey M. Friedman, Penelope Maddy, and John R. Steel. 2000. Does mathematics need new axioms? *Bulletin of Symbolic Logic* 6 (4): 401–446. http://www.jstor.org/stable/420965.

Field, Hartry. 1988. Realism, mathematics, and modality. *Philosophical Topics* 16 (1): 57–107. doi:10.5840/philtopics19881613.

Field, Hartry H. 1980. *Science without Numbers: A Defense of Nominalism*. Blackwell.

Fine, Kit. 2003. The non-identity of a material thing and its matter. *Mind* 112 (446): 195–234. doi:10.1093/mind/112.446.195.

Frege, Gottlob. 1893/1903. *Grundgesetze der Arithmetic, Band I/II*. Verlag Herman Pohle.

Frege, Gottlob. 1968 [1884]. *The Foundations of Arithmetic: A Logico-Mathematical Enquiry into the Concept of Number*. Northwestern University Press. Translation by J. L. Austin of: Die Grundlagen der Arithmetik, Breslau, Koebner, 1884. Parallel German and English text.

Frege, Gottlob. 2013. *Basic Laws of Arithmetic. Derived Using Concept-Script. Vol. I, II*. Oxford University Press. Translated and edited by Philip A. Ebert and Marcus Rossberg, with Crispin Wright.

Galileo Galilei. 1914 [1638]. *Dialogues Concerning Two New Sciences*. Macmillan. Translated from the Italian and Latin by Henry Crew and Alfonso de Salvio.

Geretschlager, Robert. 1995. Euclidean constructions and the geometry of origami. *Mathematics Magazine* 68 (5): 357–371. doi:10.2307/2690924. http://www.jstor.org/stable/2690924.

Gitman, Victoria, Joel David Hamkins, Peter Holy, Philipp Schlicht, and Kameryn Williams. 2020. The exact strength of the class forcing theorem. To appear in *Journal of Symbolic Logic*. http://wp.me/p5M0LV-1yp.

Gödel, K. 1929. Über die vollständigkeit des logikkalküls. Doctoral dissertation, University of Vienna.

Gödel, K. 1931. Über formal unentscheidbare Sätze der Principia Mathematica und verwandter Systeme I. *Monatshefte für Mathematik und Physik* 38: 173–198. doi:10.1007/BF01700692.

Gödel, K. 1986. Über formal unentscheidbare Sätze der Principia Mathematica und verwandter Systeme I. In *Kurt Gödel: Collected Works. Vol. I*, ed. Solomon Feferman, 144–195. The Clarendon Press, Oxford University Press. The original German with a facing English translation. ISBN 0-19-503964-5.

Gödel, Kurt. 1947. What is Cantor's continuum problem? *American Mathematical Monthly* 54: 515–525. doi:10.2307/2304666.

Gödel, Kurt. 1986. *Kurt Gödel: Collected Works. Vol. I*. The Clarendon Press, Oxford University Press. Publications 1929–1936.

Gödel, Kurt. 1995 [*1951]. Some basic theorems on the foundations of mathematics and their implications (*1951). In *Kurt Gödel: Collected Works Volume III*, eds. Solomon Feferman, et al., 304–323. Oxford University Press.

Goldstein, Catherine, and Georges Skandalis. 2007. Interview with A. Connes. *European Mathematical Society Newsletter* 63: 25–31.

Goucher, Adam P. 2020. Can a fixed finite-length straightedge and finite-size compass still construct all constructible points in the plane? MathOverflow answer. https://mathoverflow.net/q/365415.

Greenberg, Marvin Jay. 1993. *Euclidean and Non-Euclidean Geometries*, 3rd edn. W. H. Freeman.

Greimann, Dirk. 2003. What is Frege's Julius Caesar problem? *Dialectica* 57 (3): 261–278. http://www.jstor.org/stable/42971497.

Hamkins, Joel David. 2002. Infinite time Turing machines. *Minds and Machines* 12 (4): 521–539. Special issue devoted to hypercomputation. http://wp.me/p5M0LV-2e.

Hamkins, Joel David. 2011. A remark of Connes. MathOverflow answer. https://mathoverflow.net/q/57108.

Hamkins, Joel David. 2012. The set-theoretic multiverse. *Review of Symbolic Logic* 5: 416–449. doi:10.1017/S1755020311000359. http://jdh.hamkins.org/themultiverse.

Hamkins, Joel David. 2015. Is the dream solution of the continuum hypothesis attainable? *Notre Dame Journal of Formal Logic* 56 (1): 135–145. doi:10.1215/00294527-2835047. http://jdh.hamkins.org/dream-solution-of-ch.

Hamkins, Joel David. 2018. The modal logic of arithmetic potentialism and the universal algorithm. *ArXiv e-prints*. Under review. http://wp.me/p5M0LV-1Dh.

Hamkins, Joel David. 2019. What is the next number on the constructibility sequence? And what is the asymptotic growth? Mathematics Stack Exchange question. https://math.stackexchange.com/q/3377988.

Hamkins, Joel David. 2020a. Can a fixed finite-length straightedge and finite-size compass still construct all constructible points in the plane? MathOverflow question. https://mathoverflow.net/q/365411.

Hamkins, Joel David. 2020b. The On-Line Encyclopedia of Integer Sequences, Constructibility Sequence. http://oeis.org/A333944.

Hamkins, Joel David. 2020c. *Proof and the Art of Mathematics*. MIT Press. https://mitpress.mit.edu/books/proof-and-art-mathematics.

Hamkins, Joel David. 2020d. Who first characterized the real numbers as the unique complete ordered field? MathOverflow question. https://mathoverflow.net/q/362991.

Hamkins, Joel David, and Øystein Linnebo. 2019. The modal logic of set-theoretic potentialism and the potentialist maximality principles. *Review of Symbolic Logic*. doi:10.1017/S1755020318000242. http://wp.me/p5M0LV-1zC.

Hamkins, Joel David, and Alexei Miasnikov. 2006. The halting problem is decidable on a set of asymptotic probability one. *Notre Dame Journal of Formal Logic* 47 (4): 515–524. doi:10.1305/ndjfl/1168352664. http://jdh.hamkins.org/haltingproblemdecidable/.

Hamkins, Joel David, and Barbara Montero. 2000a. Utilitarianism in infinite worlds. *Utilitas* 12 (1): 91–96. doi:10.1017/S0953820800002648. http://jdh.hamkins.org/infiniteworlds.

Hamkins, Joel David, and Barbara Montero. 2000b. With infinite utility, more needn't be better. *Australasian Journal of Philosophy* 78 (2): 231–240. doi:10.1080/00048400012349511. http://jdh.hamkins.org/infinite-utility-more-better.

Hamkins, Joel David, and Justin Palumbo. 2012. The rigid relation principle, a new weak choice principle. *Mathematical Logic Quarterly* 58 (6): 394–398. doi:10.1002/malq.201100081. http://jdh.hamkins.org/therigidrelationprincipleanewweakacprinciple/.

Hamkins, Joel David, and Robin Solberg. 2020. Categorical extensions of ZFC_2 and the categorical large cardinals. In preparation.

Hardy, G. H. 1940. *A Mathematician's Apology*. Cambridge University Press.

Harris, Michael. 2019. Why the proof of Fermat's Last Theorem doesn't need to be enhanced. *Quanta Magazine*. https://www.quantamagazine.org/why-the-proof-of-fermats-last-theorem-doesnt-need-to-be-enhanced-20190603/.

Hart, W. D. 1991. Benacerraf's dilemma. *Crítica: Revista Hispanoamericana de Filosofía* 23 (68): 87–103. http://www.jstor.org/stable/40104655.

Heath, Thomas L., and Euclid. 1956a. *The Thirteen Books of Euclid's Elements, Translated from the Text of Heiberg, Vol. 1: Introduction and Books I, II*. USA: Dover Publications, Inc.

Heath, Thomas L., and Euclid. 1956b. *The Thirteen Books of Euclid's Elements, Translated from the Text of Heiberg, Vol. 2: Books III – IX*. USA: Dover Publications, Inc.

Heath, Thomas L., and Euclid. 1956c. *The Thirteen Books of Euclid's Elements, Translated from the Text of Heiberg, Vol. 3: Books X – XIII and Appendix*. USA: Dover Publications, Inc.

Heisenberg, Werner. 1971. *Physics and Beyond: Encounters and Conversations*. Harper and Row. Translated from the German by Arnold J. Pomerans.

Hilbert, David. 1926. Über das Unendliche. *Mathematische Annalen* 95 (1): 161–190. doi:10.1007/BF01206605.

Hilbert, David. 1930. Retirement radio address. Translated by James T. Smith, *Convergence* 2014. https://www.maa.org/press/periodicals/convergence/david-hilberts-radio-address-introduction.

Hilbert, David. 2013. *David Hilbert's Lectures on the Foundations of Arithmetic and Logic, 1917–1933*. Vol. 3 of *David Hilbert's Lectures on the Foundations of Mathematics and Physics 1891–1933*. Springer. doi:10.1007/978-3-540-69444-1. Edited by William Ewald, Wilfried Sieg, and Michael Hallett, in collaboration with Ulrich Majer and Dirk Schlimm.

Hilbert, David, and Wilhelm Ackermann. 2008 [1928]. *Grundzüge der Theoretischen Logik* (Principles of Mathematical Logic). Springer. Translation via AMS Chelsea Publishing.

Hrbacek, Karel. 1979. Nonstandard set theory. *American Mathematical Monthly* 86 (8): 659–677. doi:10.2307/2321294.

Hrbacek, Karel. 2009. Relative set theory: Internal view. *Journal of Logic and Analysis* 1: 8–108. doi:10.4115/jla.2009.1.8.

Hume, David. 1739. *A Treatise of Human Nature*. Clarendon Press. 1896 reprint from the original edition in three volumes and edited, with an analytical index, by L. A. Selby-Bigge, M.A.

Huntington, Edward V. 1903. Complete sets of postulates for the theory of real quantities. *Transactions of the American Mathematical Society* 4 (3): 358–370. http://www.jstor.org/stable/1986269.

Isaacson, Daniel. 2011. The reality of mathematics and the case of set theory. In *Truth, Reference, and Realism*, eds. Zsolt Novak and Andras Simonyi, 1–76. Central European University Press.

Jech, Thomas. 2003. *Set Theory*, 3rd edn. Springer.

Juster, Norton, and Jules Feiffer. 1961. *The Phantom Tollbooth*. Epstein & Carroll.

Kanamori, Akihiro. 2004. *The Higher Infinite*. Springer. Corrected 2nd edition.

Kant, Immanuel. 1781. *Critique of Pure Reason*. Available via Project Gutenberg at http://www.gutenberg.org/ebooks/4280.

Keisler, H. Jerome. 1977. Letter to the editor. *Notices of the American Mathematical Society* 24: 269. https://www.ams.org/journals/notices/197707/197707FullIssue.pdf.

Keisler, H. Jerome. 2000. *Elementary Calculus: An Infinitesimal Approach*. Earlier editions 1976, 1986 by Prindle, Weber, and Schmidt; free electronic edition available. https://www.math.wisc.edu/~keisler/calc.html.

Khintchin, Aleksandr. 1923. Das stetigkeitsaxiom des linearcontinuum als inductionsprinzip betrachtet. *Fundamenta Mathematicae* 4: 164–166. doi:10.4064/fm-4-1-164-166.

Kirby, Laurie, and Jeff Paris. 1982. Accessible independence results for Peano arithmetic. *Bulletin of the London Mathematical Society* 14 (4): 285–293. doi:10.1112/blms/14.4.285.

Klein, Felix. 1872. A comparative review of recent researches in geometry. English translation 1892 by Dr. M. W. Haskell. Mathematics ArXiv e-prints 2008. https://arxiv.org/abs/0807.3161.

Klement, Kevin C. 2019. Russell's Logicism, ed. Russell Wahl, 151–178. Bloomsbury Academic. Chapter 6 in The Bloomsbury Companion to Bertrand Russell.

Knuth, D. E. 1974. *Surreal Numbers*. Addison-Wesley.

Koellner, Peter. 2010. On the question of absolute undecidability. In *Kurt Gödel: Essays for His Centennial*. Vol. 33 of *Lecture Notes in Logic*, 189–225. Association of Symbolic Logic. doi:10.1017/CBO9780511750762.012.

Koellner, Peter. 2013. Hamkins on the multiverse. *Exploring the Frontiers of Incompleteness*, Harvard, August 31–September 1, 2013.

König, Julius. 1906. Sur la théorie des ensembles. *Comptes Rendus Hebdomadaires des Séances de l'Académie des Sciences* 143: 110–112.

Kreisel, G. 1967. Informal rigour and completeness proofs [with discussion]. In *Problems in the Philosophy of Mathematics*, ed. Imre Lakatos, 138–186. North-Holland.

Lakatos, Imre. 2015 [1976]. *Proofs and Refutations*, Paperback edn. Cambridge University Press. doi:10.1017/CBO9781316286425.

Laraudogoitia, J. P. 1996. A beautiful supertask. *Mind* 105 (417): 81–84.

Larson, Loren. 1985. A discrete look at $1+2+\cdots+n$. *College Mathematics Journal* 16 (5): 369–382.

Lawvere, F. W. 1963. Functorial Semantics of Algebraic Theories. PhD dissertation, Columbia University. Reprinted in: Reprints in Theory and Applications of Categories, No. 5 (2004), pp. 1–121.

Lewis, David. 1983. New work for a theory of universals. *Australasian Journal of Philosophy* 61 (4): 343–377.

Lewis, David. 1986. *On the Plurality of Worlds*. Blackwell.

Maddy, Penelope. 1991. Philosophy of mathematics: Prospects for the 1990s. *Synthese* 88 (2): 155–164. doi:10.1007/BF00567743.

Maddy, Penelope. 1992. *Realism in Mathematics*. Oxford University Press. doi:10.1093/019824035X.001.0001.

Maddy, Penelope. 1997. *Naturalism in Mathematics*. Oxford University Press.

Maddy, Penelope. 2011. *Defending the Axioms: On the Philosophical Foundations of Set Theory*. Oxford University Press.

Maddy, Penelope. 2017. Set-theoretic foundations. In *Foundations of Mathematics*, eds. Andrés Eduardo Caicedo, James Cummings, Peter Koellner, and Paul B. Larson. Vol. 690 of

Contemporary Mathematics Series, 289–322. American Mathematical Society. http://philsci-archive.pitt.edu/13027/.

Madore, David. 2020. Lord of the Rings. Twitter post, @Gro_tsen. https://twitter.com/gro_tsen/status/1211589801306206208.

Manders, Kenneth. 2008a. Diagram-based geometric practice. In *The Philosophy of Mathematical Practice*, ed. Paolo Mancosu, 65–79. Oxford University Press.

Manders, Kenneth. 2008b. The Euclidean diagram. In *The Philosophy of Mathematical Practice*, ed. Paolo Mancosu, 80–133. Oxford University Press. Widely circulated manuscript available from 1995.

Martin, Donald A. 2001. Multiple universes of sets and indeterminate truth values. *Topoi* 20 (1): 5–16. doi:10.1023/A:1010600724850.

Miller, Arnold. 1995. Introduction to Mathematical Logic. Lecture notes, available at https://www.math.wisc.edu/~miller/res/logintro.pdf.

Montero, Barbara. 1999. The body problem. *Noûs* 33 (2): 183–200. http://www.jstor.org/stable/2671911.

Montero, Barbara Gail. 2020. What numbers could be. Manuscript under review.

Moschovakis, Yiannis. 2006. *Notes on Set Theory. Undergraduate Texts in Mathematics*. Springer.

Nielsen, Pace. 2019. What is the next number on the constructibility sequence? And what is the asymptotic growth? Mathematics Stack Exchange answer. https://math.stackexchange.com/q/3418193.

Novaes, Catarina Dutilh. 2020. *The Dialogical Roots of Deduction*. Book manuscript draft.

O'Connor, Russell. 2010. Proofs without words. MathOverflow answer. http://mathoverflow.net/q/17347.

Owens, Kate. 2020. ALL the numbers are imaginary. Twitter post, @kathmath. https://twitter.com/katemath/status/1175777993979023361.

Parsons, Charles. 2017. Realism and the debate on impredicativity, 1917–1944*. In *Reflections on the Foundations of Mathematics: Essays in Honor of Solomon Feferman*, eds. Wilfried Sieg, Richard Sommer, and Carolyn Talcott. *Lecture Notes in Logic*, 372–389. Cambridge University Press. doi:10.1017/9781316755983.018.

Paseau, A. C. 2016. What's the point of complete rigour? *Mind* 125 (497): 177–207. doi:10.1093/mind/fzv140.

Paseau, Alexander. 2015. Knowledge of mathematics without proof. *British Journal for the Philosophy of Science* 66 (4): 775–799.

Peano, Giuseppe. 1889. *Arithmetices Principia, Nova Methodo Exposita* [The Principles of Arithmetic, Presented by a New Method]. Libreria Bocca Royal Bookseller. Available both in the original Latin and in parallel English translation by Vincent Verheyen. https://github.com/mdnahas/Peano_Book/blob/master/Peano.pdf.

Poincaré, Henri. 2018. *Science and Hypothesis*. Bloomsbury Academic. Translated by Mélanie Frappier, Andrea Smith, and David J. Stump.

Potter, Beatrix. 1906. *The Tale of Mr. Jeremy Fisher*. Frederick Warne and Co. https://en.wikisource.org/wiki/The_Tale_of_Mr._Jeremy_Fisher.

Potter, Beatrix. 1908. *The Tale of Samuel Whiskers or, The Roly-Poly Pudding*. Frederick Warne and Co. https://en.wikisource.org/wiki/The_Roly-Poly_Pudding.

Priest, Graham, Koji Tanaka, and Zach Weber. 2018. Paraconsistent logic, Summer 2018 edn. In *The Stanford Encyclopedia of Philosophy*, ed. Edward N. Zalta. Metaphysics Research Lab, Stanford University. https://plato.stanford.edu/archives/sum2018/entries/logic-paraconsistent.

Putnam, Hilary. 1971. *Philosophy of Logic*. Allen & Unwin.

Quarantine 'em. 2020. There's a really great joke. Twitter post, @qntm. https://twitter.com/qntm/status/1279802274584330244.

Quine, W. V. 1987. *Quiddities: An Intermittently Philosophical Dictionary*. Belknap Press of Harvard University Press.

Quine, Willard Van Orman. 1986. *Philosophy of Logic*. Harvard University Press.

Resnik, Michael D. 1988. Mathematics from the structural point of view. *Revue Internationale de Philosophie* 42 (4): 400–424.

Richards, Joan L. 1988. Bertrand Russell's "Essay on the Foundations of Geometry" and the Cambridge Mathematical Tradition. *Russell: The Journal of Bertrand Russell Studies* 8 (1): 59.

Richeson, Dave. 2018. Chessboard (GeoGebra). https://www.geogebra.org/m/wfrzeast.

Rosen, Gideon. 2018. Abstract objects, Winter 2018 edn. In *The Stanford Encyclopedia of Philosophy*, ed. Edward N. Zalta. Metaphysics Research Lab, Stanford University. https://plato.stanford.edu/archives/win2018/entries/abstract-objects.

Russell, Bertrand. 1903. *The Principles of Mathematics*. Free online editions available in various formats at https://people.umass.edu/klement/pom.

Russell, Bertrand. 1919. *Introduction to Mathematical Philosophy*. George Allean and Unwin. Corrected edition 1920. Reprinted, John G. Slater (intro.), Routledge, 1993.

Sarnecka, Barbara W., and Charles E. Wright. 2013. The idea of an exact number: Children's understanding of cardinality and equinumerosity. *Cognitive Science* 37 (8): 1493–1506. doi:10.1111/cogs.12043.

Secco, Gisele Dalva, and Luiz Carlos Pereira. 2017. Proofs versus experiments: Wittgensteinian themes surrounding the four-color theorem. In *How Colours Matter to Philosophy*, ed. Marcos Silva, 289–307. Springer. doi:10.1007/978-3-319-67398-1_17.

Shagrir, Oron. 2006. Gödel on Turing on computability. In *Church's Thesis After 70 Years*, Vol. 1, 393–419. Ontos Verlag.

Shapiro, Stewart. 1996. Mathematical structuralism. *Philosophia Mathematica* 4 (2): 81–82. doi:10.1093/philmat/4.2.81.

Shapiro, Stewart. 1997. *Philosophy of Mathematics: Structure and Ontology*. Oxford University Press.

Shapiro, Stewart. 2012. An "i" for an i: Singular terms, uniqueness, and reference. *Review of Symbolic Logic* 5 (3): 380–415.

Shelah, Saharon. 1993. The future of set theory. In *Set Theory of the Reals (Ramat Gan, 1991)*, Vol. 6, 1–12. Bar-Ilan University. https://shelah.logic.at/v1/E16/E16.html.

Shelah, Saharon. 2003. Logical dreams. *Bulletin of the American Mathematical Society* 40: 203–228.

Shulman, Mike. 2019. Why doesn't mathematics collapse down, even though humans quite often make mistakes in their proofs? MathOverflow answer. https://mathoverflow.net/q/338620.

Simpson, Stephen G. 2009. *Subsystems of Second Order Arithmetic*, 2nd edn. Cambridge University Press; Association for Symbolic Logic. doi:10.1017/CBO9780511581007.

Simpson, Stephen G. 2010. The Gödel hierarchy and reverse mathematics. In *Kurt Gödel: Essays for His Centennial, Lecture Notes in Logic*, Vol. 33, 109–127. Association of Symbolic Logic. doi:10.1017/CBO9780511750762.008.

Simpson, Stephen G. 2014 [2009]. Toward objectivity in mathematics. In *Infinity and Truth*. Vol. 25 of *Lecture Notes Series, Institute for Mathematical Sciences, National University of Singapore*, 157–169. World Scientific Publishing. doi:10.1142/9789814571043_0004. Text based on the author's talk at a conference on the philosophy of mathematics at New York University, April 3–5, 2009.

Smoryński, Craig. 1977. The incompleteness theorems. In *Handbook of Mathematical Logic*, ed. Jon Barwise, 821–865. North-Holland.

Smullyan, Raymond M. 1992. *Gödel's Incompleteness Theorems*. Vol. 19 of *Oxford Logic Guides*. Clarendon Press, Oxford University Press.

Snow, Joanne E. 2003. Views on the real numbers and the continuum. *Review of Modern Logic* 9 (1–2): 95–113. https://projecteuclid.org:443/euclid.rml/1081173837.

Soare, Robert I. 1987. *Recursively Enumerable Sets and Degrees: A Study of Computable Functions and Computably Enumerable Sets*. Springer.

Solovay, Robert M. 1970. A model of set-theory in which every set of reals is Lebesgue measurable. *Annals of Mathematics, Second Series* 92 (1): 1–56.

Stillwell, John. 2005. *The Four Pillars of Geometry*. Springer.

Suárez-Álvarez, Mariano. 2009. Proofs without words. MathOverflow question. http://mathoverflow.net/q/8846.

Tao, Terence. 2007. Ultrafilters, nonstandard analysis, and epsilon management. https://terrytao.wordpress.com/2007/06/25/ultrafilters-nonstandard-analysis-and-epsilon-management/.

Tarski, Alfred. 1951. *A Decision Method for Elementary Algebra and Geometry*, 2nd edn. University of California Press.

Tarski, Alfred. 1959. What is elementary geometry? In *The Axiomatic Method: With Special Reference to Geometry and Physics*, 16–29. North-Holland. Proceedings of an International Symposium held at the University of California, Berkeley, December 26, 1957–January 4, 1958 (edited by L. Henkin, P. Suppes, and A. Tarski).

Thurston, Bill. 2016. Thinking and Explaining. MathOverflow question. https://mathoverflow.net/q/38639.

Thurston, William P. 1994. On proof and progress in mathematics. *Bulletin of the American Mathematical Society* 30 (2): 161–177. doi:10.1090/S0273-0979-1994-00502-6.

Tourlakis, George. 2003. *Lectures in Logic and Set Theory,* Vol. 2. Cambridge University Press.

Turing, A. M. 1936. On Computable Numbers, with an Application to the Entscheidungsproblem. *Proceedings of the London Mathematical Society* 42 (3): 230–265. doi:10.1112/plms/s2-42.1.230.

Univalent Foundations Program, The. 2013. *Homotopy Type Theory: Univalent Foundations of Mathematics*. Institute for Advanced Study. https://homotopytypetheory.org/book.

Vardi, Moshe Y. 2020. Efficiency vs. resilience: What COVID-19 teaches computing. *Communications of the ACM* 63 (5): 9. doi:10.1145/3388890. https://cacm.acm.org/magazines/2020/5/244316-efficiency-vs-resilience.

Veblen, Oswald. 1904. A system of axioms for geometry. *Transactions of the American Mathematical Society* 5 (3): 343–384. http://www.jstor.org/stable/1986462.

Voevodsky, Vladimir. 2014. The origins and motivations of univalent foundations. *The Institute Letter, Institute for Advanced Study*. Summer 2014. https://www.ias.edu/publications/institute-letter/institute-letter-summer-2014.

Weaver, Nik. 2005. Predicativity beyond Γ_0. *Mathematics ArXiv e-prints*. https://arxiv.org/abs/math/0509244.

Weber, H. 1893. Leopold Kronecker. *Mathematische Annalen* 43 (1): 1–25. doi:10.1007/BF01446613.

Whitehead, A. N., and B. Russell. 1910. *Principia Mathematica. Vol. I.* Cambridge University Press.

Whitehead, A. N., and B. Russell. 1912. *Principia Mathematica. Vol. II.* Cambridge University Press.

Whitehead, A. N., and B. Russell. 1913. *Principia Mathematica. Vol. III.* Cambridge University Press.

Wikipedia. 2020. Schröder–Bernstein theorem. https://en.wikipedia.org/wiki/Schr%C3%B6der%E2%80%93Bernstein_theorem.

Williamson, Timothy. 1998. Bare possibilia. *Analytical Ontology* 48 (2/3): 257–273.

Williamson, Timothy. 2018. Alternative logics and applied mathematics. *Philosophical Issues* 28 (1): 399–424. doi:10.1111/phis.12131.

Wittgenstein, Ludwig. 1956. *Remarks on the Foundations of Mathematics*. Blackwell.

Wright, Crispin. 1983. *Frege's Conception of Numbers as Objects*, Vol. 2. Aberdeen University Press.

Zach, Richard. 2019. Hilbert's program, Fall 2019 edn. In *The Stanford Encyclopedia of Philosophy*, ed. Edward N. Zalta. Metaphysics Research Lab, Stanford University. https://plato.stanford.edu/archives/fall2019/entries/hilbert-program/.

Zeilberger, Doron. 2007. " ". Personal Journal of S. B. Ekhad and D. Zeilberger http://www.math.rutgers.edu/~zeilberg/pj.html. Transcript of a talk given at the DIMACS REU program, June 14, 2007, http://sites.math.rutgers.edu/~zeilberg/mamarim/mamarimhtml/nothing.html.

Zermelo, E. 1930. Über Grenzzahlen und Mengenbereiche. Neue Untersuchungen über die Grundlagen der Mengenlehre. (German). *Fundumenta Mathematicae* 16: 29–47. Translated in Ewald (1996).

Notation Index

Each index entry refers to the first or most relevant use in this book of a particular mathematical notation or abbreviation.

Subject Index